R. J. Mitchell
at Supermarine

R. J. Mitchell
at Supermarine

from Schneider Trophy
to Spitfire

John K. Shelton

STANDON
BOOKS

First published in Great Britain in 2008 as
Schneider Trophy to Spitfire; The design Career
of R. J. Mitchell by Haynes Publishing, Sparkford,
Yeovil, Somerset BA22 7JJ, UK

This expanded, completely updated and revised version
published in 2017 by Standon Books, The Lodge, Standon Hall,
Maer Lane, Standon, Staffordshire ST21 6QZ, UK

www.standonbooks.co.uk

A CIP catalogue record for this book
is available from the British Library

ISBN 978-0-9956781-0-1

Book and cover design by Rod Shelton

Produced by Biddles Books Ltd, King's Lynn, Norfolk

To:

Jane, Sarah
Mark and Rod

Foreword

by Julian Mitchell

If he were alive today, my Great Uncle Reginald Joseph Mitchell (or 'R.J.' as he was typically known) would no doubt attract significant attention and acclaim worldwide. After all, he has often been referred to as a 'Genius', hence it is plausible to suggest that R.J. would be photographed, interviewed, celebrated and championed ultimately as one of *the* Greatest Engineering Designers of our time.

Ironically, such contemporary 'Cult Celebrity' status and Social Media intensity could not be less suited to a man who was inherently shy, private and modest by nature. Indeed, throughout his life R.J. shied away from the intrusive spotlight of personal fame and it is well known that he consistently chose to deflect the numerous accolades attributed to *Himself*, towards his Team's collective successes instead.

Without a doubt, there is a wealth of resource available to us now, documenting Mitchell's life and the most significant stages of his career. As a young boy, I had immersed myself in an abundance of reading material recalling R.J.'s life and professional timeline. My family and I watched numerous documentaries and films celebrating his work; we visited the Science Museum in London, where I touched the Schneider Trophy, dumbstruck, in awe.

And, naturally, throughout my childhood I built numerous models of Spitfires – as have my own sons too – whilst dreaming at night that perhaps I might fly in one … some day.

Less well known however, and not fully documented until now, is the complex inner character of my Great Uncle. There is, perhaps, a generalised perception of an 'Engineering Genius' – as reflected by a linear, rational and methodical mind, embodied in a clearly defined, systematically focused personality. And fused together by brilliant, singular logic.

Yet throughout Mitchells' life – albeit one curtailed too soon – R.J. quietly repudiated this definition. And it is perhaps due to the hybrid nature of his Persona, his astonishing diversity of creative and core artistic vision – truly understood by only a few – that the essence of this great man's character has never been fully revealed.

From the outset R.J. demonstrated an intuitive, sensitive talent for creative design coupled with astonishing artistic vision. Additionally, Mitchell's diverse flexibility and extraordinary foresight would accomplish historical achievements in Aeronautical Design – despite his lack of formal training therein.

Undoubtedly, R.J.'s initial Locomotive Engineering apprenticeship served to nurture, mature and consolidate his inherent skills and erudite perception. Above all, this foundation was essential in establishing an inimitable platform for Mitchell's Pioneering greatness and his successive technical advancements.

Numerous projects have been attributed to my Great Uncle. And whilst some were truly innovative, others were more pragmatic. Certainly, however, Mitchell's ideas were transformative in terms of world-class aviation development and enhancements, impressive examples of which are substantiated by Aeronautical Design History encompassing not only the Design, Development and Application of Technical & Styling innovation, but also Supremacy in Speed, Distance & Purpose. Above all, these concepts were embodied in aspirational Beauty and Elegance – perhaps most eloquently defined by the iconic Spitfire.

It is worth noting that R.J. not only designed state of the art seaplanes and fighting machines, but he established the platform for commercial aviation today. Aviation was in its infancy when R.J. first joined Supermarine in 1916. He learnt quickly and then used his own creativity and sound engineering principle to design new craft, largely by himself.

Yet as the bounds of aeronautical knowledge were pushed further and further, it became apparent that one man simply could not be a master of all the elements that created the 'ultimate machine'. And R.J. proved himself also to be a brilliant leader. He did not seek to attract mass adoration as a Public Hero, but instead eschewed personal praise in favour of a wider Team, whom he identified as being simply 'The Best' in terms of workmanship, efficiency, and capable of producing cutting-edge designs.

One of my Great Uncle's most praiseworthy accolades is that he was fiercely loyal to, and protective of, his colleagues. Before deciding the optimal way forward, he would listen to every opinion and carefully considered each point of view, before evaluating decisionsf based on meticulous examination of hybrid scenarios. In the words of one of his design team: '*Certainly Mitchell always did the thing which should be done*'.

Perhaps the most significant example of my Great Uncle's devotion to his work is the supreme sacrifice he ultimately made: compromising his own health – and ultimately his life – in order to complete his final and most onerous work, so critical for the future of our Nation. And the successive results, thereof, which would shape World history.

His greatest legacy, one I would suggest that more than makes up for the lack of fitting National recognition that was bestowed on many of his peers, is that the name 'Spitfire' remains firmly embedded in our language and culture today as a great icon. As such, the Spitfire continues to fuel National pride and is synonymous with success, pride, courage, sacrifice and beauty.

Recently, as a testimony to R.J.'s greatness and visionary spirit, *Operation Spitfire* was founded. Based in R.J.'s Staffordshire heartland of Stoke on Trent, *Operation Spitfire*'s core mission is to restore the City's cherished Mark XVI Spitfire, currently housed in the Potteries Museum & Art Gallery. *RW388* was originally gifted by the RAF as a tribute to the People of Stoke on Trent – in formal recognition of R.J.'s place of birth and formative education.

Since inception to date, *Operation Spitfire* has attracted successful momentum. And its scope has widened exponentially to encompass a broader vision and purpose: to educate, to inspire, and to develop a new generation of Engineers, by connecting young people with local and national industries.

Operation Spitfire ultimately aspires to honour the memory of R. J. Mitchell, and to recall and validate his relevance, still, today.

I am genuinely delighted and profoundly honoured to have been invited to write this Foreword to Dr Shelton's highly commendable work. As a respected historical authority, who has researched and transcribed in fascinating detail the life and achievements of Reginald Mitchell, Dr Shelton now offers us an enlightened and fresh perspective of the Man and His Machines. He certainly frames a magnificent testimony to the enduring legacy of my Great Uncle and I conclude that in this book, R.J.'s own spirit is accurately reflected – Great Purpose fused with Passion.

Julian Mitchell, BA, MBA

Preface

'I have seen the future, and it works' – Lincoln Steffens.

R. J. Mitchell, right, with his 1929 S.6 Schneider Trophy winner.

The continued appearance of the classic and distinctive shape of R. J. Mitchell's Spitfire, still to be seen flying at air shows and commemorative events, attests to its enduring appeal – at least eighty years after the prototype's first flight in 1936, it is still in the forefront of the popular mind as the iconic aircraft of World War Two. But it is not always appreciated what a dramatic advance this machine was on the contemporary front line fighters of the RAF: from the Gloster Gamecock of 1926 to the prototype Gladiator of 1934, the maximum speed available to their pilots rose from about 150 mph to 240 mph. It might, therefore, be expected that, given the leisurely pace of British aircraft development, the Spitfire would probably be a continuation of the usual fabric-covered biplane type, limited by the drag of a fixed undercarriage and wire-and-strut bracing to a speed of about 280 mph – but soon to be opposed by the 350 mph Messerchmitt Bf 109 as war drew ever closer.

Thus the appearance of Mitchell's all-metal, stressed-skin, cantilever monoplane, with a retracting undercarriage, and also capable of 350 mph, was to be not only a vital factor in the forthcoming conflict but also a remarkable design achievement – especially as Mitchell had spent most of the previous fifteen years as Chief Designer at Supermarine creating slow-flying marine aircraft. Whilst it only requires a little thought to appreciate that the new fighter was not likely to have come from Mitchell's drawing board by some magical conceptual leap, the process by which it was achieved has not previously been systematically examined. The present work therefore draws particular attention to the most significant factor in this apparently dramatic appearance of the Spitfire – the Schneider Trophy series of competitions. As they came to an end in 1931, the British successes have now become something of a distant memory even though, at the time, Mitchell was nationally celebrated as the designer of the floatplanes which won the final three competitions.

The full story of how Mitchell came to be involved with the Schneider Trophy and the subsequent development of his famous fighter surely deserves telling and, along the way, it may come as a surprise to many to discover that he had been previously involved with the production of a series of very successful naval seaplanes and was active in the beginnings of commercial flying in Great Britain. Thus it was by no means predictable that he would go on to produce the iconic fighter so associated in the popular mind with the Battle of Britain. Indeed, he began his professional career in locomotive engineering and he never had any formal education as an aircraft designer. Yet, well before the Spitfire appeared, he had emerged as one of the most prominent designers of his time and a listing of his most significant contributions to aviation reveals promise from the very beginning:

- the Commercial Amphibian, his first independent design, won an enhanced award at the 1920 Air Ministry competition for passenger amphibian flying boats. Although this aircraft came second to the Vickers Viking, the second prize of £4000 was doubled in recognition of the promise that the aircraft had shown;

- his modification and uprating of an earlier company machine, the Sea Lion II, won the Schneider Trophy competition for Britain in 1922;

- the small fleet of his Sea Eagle flying boats constituted the first British scheduled flying-boat service, operating between Southampton and the Channel Isles between 1923 and 1928;

- his Swan of 1924 was claimed to be the world's first multi-engined amphibian passenger-carrying machine; it joined the above-mentioned flying-boat service, which was later incorporated into the newly formed Imperial Airways Ltd (later, British Airways);

- the Southampton flying boat, a military development of the Swan, was ordered, unusually, straight off the drawing-board and became the standard RAF coastal reconnaissance aircraft; indeed, *Jane's* described the machine as 'one of the most notable successes in post-war design' and it established real stability and prosperity for Supermarine;

- in the same year, Mitchell also produced his S.4 Schneider Trophy racer which revolutionised the design of virtually all the successive competition entries: he moved, in one bold step, from the usual wire-braced biplanes to a startlingly new cantilever monoplane. Compared with the top speed of 175 mph claimed for his Sea Lion III racer in 1923, the S.4 gained the World Speed Record for Seaplanes and the British Speed Record for all types with 226.75 mph only two years later;

- the increased efficiency of his metal-hulled Mk.II Southampton led to the RAF being equipped for a special Far East Flight with four of these machines. These aircraft each completed a 27,000 mile cruise between October 1927 and February 1928 to Singapore and then around Australia, which had only been visited by aircraft on four previous occasions and only circumnavigated by one earlier machine. The 62 timetabled stages were completed by all the aircraft – as Supermarine publicity said, '108,000 machine miles giving no trouble of any consequence' and as the *Daily Mail* said: 'the flight will rank as one of the greatest feats in the history of aviation';

- in 1930, Supermarine were awarded a contract (later cancelled by the Government) to build the largest wingspan flying boat in the world – greater than

the groundbreaking Dornier Do.X and only to be surpassed by the Hughes
H-4 Hercules of 1947;

- by this time, Mitchell had designed his next two Schneider Trophy racers,
  the S.5 and S.6 which, respectively, won the 1927 and 1929 contests. In the
  following event of 1931, his S.6B won the trophy outright and later went on
  the set a new Absolute Air Speed Record of 407.5 mph. This last machine
  was now of entirely metal, stressed-skin, construction and clearly looked
  forward to the Spitfire, five years later.

Whilst a full account of Mitchell's development as a designer reflects almost
the whole of aircraft history from the Blériot and Farman era until the advent of
jet engined aircraft, this present work particularly seeks to recount how his aircraft began to emerge at the end of World War One in response to current design
philosophies and how they gradually transcended these conventions. As Mitchell
had had no formal training in aircraft design, it should be doubly interesting to
see how he gradually learned his trade until he came to produce such trendsetting
and aesthetically pleasing shapes as the Southampton flying boat, the S.4 Schneider racer and, of course, the Spitfire.

It is noteworthy that, while still in his thirties, Mitchell was described in
Supermarine publicity as 'one of the leading flying-boat, amphibian and high-
speed seaplane designers in the country'; he had also been invited to give a talk
on the BBC, had been elected a Fellow of the Royal Aeronautical Society and
awarded the CBE. His reputation had been significantly advanced by his design
of the later Schneider Trophy machines which gave him unfettered opportunities
to extend the boundaries of high speed flight; as a result, the advent of the Spitfire prototype of 1936 marked a dramatic increase in speed of more than 100
mph over the most recent RAF fighter in service and led to a dramatic initial
order of 310 aircraft, three months after its first service trial. His tragically early
death at the age of 42, however, resulted in his not seeing the fighter go into
squadron service and without knowing that nearly 23,000 examples were to be
built and in a multitude of variants.

Whilst his career successes were also the result of various instances of good
fortune which must favour any career, as will be narrated, his untimely death in
1937 deprived him of direct credit for the design of a groundbreaking fighter
capable of an extraordinarily successful development throughout the war years or
for the chance to contribute other designs for his country. After the end of World
War Two, other aircraft designers were awarded knighthoods but no such posthumous acknowledgement of Mitchell's contributions was possible. And so it is
hoped that, in some way, this much expanded second edition especially will do
justice to the man and a modification to the title reflects this broadening scope: *R.
J. Mitchell at Supermarine; from Schneider Trophy to Spitfire.*

Ironically, the post-war years have seen a virtual avalanche of books on the
Spitfire and Mitchell's name is therefore still well known, but it is surprising that
there has been no comprehensive account devoted to the man and to the whole
design experience which preceded his iconic fighter. In this present edition in
particular, the author has tried to satisfy this need by fully recording the whole
evolution of his design output, in chronological order, with photographs and
drawings of every main type which was completed – as well as by bringing
together salient information, often no longer readily available to the general
reader, concerning Mitchell's output and the factors affecting it.

Recent writings have provided fresh insights which are taken account of,
with the aim of not only bringing the previous edition completely up-to-date, but

also of presenting the fullest possible account of our subject. In particular, there is a detailed consideration of the fatal crash of Kinkhead's S.5 and of the wooden hull construction techniques which were so important to Mitchell's early designs. Additionally, the emergence and importance of Rolls-Royce engine technology is more fully described and the dramatic advance in performance of the Spitfire is now more fully illustrated – with reference to the earlier conventions of British fighter design and to the other aircraft which also attempted to fulfil the F.7/30 Air Ministry requirement (which lead to the appearance of the Spitfire). Thus the unique elliptical and thin wing, which was such a vital feature of Mitchell's fighter, is also more fully documented. And it was felt that more space should be devoted to the development of the Spitfire after Mitchell's death and to the uncertainties surrounding the acceptance of this aircraft into squadron service. The vital work of Joe Smith in fully realising the potentiality of his predecessor's concept is credited, together with an account of the emergence and continuation of the Spitfire as a national icon.

Acknowledgements
The excellent Putnam book, *Supermarine Aircraft since 1914*, is still the main reference source, but its scope was not concerned directly with Mitchell's development during his two decades as Chief Designer – which must be inferred from its survey of the company's aircraft, often arranged by type rather than chronologically. The fine *Spitfire Story* by Alfred Price is very informative about the genesis and development of Mitchell's fighter but, of course, it is not concerned with his whole career. One thus turns to the work of his son, Gordon Mitchell, which does give considerable biographical information and a chronological survey of his father's whole design output. But whilst Dr Mitchell performed an important service by raising the profile of his father, quite a few aircraft are only briefly dealt, and there are no three-view drawings of the machines and no comprehensive set of photographs.

The annual publications of *Jane's* have been invaluable sources of information, as have the the books by Barker, Eves, James, Mondey, and Pegram, on the Schneider Trophy contests. It is also hoped that the extracts from the reminiscences of various pilots will help restore memories of their achievements and contributions especially to Supermarine aviation between the two World Wars. Such fliers as Atcherley, Biard, Grieg, Henshaw, Livock, Nichol, Orlebar, Penrose, Quill, Schofield, Snaith, and Waghorn supply much information about that essential factor in any designer's output – what it was like to fly the machines.

They also give revealing insights into the character of the designer, as do several Supermarine employees – Black, Griffiths, Shenstone, Smith, and Webb – whose accounts also make it possible to discover much about the complex nature of the man behind the machines. Additionally, the works of Bulman, Penrose, and Viscount Templewood have been valuable sources of information concerning the economic and political factors affecting British aircraft development (and therefore Mitchell) during this period; the recent works of Sinnot and McKinstry have also been very helpful in filling out this background to Mitchell's working life. Cozens' unpublished MS concerning the history of South Hampshire aviation, helpfully copied to me by Solent Sky, also gave many insights into Mitchell's career at Supermarine. These authors quoted from and/or referred to in the main text are detailed in the Bibliography – which has been considerably augmented in this edition in order to provide extensive opportunities for further reading in related areas.

Every effort has been made to obtain permissions to reproduce material from these authors but, if there have been any omissions, credit will be given in subsequent printings or editions. Meanwhile, I am grateful for permission to quote from the following authors and publishers (listed in the Bibliography): G. P. Bulman, Alec Henshaw, Julian Lewis, Alfred Price, *Jane's*, Crecy, Robert Hale, Royal Aeronautical Society, Putnam, *The Aeroplane*, and *Flight International*; I have been unable to trace Seeley Service and Co. (A. H. Orlebar), although his daughter, Mrs B. Findlay, is agreeable to my quoting from his book, nor have I been able to contact Hurst and Blackett (Biard), G. T. Foulis and Co. (Walrus), United Writers (Griffiths) or J and K. H. Publishing (Webb). I must also thank the head teacher of the Reginald Mitchell Primary School, Mr A. V. Stancliffe (Arthur Black) and I am particularly appreciative of the kind assistance I have received from staff at the Royal Air Force Museum, Cambridge University Department of Manuscripts and Archives, The Royal Aeronautical Society, including the Yeovil and Southampton branches, the Rolls-Royce Heritage Trust, and the staff at Southampton Solent-Sky Museum. More recently I have been grateful for the exchanges of information with the Rt Hon. Dr Julian Lewis, MP (*Racing Ace*) and Ralph Pegram (*The Schneider Trophy Seaplanes and Flying Boats*).

The author's own general arrangement drawings are based on those in *Supermarine Aircraft since 1914* (C. F. Andrews and E. B. Morgan), *The Schneider Trophy Seaplanes and Flying Boats* (R. Pegram), *The Spitfire Story*, (A. Price) and various MAP drawings. The drawings have also incorporated additional information obtained from my own collection of photographs and from those now available online.

The 'Chronicles' which appear in the first two chapters of this book represent the author's version of how press reports at the time might have looked.

And whilst I must take ultimate responsibility for the final form of this book, I must pay special tribute to my son Rod: in all IT matters 'the child is father to the man' but the assistance I have received has been far more than just technical: not only have I been spared interpreting the minutiae of *New Hart Rules* and *Fowler's Modern English Usage*, I have benefited greatly from his professional viewpoint on layout and other publishing conventions.

Photo credits
(Letters in brackets indicate the placement of the image on the page.)
BAE Systems 32, 43(a–c), 45, 84(b), 112, 135(a–b), 136, 168(b), 190, 193, 219(a), 231, 236, 256, 259, 262, 267, 268, 271, 272(a), 273, 278, 283, 285(a–b), 293(b), 290(a–c), 291(a–c), 292(a–b); Dominic Winters, Auctioneers 21; P. Jarrett 18, 27(a), 28, 30, 38, 61, 66(a), 69, 71(a), 77, 78, 83, 84(c), 90(b), 103, 109(a–b), 122, 126(a–b), 147, 160, 177, 195(a–b), 207(b), 208(b), 210, 232, 258(a–b), 261, 293(a), 308(b), 315; E. B. Morgan 17, 43(a–c), 50(a), 64, 68, 86, 90(a), 98, 107, 126(a–b), 135(a–b), 150, 158(b), 175, 206, 303, 306, 308(a), 314; Royal Air Force Museum 3, 102, 133, 211(b), 217, 218, 219(b), 220(a–b), 222, 270, 272(b), 291(a); *The Sentinel*, Stoke-on-Trent 84(a); Solent-Sky v, 7, 9, 10, 16, 27(a), 31, 50(b), 35, 52, 57, 66(b), 71(b), 72, 74, 76, 88, 91, 94, 111, 120, 127, 128, 142(a–b), 149, 159, 164(a–b), 166, 168(a), 176(a–b), 178, 186, 200, 201, 203, 207(a), 208(b), 211(a), 254, 268, 275, 304, 305, 310, 316, 337. Other photographs are from the author's own collection.

Contents

List of General Arrangements

Permission to reproduce any of these 3-view drawings should be obtained from the publisher.

R. J. Mitchell at his drawing-board, c. 1930.

— Introduction —

The Designer
and his Aircraft

In the chapters which follow, the various aircraft that came from the drawing-board of Supermarine's Chief Designer will be described in chronological order, as established by their first flights. Such a method is intended to give an accurate picture of the varied and contrasting design problems that confronted Mitchell at any one particular time – in fact, this aspect of his design career was singled out by *Flight* magazine in his obituary:

> His versatility will be appreciated when it is pointed out that his productions ranged from heavy, long-range flying boats to tiny single-seat landplane fighters and on more than one occasion he had two or three very different types of aircraft passing through the design stage at the same time, so that he frequently had to switch his mind from one problem to another of a totally different character.

The earlier listing of Mitchell's successes (and there were also some failures) indicated a considerable output in his relatively short working life and this is obviously not simply attributable to 'versatility' or talent. A capacity for hard, concentrated work was also clearly involved, especially when one comes to consider the variety of aircraft he was called upon to design: for example, between 1920 and 1922, the newly appointed Chief Designer, with only three previous years' experience in the aircraft industry, was responsible for the design of a passenger-carrying prototype (the Commercial Amphibian) which required the innovation of a retracting undercarriage design, a fleet spotter (the Seal) with the added complexity of folding wings, beginning the design for a replacement for the large World War One Felixstowe coastal reconnaissance flying boat (the Scylla), and the modification of an earlier company machine for the 1922 Schneider Trophy contest (the Sea Lion II).

Additionally, the present chronological approach (instead of a neater tracing of the development of certain basic Supermarine groups of machines) also reveals how the Chief Designer moved, by no means in a straight line, from a dependency on the traditions of aircraft design and construction that he inherited to the boldness and originality of much of his later work.

In this respect, the year 1925 can be seen as the point where Mitchell's designs revealed a man with full confidence in his own ideas and with the ability to create new shapes – shapes which began to influence the less traditional sort of aeroplane and which found particular fulfilment in the Spitfire. Indeed, the S.4 design, for the Schneider contest of 1925 (*see* photo overleaf), has been described by Alan Clifton (*see* below) as having 'breathtakingly clean lines, which caused a sensation when photos were released'.

This aircraft also illustrates how the demands of these competitions brought out the originality of our designer: whilst Mitchell has gained popular fame for his Spitfire, it should be remembered that his success with these racers had already placed him foremost among aircraft designers in the view of the aviation fraternity – for example, when Jeffrey Quill joined Vickers in January, 1936, as a young test pilot, before the Spitfire had flown, one of his most memorable early incidents was meeting 'with the then *already legendary* R. J. Mitchell' [my italics].

Mitchell's Main Aircraft Types

Most commentators are not just impressed by how soon Mitchell established his reputation as a major designer (being awarded the CBE when he was only 36) but also by the range of his designs – and this breadth will become obvious in the following chapters. But first, it will be useful if the main types of his aircraft are distinguished.

The S.4.

Whilst the S.4 marked the beginning of one of Mitchell's main design types, the racing floatplane, 1925 also saw the first flight of another of the designs of his early maturity, the Southampton:

The Southampton Mark I.

This machine, often regarded as one of the most graceful of all his biplanes, represents the second main category of his aircraft: the larger flying boats. The special requirements of these two types were a continual challenge to the ingenuity of Mitchell and the team he began to collect around him.

The third main type with which he was associated was the medium-sized amphibian. Here the particular requirements of the type did not require major departures from current practice and Mitchell obviously saw no reason to make changes for their own sake – as exemplified by the Walrus as late as 1936:

Walrus prototype over Gibraltar.

The following list of twenty-one distinct types of aircraft completed indicates the preponderance of these three types, as well as other different aircraft (including the Spitfire) which can be seen appearing from time to time:

Medium sized flying boats	Larger flying boats	Schneider Trophy aircraft	Landplanes
Commercial Amphibian	Swan	Sea Lion II/III	Sparrow I/II
Seal II	Southampton I–IV	S.4	Type 224
Sea King II	Nanok/Solent	S.5	Spitfire
Seagull II–IV	Air Yacht	S.6/6A/6B	
Sea Eagle	Scapa		
Scarab/Sheldrake	Stranraer		
Seamew			
Seagull V/Walrus			

It will be shown that Mitchell's almost twenty productive years were mainly concerned with the improvement of marine reconnaissance and passenger aircraft before the advent of the Spitfire. But while these developments constituted the mainstay of Mitchell's design activity, he was respected in the industry for his work in the very different sphere of high-speed aircraft – their successes established Mitchell as the foremost designer of high-speed aircraft, each of which had successively improved not just upon the seaplane speed record but also, in the case of his last two machines, upon the World Absolute Air Speed record. As Mitchell's main design work was concerned with passenger-carrying amphibians, naval reconnaissance flying boats, and ship-based fleet spotter planes, where speed was by no means the main criterion, the Schneider Trophy machines were obviously a far more significant influence on the emergence of the Spitfire – as Quill said:

At the time the Spitfire was designed, Mitchell's design team, because
of its previous involvement with the S.4, S.5, S.6 and S.6B Schneider
Trophy racing seaplanes ... had more practical knowledge of high speed
aeronautics than any other design team in the world. They were mentally
adjusted to, and dedicated to, the search for the ultimate in aerodynamic
efficiency and the achievement of the highest possible speeds. They were
not going to allow themselves to be constrained by convention or other
extraneous considerations from achieving these aims. They were young
and, I believe, very single-minded ... Members of his team still retained
the basic attitudes acquired from racing seaplane programmes throughout
the life of the Spitfire.

It was from this experience that the otherwise unlikely development of a
land-based fighter, the Spitfire, emerged from the drawing-board of a designer
whose only other land-based aircraft (apart from the immediate predecessor of
the Spitfire, Type 224) was a one-off response to an Air Ministry light plane
competition ten years before. However, the link between Mitchell's specialised
high speed Schneider Trophy designs and the Spitfire was not a direct one, as is
sometimes stated or assumed, and so it is also the purpose of this book to trace
the history of the Schneider contests and their aircraft, in order to arrive at a
more accurate assessment of the impact of these events upon Mitchell's career
and its culmination in the Spitfire.

Nor was Supermarine's winning of the Schneider Contests (four times out
of five in the eleven events) *entirely* due to the necessarily superior qualities of
the Mitchell designs. Chance played a considerable part in the Schneider Trophy
saga and, in particular, the state of various nations' aero engine industries at
vital moments. It was also fortunate that the major international aeroplane
competition during the main part of Mitchell's design career turned out to be
specifically for water-based aircraft that were, at least, closer to the main
concerns of his company than a competition for land-based racers would have
been. And had not America made a sporting gesture in 1924 and had Mussolini
not decided that his dictatorship would be well served by an Italian victory in
1925 (*see* pp. 116–117), the competition might very well have already been won
outright and there would have been no further contests to initiate the three high-
speed designs which so established Mitchell's high reputation in the aircraft
industry and beyond; the Walrus fleet spotter and perhaps the very promising
bomber that he was working on at the time of his death would have been his
wartime legacies, not the Spitfire.

Mitchell's Work Ethic and Temperament

At the age of 36, Mitchell was described in Supermarine publicity as 'one of the
leading flying boat, amphibian and high-speed seaplane designers in the country'; he had also been invited to give a talk on the BBC, had been elected a
Fellow of the Royal Aeronautical Society and awarded the CBE.

These successes, which were achieved before his early death at the age
of 42, clearly suggested that he might well be entrusted with the design of an
outstanding fighter when the need arose but it was especially fortunate that he
had become involved with the design of the later Schneider Trophy machines
which gave him *unfettered* opportunities to extend the boundaries of high speed
flight. As a result, the advent of the Spitfire prototype of 1936 marked a dramatic
increase of more than 100 mph over the most recent RAF fighter in service
and led to an even more dramatic and unprecedented initial order of 310, three
months after its first service trial. He died, however, without seeing the fighter go
into squadron service and without knowing that nearly 23,000 examples were to
be built and in a multitude of variants.

In the pages which follow, these progressive stages in the remarkable career of R. J. Mitchell will be described. But, first, it will appropriate to describe something of the character and capacity for hard work which produced these significant landmarks and this can be especially appreciated when one notes other makers' less successful responses to the challenge of the Schneider Trophy. The successes of his Trophy aircraft indicate his attention to design detail as well as his ability to produce, to a very tight schedule, effective solutions to the many problems which resulted from engineering at the very forefront of aircraft technology. After he became both Chief Designer and Chief Engineer at Supermarine, he was responsible for the appointing of virtually all the design team which had been built up between 1920 and 1935; thus the team that he had collected around him and his own hardworking and conscientious example were the most significant factors producing the reliability of his aircraft.

The previous list of aircraft for which Mitchell and his design team were responsible did not include a multitude of projects which never left the drawing board or the three very large uncompleted projects which occupied a considerable amount of design time but which never flew: the Scylla, cancelled after the hull was completed, the Type 179 Giant, cancelled before the hull was completed, and the prototype Bomber whose design was overseen by Mitchell and later abandoned after the fuselages were destroyed by enemy action.

If these three uncompleted projects are added to the list of distinct types which actually flew, it is worth noting that, in the seventeen years that Mitchell was active as Chief Designer, he had had overall responsibility for twenty-four different aircraft types. Joe Smith (who took over as Chief Designer after Mitchell's death) summarised his predecessor's professional capability and output in the following way: 'Thinking back, I have realised that no other man of my experience has produced anything like the number of new and practical fundamental ideas that he did during his relatively short span of working life.'

This varied output, which was extended to landplanes in 1924 with the Sparrow, was described by Arthur Black, Mitchell's Chief Metallurgist, as follows:

> In the sixteen years after he became Chief Designer at the age of twenty four, he designed the incredible number [of twenty-four machines] ranging from large flying boats and amphibians to light aircraft and from racing planes and fighters to a four engined bomber. This diversity of effort and its amount marks R. J. Mitchell for the genius he was.

Even to the casual observer, it must be clear that such an output suggests that a capacity for hard work was one of Mitchell's character traits. Alan Clifton, who knew Mitchell from when he joined him at Supermarine in 1923, has left an appreciation of Mitchell which attests to this aspect of the man but which also indicates other qualities that were necessary to the success of his career: he recorded how Mitchell would visit the drawing office and study someone's detailed drawing, head on hands, thinking rather than speaking. In reply to questions, a small group would gradually gather round until some conclusion was reached; Mitchell would then move on to another board to repeat the process.

Harry Griffiths, who joined Supermarine as a laboratory assistant in 1929, has left the following observation:

> When a problem was being discussed in the drawing office he would stand by the drawing board listening to all the arguments as to what should be done – on these occasions he had the habit of rolling a pencil back and forth on his hand (it was always a very black pencil!) – and when he had heard enough he would push everyone aside, draw a few lines on top of the existing drawing saying, 'This is what you will do,' throw the pencil down and march back to his office.

Ernest Mansbridge, who joined Clifton in 1924 to work on stressing, remembered Mitchell for a similar method of dealing with a problem by calling in the leaders of the relevant groups and getting them arguing among themselves. He would listen carefully, making sure that everyone had said what he wanted to, and then either make a decision or go home and sleep on it. Joe Smith put the matter in this way: 'his work was never far from his mind and I can remember many occasions when he arrived at the office with the complete solution of a particularly knotty problem which had baffled us all the night before'. In fact, Mansbridge expressed the suspicion that, with many problems, Mitchell's discussions were essentially a means of ensuring that he had not overlooked anything and that, otherwise, he had already reached a decision beforehand.

He had also been appointed Chief Engineer in 1920 and Arthur Black has given a glimpse of his daily inspection also of the manufacturing process:

> I remember how R.J.'s well built figure, medium height with fair colouring, could be seen in the workshop each morning, studying with complete concentration the developing shape of the aircraft being built. He would walk round it and study it from all angles, now and then examining a detail minutely. I sometimes wondered if he was aware how closely he was watched for some clue as to what his reactions were going to be. If he was satisfied, then he would pass on to the next job; but if he was not satisfied, then much of the design work and manufacture might well have to be done again. But his outlook was strictly practical and having discussed the matter with those concerned, a satisfactory compromise was usually arrived at.

Griffiths has also left the following anecdote from 1927 concerning attention to detail which is also indicative of why Mitchell was respected in the firm:

> In the S.6 the fuel was carried in the floats and was pumped up through the struts to the engine. In level flight this would have been O K but during the race the aircraft was banked through 80 degrees in order to negotiate the sharp bends of the course and this created such high centrifugal force that the fuel supply would have been cut off. Thus a small header tank was located in front of the engine to hold a reserve of fuel sufficient to maintain a supply during turns, and the pumps were arranged to deliver an excess of fuel. This meant that on the straight part of the course some fuel had to be returned to the float tanks.
>
> A valve on the front of the header tank had two spring loaded ports which were supposed to split the overflow into equal parts for return to the floats, but inevitably it all went down one pipe resulting in a potential out of balance …
>
> We tried all sorts of combinations of spring loaded valve flaps, differing pipe sizes and other devices to equalise the flow without success and the race was getting nearer every day.
>
> One Sunday morning [n.b.], near to exasperation, we were fitting yet another variation when Mitchell came along and stopped to have a look. At the top of the valve housing there was a small hole leading into the tank which was intended to allow air to escape as fuel went in.
>
> He pointed to the hole and asked, 'Why is that there?' and hearing that it was an air bleed was quiet for a few moments. He then said, 'Stuff it up'. I was sent to the stores to get an aluminium rivet of the right size and we hammered it in. We then reassembled the valve in its original form and switched on the pump for a test run.
>
> Eureka – no matter what we did the fuel split into two equal parts!

A fortuitous result of Mitchell's beginning at Supermarine with no prior knowledge of full-size aircraft design or manufacture was that he had to learn the trade from the bottom up and, with only seven drawing office staff when he became Chief Designer (and there were no more than double that number by 1923), the firm was small enough during these formative years of Supermarine that he himself could ensure that his high standards were applied at all stages of the production of aircraft.

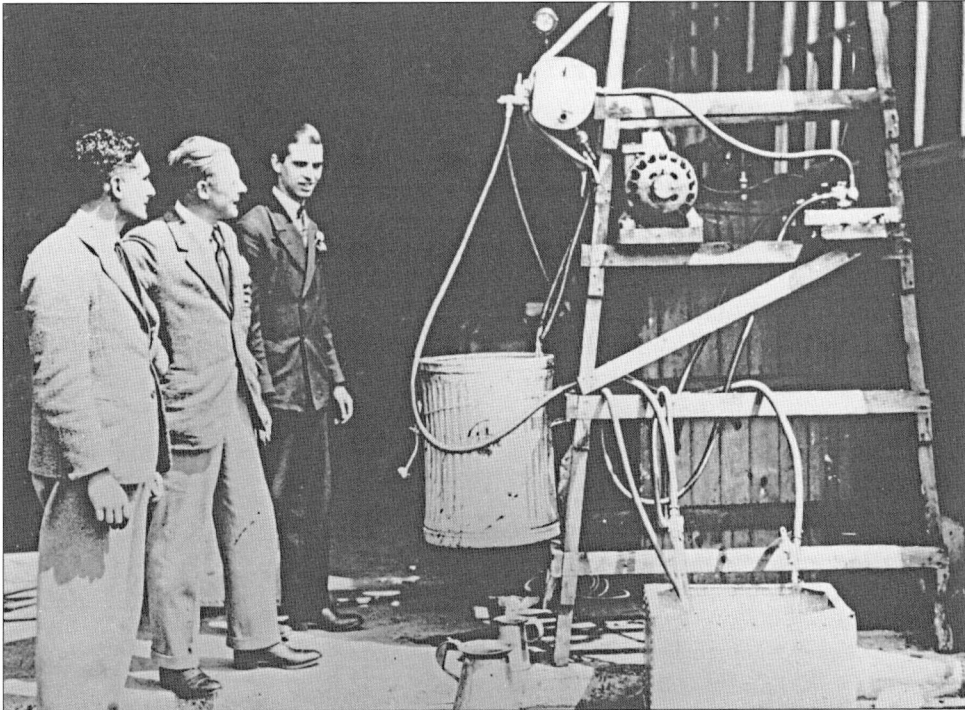

Mitchell (centre) with Arthur Black (Chief Metallurgist, right), with the typically improvised fuel system test rig mentioned opposite– the header tank is top centre, the dustbin simulated the engine and the two square tanks the floats. It had been moved outside because of the fire risk. – the motor, apparently, was not flameproof.

Despite the dominant position he established at Supermarine, and in those more authoritarian days, most accounts of Mitchell's personality agree on his ability to listen rather than parade his own opinions, on his basic shyness (he had a slight stammer), and on his lack of pomposity. One member of the 1931 RAF High Speed team, Grp Capt. Snaith, has described Mitchell's demeanor when one of the racing seaplanes developed a nearly fatal rudder flutter – which resulted in buckled rear fuselage plates and which raised serious doubts about the current Supermarine effort to compete safely in that year's Schneider Trophy event. Practice flying had been stopped and all the experts called in; during the panic and hubbub that ensued, Mitchell sat in a corner hardly saying a word. But it was he who came up with the solution soon afterwards of adding balancing weights to the rudder (*see* pp. 174–175).

Another member of the High Speed team, Flg Off. Atcherley, has also given a similar assessment of Mitchell: 'He was always keen to listen to pilots' opinions and never pressed his own views against theirs … He set his sights deliberately high, for he had little use for "second-bests". Yet he was the most unpompous man I ever met'. And a further example of his open-mindedness – which was to have very far-reaching results – was to be seen during the very early stages of the development of the Spitfire. At this time, when there was no formalised liaison between manufacturers and the air force, Mitchell got the Vickers Chief Test Pilot, 'Mutt' Summers, to arrange a special visit to the Aeroplane and Armament Experimental Establishment at Martlesham Heath so that he could hear the views of the RAF test pilots on the merits and shortcomings of the current fighters in service.

This self-effacing willingness to listen to all points of view, however, was not matched by a readiness to bear fools gladly. Most accounts mention his shortness with those who did not get his message quickly enough. For example, Joe Smith said that 'R.J. was an essentially friendly person, and normally even-tempered, and although he occasionally let rip with us when he was dissatisfied with our work, the storms were of short duration and forgotten by him almost immediately – provided you put the job right'.

Mitchell's condition after his operation for bowel cancer in 1933 exacerbated his testiness but, unfortunately for those working with him, 'none of us knew at the time' (reported Denis Le P. Webb, who began an apprenticeship with Super-marine in 1926); Lovell-Cooper (who joined two years earlier and eventually became chief draughtsman) recorded that 'he used to have terrible tempers as you can imagine. He was in shocking distress a lot of the time'. But even before then, it was not unknown for him to contemptuously flick aside a drawing that did not satisfy him and even to tear it into shreds if it particularly displeased him; and his secretary, Miss Vera Cross, reported that he had no time for those who did not measure up to his standards.

Nor did he encourage interruptions when deep in thought at his drawing-board, as Joe Smith has recalled:

> A mental picture which always springs to my mind when remembering him, is R.J. leaning over a drawing, chin in hand, thinking hard. A great deal of his working life was spent in this attitude, and the results of this thinking made his reputation. His genius undoubtedly lay in his ability not only to appreciate clearly the ideal solution to a given problem, but also the difficulties and, by careful consideration, to arrive at an efficient compromise.
>
> One result of his habit of deep concentration was that he naturally objected to having his train of thought interrupted. His staff soon learned that life became easier if they avoided such interruption ... If you went into his office and found that you could only see R.J.'s back bending over a drawing, you took a hasty look at the back of his neck. If this was normal, you waited for him to speak, but if it rapidly became red, you beat a hasty retreat!

The following chapters will show how often his perfectionism resulted in either a new type of aircraft or a new piece of mechanism for an aircraft functioning satisfactorily virtually from the very start – by no means to be expected in these early days of aviation development. Perhaps the best examples of reliability would be the Schneider Trophy racers, 1927 to 1931, where Mitchell was working under considerable time pressures and at the limits of technical know-ledge and yet produced machines which, unlike most of the competition, were not seriously affected by malfunctions. Certainly there were failures of airframe structures, notably with the S.4 and perhaps with Kinkead's S.5 (see Appendix Five), but this was, essentially, attributable to the limitations of contemporary understanding rather than of any neglect of practical matters. Joe Smith describes this application to detail:

> When in the throes of a new design, the arrangement of which had been decided, he would spend almost all his time in the drawing office on the various boards. Here he would argue out the details with the draughtsmen concerned, and show a complete grasp of the whole aircraft ... Construction of the machine having begun, he would spend some time each day examining and assessing the result. If he was not satisfied with the way something had turned out, he would go back to the drawing-office and, having discussed the matter with the people concerned, either modify it or leave it, as the case might be. And always the practical aspects of the proposed alterations would be borne in mind in relation to the state of the aircraft, and the ability of the works to make the change.

However wide-ranging was Mitchell's own involvement in aircraft production, the early appointments that Mitchell made to his design team (*see* details p. 15) indicate the expanding activity of his offices and also the sorts of expertise that the Chief Designer needed to support him as the work of building aircraft became increasingly complex. It is worth noting how these men rose 'through the ranks' to become the leaders of the various design and construction teams; clearly, they had measured up to Mitchell's standards.

At this point, Supermarine's long-serving test pilot, Capt. Henri C. Biard, ought also to be mentioned for, after all, he was the only man in the company who could tell Mitchell how his designs actually worked in the air. He had been a pilot in World War One, became an instructor, and then first came to the notice of Supermarine as a pilot on their Channel services (*see* p. 31). Thus he had had a wide experience of different aircraft types and Mitchell must have listened to his comments carefully about aircraft performance. There is some confirmation of the influence of Biard from a neighbour of his, G. A. Cozens: in the course of his observations on the early medium-sized amphibians, he wrote, 'Once more it was Captain Biard who put the Seal through its trials and [it was] his suggestions and advice that R. J. Mitchell carried out in establishing this aircraft as the first of the long line of Seagulls'.

Capt. Henri C. Biard.

Horseplay and Expletives
Biard also had a reputation as a practical joker and this reminds one of Webb's observation which gives another view of Mitchell's team in the earlier days:

> They were a very high-spirited crowd and given to rather boyish pranks but after all they were youngish and very hard working and so played hard as well. The advent of Vickers [takeover, December 1928] sobered them up a bit – alas – or were they just getting older?
>
> Marsh-Hume [the business manager] was a bit pompous and so he was the inevitable target for the other bright young men in the form of Wilf Elliot [works manager], Henri Biard, Charles Grey [secretary] and R. J. Mitchell, who had been known on more than one occasion to congregate outside Hume's bungalow and serenade him in the early hours in a raucous and unmelodious manner to the discomfort and embarrassment of Hume and his missus and the fury of Hume's neighbours.

No doubt maturity did bring about a more serious attitude but there are various later accounts of practical jokes, including Mitchell's dismantling of a colleague's bed when staying at a hotel and his setting fire to another's notes during his speech. It is also recorded that, when taking his brother out for a drink, he chose a pub frequented by Supermarine workers, who were not at all disconcerted by his arrival. His lack of 'side' and, outside work hours, his readiness to be 'one of the boys' was complemented by his taking an active part in the firm's sporting activities – particularly tennis and cricket. Nevertheless, only the breezy RAF High Speed Flight Schneider Trophy pilots called him 'Mitch'; in the works, 'R.J.' was the limit of familiarity.

On the other hand, Major G. P. Bulman (*see* index) knew Mitchell very well and the following anecdote shows him employing the name 'Reg' used within the family. It also hints at some possibly more earthy epithets used by our designer:

> We happened to run into each other at the Paris [Aero] Salon, in 1930, probably, and together spent the evening by visiting the famous Folies-Bergère (as all good Englishmen do). Having a drink in the bar during the interval Reg (a most shy man) was confronted by a would-be alluring damsel whose attentions he repelled with a note taken from his pocket, begging her to resume her promenade. At the end of the show we went to the cloakroom to regain our overcoats. ... 'My God!', he suddenly exclaimed, 'I must have given my cloakroom ticket to that ruddy girl in the interval instead of the cash I thought I had.' He then spotted the coat still on its peg with huge relief, but the French attendant with typical Gallic logic replied, 'No ticket, no coat,' and was adamant ... After a word between us I started to assail the vigilant Frenchman with a torrent of terrible French and worked him in his gathering fury towards the far end of the counter away from the precious coat ... Mitch leapt over the counter, grabbed his coat and took to his heels. I followed with all speed, the two of us tearing round a corner outside, choking with laughter – one of the world's most famous aircraft designers and a British Government representative [Asssistant Director (Engines) in the Joint Directorate of Scientific Research and Technical Development, Air Ministry].

The film about Mitchell, *The First of the Few* (1942), gives an anodyne view of the designer and his workplace, not only because film dialogue was more

HRH Edward, Prince of Wales in front of the Swan, 27th June,1924. On the Prince's left is Sqd. Cmdr James Bird, director. Next to him, is R. J. Mitchell; Biard is behind him; on far left is Charles Grey, secretary, and next to him in front, is W. Elliot, works manager – the 'singers' mentioned above.

decorous in those days but also because it was made with an eye to raising the morale of its wartime audiences. However, those who worked with him often reported on his blunt, down-to-earth remarks. Apart from the well known remark on the naming of his Type 300 design 'Spitfire' – 'It's the sort of bloody silly name they would give it' – there is his reported comment to Shenstone when the elliptical shape of the Spitfire wing was being discussed – 'I don't give a bugger whether it's elliptical or not, so long as it covers the guns'. His reaction to the Schneider Trophy engines' cutting out, due to excess sealant in the fuel piping being washed loose, was equally blunt – 'You'll just have to bloody well fly them until all this stuff comes out' – and, when Jeffrey Quill had confessed to being overawed by the Supermarine boffins with slide-rules sticking out of their pockets, Mitchell's advice was not for the polite circles of his day –'If anybody ever tells you anything about aeroplanes which is so bloody complicated you can't understand it, take it from me it's all balls'.

These quotations bear out Gordon Mitchell's report that his father's language in men's company 'was sometimes colourful' and no doubt it reflected his early apprenticeship at the Kerr Stuart locomotive works immediately after he left school. On the other hand, the son narrated how, when chairing a meeting which included high-ranking Air Ministry officials and when some 'forceful language' was expressed in front of his secretary, Vera Cross, Mitchell was swift to rebuke the use of such language in her presence.

The First of the Few unfortunately shows little of Mitchell's workday environment; nor does it give any insight into the mind of a practical engineer – showing him gazing at birds for inspiration, despite the fact that the principles of flight had already been established by observing birds gliding at the end of the 18th century by George Cayley. Jeffrey Quill supplied a necessary corrective when he wrote that Mitchell had better things to do other than 'looking at bloody seagulls'. This being said, one often suspects that Mitchell's down-to-earth remarks often disguised the aesthetic considerations behind his designs which later chapters describe.

Eulogy or Otherwise

It has not been recorded how, as a young man, he comported himself during his early years as Chief Designer and Engineer but his coming to terms with the established and unfamiliar world of boat building techniques and their planing characteristics must have been facilitated by a lack of 'side' at a critical period in the fortunes of Supermarine. Thereafter, Mitchell's continued willingness to listen to the growing number of assistants must also have been an encouragement for them to stay with Supermarine and, as the demands of the aviation industry grew more complex, it became increasingly important to take note of their expertise. Arthur Black again: 'he was always ready to listen to a technical argument and remained always open to conviction'.

Almost every one of these men remained with Supermarine for the whole of Mitchell's working life. The effect of working together to produce a very tangible end product and the constant pressure to deliver this product on time, must have been important factors uniting them. Admittedly, their specialist marine field and the years of economic depression during this period would also have reduced their chances of moving elsewhere, but Mitchell's leadership and his less serious moments were no doubt contributory factors in their remaining with Supermarine for the whole of their careers – despite (or, perhaps, because of) Mitchell's demanding nature. Black speaks of 'the wonderful experience of working in an atmosphere of continual achievement' and of being 'grateful to R. J. Mitchell for the technical standards he so ably established'.

On the other hand, it must be recognised that almost all the memories of Mitchell were recorded by men who had been appointed by Mitchell in the early days of Supermarine and promoted by him. They were, thus, always instinctively subordinate to him: when Jeffrey Quill came to Supermarine as test pilot in 1936, he noticed how Mitchell, in spite of having had no formal training in aeronautics, was 'held in some awe by his staff'. Joe Smith, although eulogising somewhat from the distance of seventeen years after his predecessor's death, has indicated what qualities he considered contributed to Mitchell's leadership:

> He never shirked full responsibility, and his technical integrity was un-questioned. He won the complete respect and the confidence of his staff, in whom he created a continuous sense of achievement. He placed himself firmly at the helm, and having made decisions, expected and obtained full cooperation of all concerned. But, in spite of being the unquestioned leader, he was always ready to listen to and to consider another point of view, or to modify his ideas to meet any technical criticism which he thought justified … The effect of this attitude on the team of young and keen engineers which he collected around him can well be imagined.

Today, design complexities are such that many might have reservations about the idea of one man completely dominating the design output of a company, but it should be borne in mind that, during Mitchell's lifetime, it was still possible for one man to have a complete grasp of all the detail that went into the making of the aircraft of the 1920s and 1930s. As Alan Clifton has said: 'R.J. was widely considered the greatest aeroplane designer of his time when one man's brain could carry every detail of a design'.

However, a special mention ought to be made of the input of Beverley Shenstone after 1931. Whilst he had startled Mitchell at his interview by presum-ing to suggest significant alterations to the wing of the Giant flying boat cur-rently under consideration, he was taken on two weeks later on a temporary basis and offered a permanent position soon afterwards. Unlike others in the design team, Shenstone was eleven years Mitchell's junior, being aged 25 when he joined Supermarine, but bringing with him experience of German aerodynamic theory that was well in advance of contemporary British practice. It would seem that Mitchell soon recognised that the new man might well allow Supermarine to progress beyond their already acknowledged lead in flying boat design and was not so flushed by his own successes that he did not appreciate what the younger man Shenstone might contribute. It is most likely that the revised scheme for the proposed Giant (*see* pp. 198–204), the then revolutionary move to the thinner wings of the Stranraer and the Spitfire, and the renewed attention to drag redu-cing measures, owed much to his input. His interest in elliptical wing shapes was highly relevant to the developing outline of the Spitfire wing and his interest in flying wings was, in all probability, behind the move to a deltoid wing shape for the later bomber proposal (*see* Chapter Nine).

Nevertheless, Shenstone always acknowledged that design decisions were thrashed out with Mitchell and the others, and thus it is still correct to regard the later, more complex, aircraft designs as deriving essentially from Mitchell him-self. He had to be convinced of the rightness of the arguments of others but it would appear that he retained his open-mindedness to the end. Thus, whilst the size of his design team gradually expanded, Mitchell was always able to make the main conceptual decisions and also to oversee and influence the detailed work to such an extent that it is quite justifiable to speak without undue over-simplification, in the chapters which follow, of *his* designs.

Some of the last words on the matter of Mitchell's influence might well be left to the record-breaking pilot, Alex Henshaw, who gave the following brief account of his first impressions of Supermarine when he joined them shortly after Mitchell's death: 'To start with, it was on a smaller scale and less affluent and ostentatious in its general mode of work and most of the operatives were ordinary people who had been in amphibious aviation for a long time and were dedicated to their work'.

But perhaps the most telling example of the way Mitchell built up this team to implement the requirements of Supermarine and its customers was his reaction to the arrival in the design office of the Vickers designer, Barnes Wallis, in his projected role as Chief Assistant to Mitchell. This transfer had come about as part of the reorganisation that took place when Vickers acquired Supermarine in 1928; by then the passion of Mitchell and his team for engineering detail, their ability to get their aircraft into the air on time and to have won the Schneider Trophy Competition in Venice was an obvious attraction to the management of Vickers when looking for expansion in the aviation field. Accordingly, Mitchell's contract to remain with Supermarine until 1933 was retained and his design team was kept in the new organisation as an entity, with the exception of the proposed addition of Wallis who already had a good reputation in connection with the design of the R100 airship.

Mitchell is reported to have suggested to his staff not to make Wallis too comfortable and, when Wallis began to interview the various members of the design staff in order to form an assessment of those he would be dealing with, Mitchell disappeared from the works and was not seen again for nearly two weeks. His reappearance after the Vickers board recalled Wallis to their Weybridge factory might tell us something about the prima donna nature of chief designers; it certainly indicates the status of Mitchell in the expanded company set-up and, in the present context, suggests a great deal about Mitchell's faith in the integrity of the design team which he had built up and shaped over the previous nine years. (An interesting sideline to the Wallis interlude was reported by Shenstone in his diary: when asked by Mitchell, about three years later, to obtain from Barnes Wallis some information about strength testing that has just been carried out on a Vickers aircraft, the answer received was a blunt 'No'.)

In another incident, the chief designer threatened the works superintendent, Trevor Westbrook, that he would walk out when Westbrook refused to stop the work of pneumatic drills just below Mitchell's office during normal working hours; he said that he would be at home if the Managing Director asked where he was. At this, Westbrook, who had also been sent down from Weybridge at the time of the Vickers takeover and was a formidable personality in his own right, capitulated.

It might be added that, despite his willingness to listen to others, to lead by example, and to associate with the workforce outside working hours, his family background must have played a vital role in his success. His father had been a headmaster, he then became managing director of his own printing firm, and rose to a prominent position in the local Freemason fraternity. Victoria Cottage, Normacot, in which Reginald grew up, was detached and serviced by a maid and so it is very likely that the son did not find it too difficult to give orders and to expect them to be carried out even though he was appointed Chief Designer and Chief Engineer when he was only twenty-five.

Perhaps his being in a position of authority in the workplace carried over into his domestic life; Mitchell's son remembers that his father was 'damned difficult to live with' and that there were occasionally 'some pretty awful rows'. On the other hand, Webb also recalled a different side of Mitchell's character: when still a very junior apprentice he found that Mitchell 'was friendly and pleasant' and that he put him completely at ease: 'It was quite obvious that R.J.'s successes had not gone to his head and never did. Later, if he saw me footslogging over to Southampton and he was making his stately way in his Rolls-Royce, he would not be above offering me a lift'. The Schneider Trophy pilot, Fl. Off. R. L. R. Atcherley, spoke in somewhat similar terms: 'When I first met him, I was struck by his young and carefree looks. He was a man with an alert and inquisitive mind, and in spite of his very considerable attainments in the world of aircraft design he was always ready to crack a joke or take on anyone on his own wavelength.'

In the presence of strangers or women (he left interviews of female staff to others), a certain remoteness was the result of shyness and a slight stammer which increased his dislike of public speaking – something that was more demanded of him as his success as a designer increased. Nevertheless, Joe Smith remembered that he could also be charming with 'an engaging smile which was often in evidence and which transformed *his habitual expression of concentration*' [my italics]. Mitchell's son has also left a boyhood anecdote which relates to this 'concentration': having been shown round his father's workplace he was asked how he had got on; to the reply that he had enjoyed it, the father rejoined, 'I don't care a damn whether you enjoyed it, I want to know what you learned'.

Harry Griffiths has also supplied a reminiscence of Mitchell which encapsulates some of his apparently contradictory character traits and foibles:

> 'R.J.' was human like the rest of us – he could be moody but in general he had a pleasant personality and I always had the impression that he was somewhat retiring yet he was decisive and when necessary could be very firm.
>
> He had a small personal staff consisting of a clerk, two typists and an office boy – they were all loyal to him and understood his moods. When any unwanted visitor asked to see him he would tell his staff, 'I'll see him in ten minutes' and they knew that this meant 'Get rid of him!' It worked well until a new typist arrived and the visitor was ushered in after precisely ten minutes! It only happened once.
>
> I've already said that his office was immediately over the laboratory and occasionally he would come downstairs to see Arthur [Black] and would always stop and ask how I was getting on. Sometimes these visits would be to ask the boss if he fancied a game of golf and off they would go for the afternoon. On another occasion he came and played merry hell because the office was untidy, although in fact it was no worse than usual …

According to his son, practical matters such as money were left to his wife and she would hand out cash for his personal requirements and replace it when required. She also was accustomed to his talking at one moment and to his being miles away the next; and she also soon learned to contact his personal secretary when preoccupation with some design problem led to the evening meal at home going cold. And as Miss Vera Cross grew into her job as his secretary, she soon organised his very imperfect filing system, learned how to prevent interruptions and relieved him of the main burden of correspondence, which he hated – although she had often to wait beyond office hours before letters were signed.

Sir Robert McLean, the managing director of Vickers (Aviation) during the whole of Mitchell's working life, made an interesting observation on Mitchell's character: 'He was a curious mixture of dreams and common sense'. The latter trait has already been fully described but the mind which produced the startling advances and aesthetics of the revolutionary S.4 and of the Spitfire must also be attributed to this otherworldly side of his nature. Apart from seeking a mental break from the inevitable minutiae of aircraft design or the later demands of becoming a director of the firm by taking time off for golf, he would, on a nice afternoon, also slip away for a few hours' sailing. This absenting himself is not necessarily at variance with previous accounts of his fierce work ethic but must surely have been a necessary part of the other-worldliness that Sir Robert spoke of: bearing in mind his well-known concentrated stance at his drawing-board, as described by Smith earlier, and the intuitions that prefigured his many groundbreaking designs, Yeats' lines about prominent persons in history put the matter rather well:

> His eyes fixed upon nothing,
> A hand under his head.
> Like a long-legged fly upon the stream.
> His mind moves upon silence.

* * * * *

Mitchell's Senior Design Team

Arthur Shirvall joined as an apprentice in 1918 and was later attached to a qualified naval designer who had been seconded to help Supermarine with hull design; Mitchell greatly admired his drawing of hull lines ('Shirvall always makes things look very nice') and he was eventually put in charge of Hydrodynamic Hull Design and Tank Testing;
Harold Smith was taken on as an apprentice at about the same time and rose to become Chief Structural Engineer;
Jack Rice was apprenticed in 1922 and eventually became Head of Electrical Design;
Joe Smith, after an apprenticeship with Austin, joined in 1922 as a draughtsman; Mitchell made him Chief Draughtsman only five years later; after Mitchell's death he became Chief Designer and was responsible for the development of all the Spitfires after Mark I;
Alan Clifton had replied in 1922 to an advertisement for a 'mathematician for strength calculations'; Mitchell, apparently, preferred to carry on doing all his own stress calculations rather than face the appointment procedure but he did see Clifton the following year, by which time he had a degree in engineering; he later became the Principal Assistant to Joe Smith, then Head of the Technical Office of the Design Department, and finally Chief Designer after Smith's death;
Eric Lovell-Cooper joined in 1924 as a draughtsman, coming from the Boulton Paul aviation company; he later became Chief Draughtsman;
Ernest Mansbridge also came in 1924, with an engineering degree; he worked with Clifton on stressing but later was directly responsible to Mitchell for aerodynamics, performance, weights and flight testing; his later duties were predominantly Flight-Testing and Performance Estimation;
Jack Davis joined as apprentice in 1925 and later went for experience to the aircraft manufacturers Westland and Boulton Paul for two years; he later became Senior Design Draughtsman;
Arthur Black was appointed metallurgist in 1926; at this time much aircraft construction was changing from wood to metal and Supermarine was one of the first firms to employ such an expert; Black later became Chief Metallurgist;
Beverley Shenstone, joined at the end of 1931 and was the first appointee to hold a degree in aeronautical engineering; he rose to the position of Chief Aerodynamicist.

R. J. Mitchell at the time of his wedding, 22 July, 1918.

— *Chapter One* —

1895 to 1919
Early Days

Reginald Joseph Mitchell was born on the 20th of May, 1895 at 115 Congleton Road in the Butt Lane district of Kidsgrove; in the same year, the family moved a few miles to Longton, one of the the 'six towns' soon to be constituted as Stoke-on-Trent: first to 87 Chaplin Road, in the Normacot district, and then to the nearby Victoria Cottage at 1 Meir Road, Dresden, where Reginald grew up. He died only forty-two years later on 11 June 1937; nevertheless, from 1919, when he became Chief Designer at Supermarine Aviation, Southampton, his relatively brief career spans the whole development of aviation since the pioneering days until just before the beginning of the jet era.

During this time, the performances of aircraft and the expectations of those using them changed so rapidly that any designer, looking back over a career during these years, must have felt privileged to have been part of an industry at the most crucial part of its development. A simple illustration of this rapid advance in aircraft technology might be a comparison of the technology and aerodynamic practice that Mitchell inherited when he modified an aircraft for his first Schneider Trophy winner in 1922 with that which had been developed for his winning design in 1931.

The first of these Schneider machines, the Sea Lion II, had a wooden hull that was a fine example of the boat builder's craft:

Sea Lion II.

It had the usual biplane arrangement of the wings, separated by struts and braced by wires, as this approach was the almost universal way of achieving the necessary wing area for the lightest structure; the stressing considerations involved at that time being akin to those of the bridge builder. The flying surfaces were fabric covered, again for lightness, and the power available to drive them through the air was 450 hp; it was, nevertheless, able to reach a maximum speed of

160 mph and to establish the first World Speed Record in the maritime aircraft class, in the course of its 1922 Schneider Trophy win.

By contrast, Mitchell's S.6B, which won the Schneider Trophy outright in 1931, was of an all-metal, cantilever, monoplane structure with metal skinning to the flying surfaces that owed little directly to other industrial practices and was the result of exhaustive wind tunnel testing. The power available to this aircraft was now 2,350 hp and it gained the World Absolute Speed Record at 407.3 mph. Over this very brief span of only nine years, the increase in maximum speed and power can be averaged out at 26 mph and 211 hp per year (The intensive wartime development of the Spitfire over a similar period was actually less – 11 mph and 144 hp per year.)

Supermarine S.6B.

Schooldays

It is worth recording that, as recently as the 17th December, 1903, the year that Mitchell began elementary school, the Wright brothers made the first powered aircraft flights, lasting less than one minute in duration. Only six years later, progress was such that Blériot made a stir by flying across the Channel and aerial activity came to Britain quite soon afterwards, with Alliot Vernon-Roe being credited with making the first flight by a British-designed, -built and -powered aircraft on July 12, 1909.

In the same year, air shows were organised in Doncaster and Blackpool and, in 1910, the young Mitchell would be caught up in the local interest in flying, as crowds flocked to aviation meetings which took place closer to home at Wolverhampton and Burton-upon-Trent. Even nearer, a Wright machine was put on display at the Hanley (Stoke-on-Trent) Park Fete in the same year. By this time, Mitchell was just turning fifteen and his enthusiasm for the air would have been further stimulated by the flight of Louis Paulham, in a Farman biplane, which passed no more than twelve miles west of the family home on the way to winning the *Daily Mail* London-to-Manchester competition. Two years later, another early aviator, Gustav Hamel, came to nearby Stafford and to Stone, to which special trains were organised, and he also came as near as Longton for the Whitsuntide fete.

The young Mitchell was known to have had a passion for building model aircraft, no doubt informed by press photographs and the displays of the very earliest aircraft, particularly the successful machines of the Wright brothers, Farman, and Blériot, which were prominent in the events organised in his area. But an account that Mitchell's models 'swooped and dipped' (which full-size machines seldom do outside airshows) would seem at variance with the approach of a lad who, at the age of sixteen, made his own lathe and, later, a dynamo. It would seem more likely that his obsession with model aircraft would be directed more technically towards understanding the principles of aerodynamics, as exemplified in straight, level flight – a good preparation for his first Supermarine aircraft which had to satisfy the Air Ministry inspectors that it could 'fly itself' for at least three minutes (*see* p. 40).

Sidney Camm, designer of the future Hurricane, showing a similar interest in model flight.

There was a short-lived aircraft firm nearby at Wolverhampton (the Star Aeroplane Company, which, in 1910 offered a monoplane based on the Antoinette aircraft and a biplane based on the Farman type) but there is no evidence that the young Mitchell ever visited it. And, even if this had been countenanced by the family, any such precocious visits to manufacturers with longer-term prospects, such as Sopwith or Shorts, would have involved travelling considerable distances from Stoke-on-Trent – at a time when even motor transport was in its early stages: the year of Mitchell's birth saw the very first car journey in Britain and Herbert Austin began car building in Britain ten years later when Mitchell was about to go to Hanley High School.

But, at least, the young boy had the advantage of a good educational background as his father, Herbert Mitchell, a Yorkshireman, had moved to the Potteries to become a headmaster at Longton. The son also had the advantage of a more practical education as his father subsequently became a master printer and, eventually, a managing director of a printing company. In view of the aesthetic aspects of his design work, mentioned later, it should be mentioned that Reginald's younger brother, Billy, was to set up his own business designing patterns for the local pottery industry and that his nephew, Jim, became an artist whose aviation prints of his uncle's Spitfires sold nationally.

It is also noteworthy that, having decided to read all the novels of Walter Scott, it is reported that – whilst still a schoolboy – he persevered to the bitter end. It is not recorded if his father had anything to do with this perseverance but he would certainly have approved of Reginald's sticking to the task as he was known to demand high standards of conduct and application.

Apprenticeship

Similar perseverance was required when the son took his first major step away from his father's world and towards his own future career by being apprenticed in 1911 to the locomotive engineering firm of Kerr, Stuart and Co., in Fenton, another of the Potteries towns. Shipbuilding, bridge building, or textiles might just as well have provided an engineering introduction to the world of industry at that time: the local locomotive maker would have similarly seemed to offer just as good and as stable a career beginning: by the time Mitchell left the company, their narrow and standard gauge engines had been sold as far afield as California,

Chile, Mexico, China and India – and examples have been maintained at the Talyllyn, West Lancashire and Leighton Buzzard Railways. Reginald's early training in the workshops must have been a culture shock to a lad brought up in a middle class environment (including the works of Walter Scott); returning each day covered with the oil and grime of the engine sheds was not to his liking. He was all for abandoning the apprenticeship but his father would have none of it and Reginald stayed on. At least he won one minor victory when his foreman likened the tea that Mitchell had made to urine (or words to that effect). The foreman was much better pleased with a second mug which Reginald had personally doctored accordingly.

Looking back from Mitchell's later appointment as assistant works manager at Supermarine's with its nononsense working practices (*see* later), these early days were not wasted; more important, however, was his move to the Kerr Stuart drawing offices and his attending the Wedgwood Burslem Technical School for evening classes in engineering drawing, mathematics, and mechanics. This more cerebral work clearly matched his potential better as he was awarded one of three special prizes presented by the Midland Counties Union.

By the time he was twenty-one, he had completed his apprenticeship and the First World War had been raging for two years. He made attempts to join the forces but his engineering training was considered more useful in civilian life – where he undertook some part-time teaching at the Fenton Technical School. Perhaps his wish to join the war effort, and certainly his interest in flight, which had also been expressed by the keeping of homing pigeons, was soon to become a reality when he then took the fateful decision to apply for the post of Personal Assistant to the Managing Director of the Pemberton Billiang aviation works at Woolston, Southampton.

This small company was fully engaged in the war effort and their main interest lay in marine aircraft, so Mitchell was not only applying for employment in a relatively esoteric form of engineering but also in the doubly remote one of seagoing aircraft. One might say that this type of product was triply remote as 'hydro-aeroplanes', as they were then called, were less developed than the early landplanes: it was only on 28 March 1910, that Henri Fabre made the first take-off from water by a powered aircraft and in the January of the next year that Glenn Curtis took off from water in San Diego with a more practical hydro-aeroplane. The first British aquatic events took place in November, 1912, and involved an own-design machine on Lake Windermere and a converted Avro landplane at Barrow-in-Furness. An eyewitness account of the latter's first flight, on the 18th, was reported in *The Aeroplane*: 'The machine left the water several times, just rising clear, so that Commander Masterman saw daylight under the floats. He computes the distance travelled to be 50 or 60 yards in these skips.' The hydro-aeroplane then proceeded to capsize. It was repaired, however, and flown successfully with the newly formed Lake Flying Company of Winder-mere, which was contracted by the Royal Naval Air Service as the primary developer of seaplanes. (Much to the annoyance of Beatrix Potter and Canon Rawnsley, founder of the National Trust.)

The Advent of the Schneider Trophy
By this time, landplanes were beginning to emerge as reasonably reliable and efficient and so, on September 5, also in 1912, at the Aéroclub de France, Jacques Schneider (*see* Appendix Four) announced a trophy contest that was designed to develop aircraft which could also operate efficiently from water. As there was no airport infrastructure for landplanes but plenty of sheltered shipping ports with (flat) areas of water from which to operate, it would seem only

sensible to consider developing seagoing aircraft for the next generation of transport vehicles.

Thus the Schneider Trophy, as it soon came to be known, was envisaged not just as a speed event but as a competition to find the best practical marine machines – hence the flotation and water navigability tests that had to be successfully completed prior to the main flying event. The contest was to be held annually, to take place over open sea and over at least 150 nautical miles. Entries were to be limited to three aircraft selected by each country's aero club with the winning club to organise the following contest; the country gaining three wins in five years became the outright holder of the trophy and a substantial cash prize of 25,000 francs was offered to the winning pilot of each of the first three events. Competing aircraft had to be seaworthy, as demonstrated by various tests speci-fied by the organising club, such as navigability on water and taking off and landing from water. As the competition was never to be just a straightforward race, the competitors were to start at intervals, decided by lot, to avoid bunching at turn points. Nevertheless, the main flying part was a contest of relative speeds, as suggested by the form of the trophy, a sculpture of a winged figure kissing a zephyr recumbent on a breaking wave. The original trophy (now in the Science Museum, London, beside the winning S.6B floatplane) is a large and handsome art nouveau creation in silver and bronze, the wave over 22 inches wide, mount-ed on a pedestal of dark-veined marble:

A small scale replica of the Schneider Trophy presented to Mitchell after his designs won the three consecutive competitions, 1927, 1929 and 1931; small plaques on the sides of the replica record the various Supermarine wins and the pilots involved. [This replica sold in 2010 for £27,000.]

The first competition for the new trophy, on April 16, 1913, saw only four contestants, three representing the Aero Club of France and an American. Their aircraft all betrayed their landplane origins, especially the Deperdussin, entered by Maurice Prévost, which had a relatively streamlined monocoque fuselage but was equipped with an improvised float undercarriage supported by a maze of bracing wires and struts; Dr Gabriel Espanet and the American, Charles Wey-man, were both in Nieuports which also had very experimental-looking floats.

(The more efficient vee-bottomed floats, with their inherent shock-absorbing characteristics, did not appear until much later.) In view of a preponderance of biplanes in the next few competitions, it is interesting that the two aircraft types which completed the race were monoplanes, Garros being hardly slower than Prévost, even though his Gnome rotary engine was only half as powerful as the other's 160 hp model – the four turns on each of the 28 laps tended to equal out the power differences, especially as the indifferent handling of these early aircraft with cumbersome floats made cornering at speed difficult. All these aircraft had a third float under the tail.

French domination of the first Schneider competition was followed by a similar result in the next Gordon Bennett Cup event at Rheims but such winning ways came to an end at the second Schneider Trophy contest. It was held on April 20, 1914 over the same restrictive course as on the previous occasion although there were now no taxiing requirements, merely two touchdowns and take-offs within a specified distance to serve as the seaworthiness tests. The number of participants was higher, with a team of three French and a Swiss, who entered with the first of the many future Schneider Trophy flying boat machines.

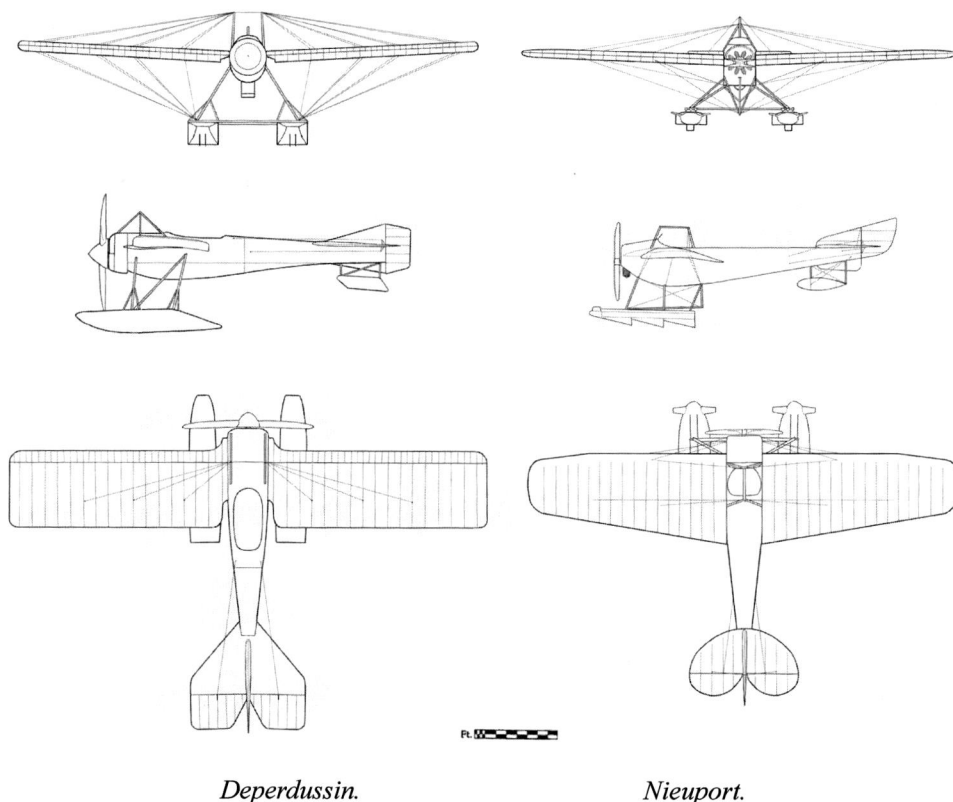

Deperdussin.   *Nieuport.*

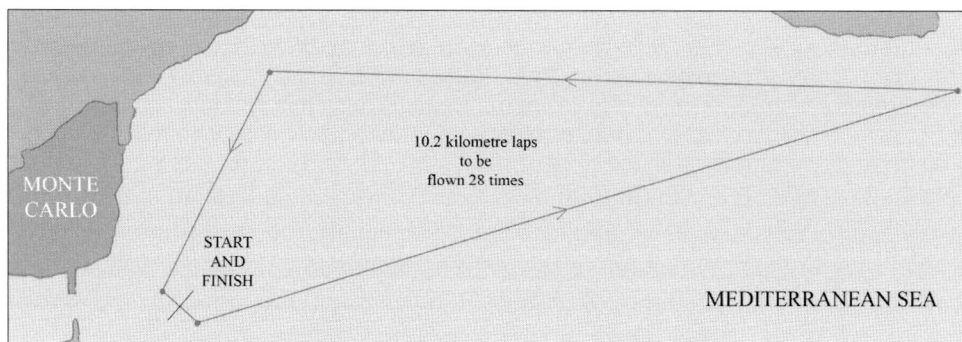

Schneider Trophy Course, 1913 and 1914.

New Schneider Trophy Attracts Aviators. Monaco, April 16, 1913

The two weeks of flying competitions concluded today with the contest for the new hydro-aeroplane cup and first prize of FF25,000, offered by Jacques Schneider, son of a wealthy arms manufacturer.

Contestants had to demonstrate their mounts' seaworthiness by taxiing for half a lap and then fly the 173 miles of a 28 lap course over Roquebrune Bay. Taking part were three French contestants and an American.

Maurice Prévost went first in a Deperdussin; Roland Garros started second in a Morane-Saulnier but had difficulty in taking-off, drenched his engine and had to be towed back to harbour; Gabriel Espanet in a Nieuport got away third, followed by the American, Charles Weyman, also in a Nieuport.

Espanet had to drop out with a broken fuel pipe; Prévost completed the course but he had taxied across the finish line when the rules decreed that he should have flown across. With 24 of the laps completed, Weyman was forced to alight with a burst oil pipe which put the engine beyond the possibility of repair that day.

Meanwhile, Garros had dried out his engine and rejoined the race but he was beaten into second place by Prévost who had taken off again and completed another lap, this time flying across the finish line; his total time produced a competition average speed of 45.75 mph, although his actual air time gave him 61 mph.

Briton Wins Second Schneider Trophy Contest. Monaco, April 20, 1914

Today, Howard Pixton added the Schneider Trophy to his other successes although his Sopwith Tabloid was the smallest and lowest-powered aircraft in the race. The aircraft had clumsy makeshift floats and a peculiar 'sit' on the water but in practice its take-off was seen to be smooth and swift compared with the sluggish performance of the French monoplanes.

The two French Nieuports, flown by Pierre Levasseur and Gabriel Espanet were first away, followed by Ernest Burri in an FBA flying boat flying for Switzerland. The FBA excited the crowd with a long, porpoising take-off, beginning with a series of hops and finally bounding and ricocheting into the air. Pixton took off a quarter of an hour after the two Nieuports and the Sopwith biplane was obviously faster and more manoeuvrable than the others.

After 10 of the 28 laps, Pixton was over 10 minutes ahead of the French Nieuports which were now having engine trouble from the strain of chasing the British plane and both ended up with seized pistons, leaving the race to Pixton and Burri. On his 15th lap, Pixton now began to suffer misfires in one of the nine cylinders of his 100 hp Gnome Monosoupape's rotary engine. Another Briton, Lord Carbery had now joined the contest in a borrowed Deperdussin but he soon retired, again leaving only Burri in contention. Then Prévost re-entered the race but suffered engine failure almost at once; Lavasseur also re-joined the competition on a borrowed machine but he was let down again by his engine after 9 laps. Then Burri was seen to stop on his twenty-third lap but was able to take off again, having taken on more fuel.

One by one Pixton ripped out the drawing pins which were his crude lap counter as his engine settled down on its eight good cylinders and the lap times improved again with Pixton nursing his engine and concentrating upon very accurate flying. In this manner, he managed to complete the course, beating Burri into second place. With an average speed of over 86 miles an hour he had beaten Prévost's average of the previous year by 25 miles an hour.

There was also a Briton, flying a French machine and an all-British interest: C. Howard Pixton, previously a firm test pilot, was entered in the place of their unavailable test pilot, Harry Hawker, flying a special floatplane version of the Sopwith Tabloid, specifically in the hope of publicity for this production-line machine. It was a biplane which T. O. M. Sopwith believed could be made smaller and lighter than an intrinsically cleaner monoplane. The result was a 25 foot span machine with only one set of inter-plane struts on either side and a neatly cowled engine; as it was smaller and powered by an engine considerably lower in power than that of all the other aircraft, it was not considered to be in serious contention. and its 'sit' in the water was even more pronounced and ungainly than that of the previous three float conversion types. Nevertheless, the British designer was vindicated when Pixton's biplane was the winner.

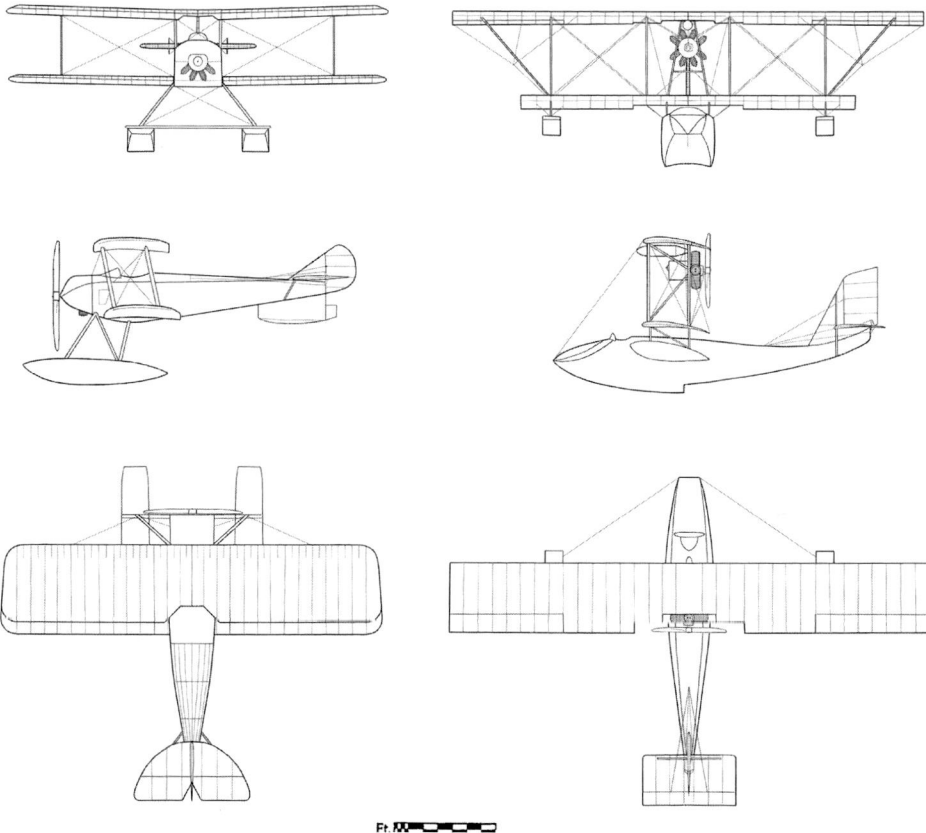

Sopwith Schneider.   *FBA Flying boat.*

Woolston: Nighthawk and other early aircraft

At the time of this second contest, Mitchell was exactly one month from his nineteenth birthday and still an apprentice to the locomotive engineering firm in Fenton; if his mind had already been turning to aviation as a career, reports of the first two Schneider Trophy contests could hardly have been encouraging to his family – of those aircraft which managed to cross the start line only four out of nine completed the course.

World War One then brought such civilian competitions to an end but, at least, it saw a significant increase in the development of aircraft and of water-cooled British aero-engines. Aeroplanes were now becoming sufficiently reliable to play a significant part in warfare, mainly in fighter, reconnaissance, target and gunnery spotting duties, so the demand for aircraft for the war effort now made a career in aviation at least something of a prospect. Nevertheless, it was a doubly bold decision at that time for a provincial lad to go to the (then) remote Woolston, about to be incorporated into the Borough of Southampton for an interview at the Pemberton Billing flying boat firm, engaged since 1912 in making vehicles even more unfamiliar than the newfangled motor omnibus.

As his closest working colleagues later testified to his apparently intuitive feel for what sort of aerodynamic shape would work, his early design of model aircraft must have somehow generated an instinct to head for the unknown world of aviation in the same way that an exceptional person fifty years later, with all the insouciance of youth, would have struck out for a place in the space industry.

On being offered the position of personal assistant to the General Manager, Scott-Paine, it is reported by his son that he straightway asked for his belongings to be sent down to Southampton. No doubt his keenness to involve himself in this new industry and his youthfulness overcame any misgivings he might have had when he saw what a small-time and under-funded operation Supermarine was at that time. The following extract from G. A. Cozens' manuscript, 'Concerning the Aircraft Industry in South Hampshire', describes the humble beginnings (1913) of the firm at Oakbank Wharf, Woolston, which Reginald Mitchell joined three years later:

Extracts from G. A. Cozens' Manuscript
concerning the Aircraft Industry in South Hampshire.

Supermarine seems to have begun almost by accident and in the early stages the unpredictable nature of the firm's founder [Pemberton Billing] and his equally colourful general manager [Scott-Paine] might have diverted the destiny of Supermarine in any one of several directions. The factory was in a part of Mr Kemp's boatyard just above the Floating Bridge on the Woolston side of the River Itchen, and a number of strange contrivances were built there. Mr Kemp often said that it was he who kept the little firm going, and indeed the works facilities like the sawmill were very useful and the workforce, who were largely the Kemp boatyard men at the start, were versatile and able to carry out some unusual projects.

[*G. A. Cozens lived close to the Supermarine works in the early days of the company and was a school friend of one of their workers as well as a neighbour of Biard, their test pilot. In his collection of information about the local aircraft industry, he has left some fascinating, often anecdotal, information concerning its formative years, particularly on the subject of Supermarine. This material reveals how important Supermarine was to this early development of what was a relatively esoteric branch of aviation, as well as how basic were design, building and employment practices – at about the time that R. J. Mitchell joined the firm. The author believes that this extract, and others following, whilst sometimes inaccurate, deserve to be more widely known. He is especially grateful to the Archive Section of Solent-Sky, Southampton, for making available a copy of this Manuscript.*]

It is often stated that Mitchell had arrived at Woolston in 1917; however, the drawing reproduced overleaf, relating to the central nacelle, its gun mounting, and the various cable runs of the P.B.31 Nighthawk and initialled 'R.J.M.', is dated 18 September, 1916. C. G. Grey has written that the young Mitchell 'had been discovered by Mr Pemberton Billing as a competent draughtsman' and the interview must have occurred before September 20, 1916, when the company was sold by Noel Pemberton Billing to Hubert Scott-Paine. The P.B. aircraft in question was completed and first flew (presumably with Mitchell's control cables now installed) in February 1917. The *Flight* and *Aeroplane* obituaries to Mitchell in 1937 both stated that he joined the company in 1916.

Detail from Works Drawing
initialled by Mitchell

The first version of this Nighthawk machine was the P.B.29E and had been created in response to the 1915 bombing raids by German airships. In order to reach the heights attainable by these invaders, an aircraft with a large wing area was required; the usual biplane principle for lightness of construction was applied with a vengeance and resulted in its quadruplane configuration. It crashed soon after its first flight after delivery to naval pilots for handling trials.

Despite the generally unprepossessing nature of the Supermarine establishment, the young Mitchell – builder of small model aeroplanes – must have been

Drawing of the Nighthawk initialled by R. J. Mitchell and dated 1916.

excited to be working on its successor, the new 60 foot span monster quadruplane, the P.B.31, which stood nearly 18 feet high. Just as Mitchell's career ended with, and was brought to the attention of the widest public by the creation of the Spitfire, so his design career that was devoted mainly to marine aircraft started with another land-based fighting system: the large wing area of the P.B.31 was designed to support the weight of a Lewis gun in the nose and, unusually, a non-recoil 1.5 pounder cannon mounted in the top wing pylon, together with another Lewis gun. Equally unusually, there was also a 5 hp engine and generator installed to power a movable searchlight at the very front of the aircraft for searching out airships at night.

P.B.31 Nighthawk under construction. R. J. Mitchell fourth from right?

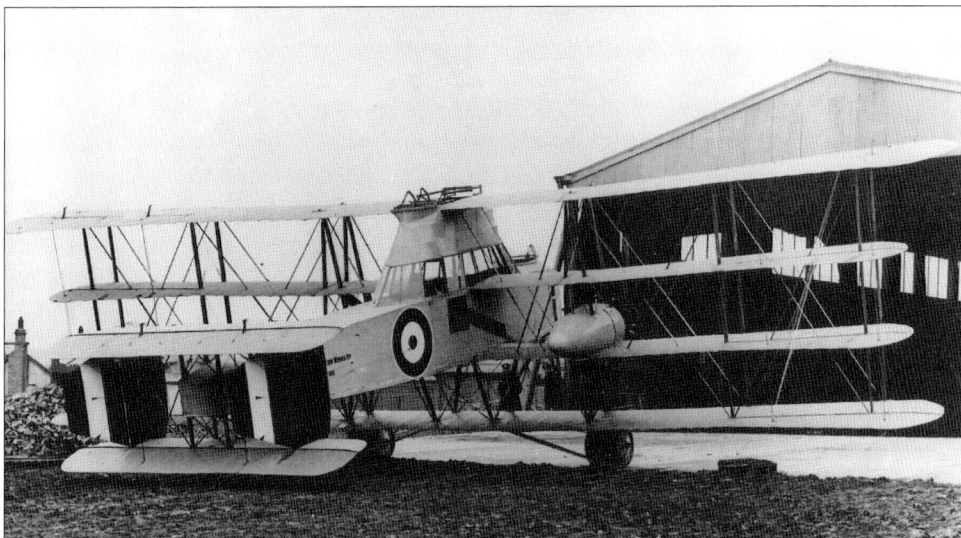
The completed P.B.31 Nighthawk.

Under a well-known American test pilot of the day, Clifford B. Prodger, the aircraft was able to reach 75 mph, with a landing speed of only 35 mph but, underpowered by two 100 hp Anzani engines, it took one hour to reach 10,000 feet (a climb rate of about 3 ft per second). Our embryo designer would have learned immediately how aircraft designs were, more than anything else, at the mercy of engine development and how the need for slow landing speeds had inhibited overall performance when there were none of the modern aids such as prepared

runways or night landing facilities available. Meanwhile, the problem of successfully combating the German Zeppelins had been solved by 1917: the development of an explosive bullet enabled sufficient oxygen to enter the airship's
hydrogen bags for them to be ignited by an incendiary bullet. As both these
bullets could be loaded into the gun magazines of a conventional aircraft, the
development of the heavily armed Nighthawk was not required.

An earlier Pemberton Billing design, the P.B.25, had fared better with an
order for twenty machines and these had a configuration that was to become very
familiar to the Southampton firm: the small biplane with a pusher engine. Like
the P.B.31, it was also a landplane as World War One mainly encouraged the
development of aircraft to operate over the battlefields of France and Belgium.
The embryo company also gained further valuable design and structural
information as the wartime effort involved repairing other firms' aircraft and the
building of others – in particular, twelve Short S.38 and twenty-five Norman
Thompson N.T.2B seaplanes.

Early photograph of Supermarine workers, with tools of their trade,
in front of a Norman Thompson N.T.2B flying boat centre-section. Mitchell is standing,
third from right.

Design experience of a more direct sort came the way of Mitchell for, when
he joined the company, it was completing an order to build some Short Type 184
floatplanes. The association with this firm was particularly important as firms
licenced to manufacture Short's products were supplied with full sets of blue-
prints and had to send their staffs to the parent company for instruction. Other
budding aviation firms such as Fairey, Westland, English Electric and Parnall
also benefited from this arrangement – not to mention Mitchell himself.

The company had also just finished two flying boats specified by another
pioneer aviation group, the Admiralty design team. This unit, from the Royal
Naval Air Station at Eastchurch, together with the one at the Royal Aircraft
Factory at Farnborough, represented much of the contemporary ability to tackle
aerodynamic and structural problems in a scientific manner. This first group of
people was responsible for designing for the war effort but not for construction
and, as a result, they were instigators of the Handley Page heavy bomber and
a range of Sopwith aircraft that were originally intended for naval use. Most of
the early theoretical work, in particular the seminal *Handbook of Stress Calcul-*
ations, came from this source and so it was fortunate for the new company that
some of this leading team was sent down to the works at Woolston to draft out
the details of new naval machines.

The first two aircraft completed to Admiralty designs were accordingly known as A.D. Boats and again exemplified the pusher biplane configuration of the P.B.25. With the addition of the P.B. sort of wing superstructure to the new boat-like hull, there begins to appear the general flying boat formula that was to inform Mitchell's Commercial Amphibian, Sea Eagle, Sheldrake, and which led up to his well-known Walrus. Of especial importance, the particular details of flying boat hull construction also came to the new company: F. Cowlin, the Technical Supervisor at the Royal Naval Air Station has recorded how he went down to the then Pemberton Billing firm and 'learned a great deal about hull design from Linton Hope, who joined the section for a time while we were engaged on the A.D. Boats'. The lines and structure thus laid down by this well-known yacht designer became the basis of all the wooden flying boat structures that Mitchell subsequently utilised and were a considerable advance on those employed by the current naval flying boat, the Felixstowe.

By 1917, the first of the twenty-four production machines to be ordered was undergoing acceptance trials for the Navy and so Mitchell, who obviously knew much more about heavy locomotives than lightweight wooden aircraft, had ample opportunity to see the constructional techniques employed as the rest of the aircraft were built. As Penrose wrote: 'the Admiralty had found floatplanes too dependant on smooth water; they were interested in the far heavier flying boat hull which in the Linton Hope approach consisted of a double skin of mahogany planking with fabric in between, with rock elm strips forming almost circular ribs, longitudinally stiffened by closely spaced stringers'.

Aside from structural matters, Mitchell would also have discovered that the hydrodynamic aspects of hull design were in their infancy. The early test pilot, John Lankester Parker, described his first acquaintance with the A.D. Boat as follows: 'Not only did it develop a formidable porpoise at a very low speed, but nothing I could do would prevent it turning in ever smaller circles to the right, despite the fact that my passenger went out on the port wing-tip [!] to keep one float well and truly in the water.' The young landlubber Mitchell would also have discovered that it was one thing to design an efficient yacht hull but quite another to produce one that would easily plane over the water and break free from its suction in a controllable manner – on the one hand, a calm sea and no wind might prevent the currently low-powered aircraft from being able to 'unstick'; on the other, it might 'porpoise' in a series of bounces over waves until the right flying speed and angle might be achieved. (The standard World War One Felixstowe flying boat embodied an alternative type of hull which had a flying boat forebody married to a conventional landplane fuselage of longerons and struts but it was found to need additional planking which affected its performance as well as its flying trim. It was no better a performer on water than the A.D. Boat.)

A second Admiralty design, the A.D.Navyplane was another biplane pusher type of seaplane but this time it had twin floats instead of a flying-boat hull. Its detail design and construction were left to the Woolston firm but the plane did not go into production as its intended powerplant, the American Smith air-cooled radial, did not live up to expectations. This Navyplane was still undergoing trials at the time that Mitchell joined the company and there was soon work for him to do drawing up a specification for an improved version. However, as the war was almost at an end and as the Short 184 was performing adequately enough the duties for which the Navyplane was intended, nothing came of this effort.

There was also a further requirement at this time – for a single-seat seaplane or flying-boat fighter – issued by the Air Department under specification N.1B. As the Supermarine design for this aircraft flew later than the other designs

A.D. Flying boat.

mentioned above, and as Mitchell's involvement with it was the greater, it will be described in the next chapter.

Marriage and Early Hopes

The unlikely development of a lad from the middle of England, trained in locomotive engineering, into the designer of sea-based aircraft was only possible because the managing director must have recognised something about the applicant for the post of Personal Assistant when he stood before him. Nor was his locomotive engineering background necessarily a drawback: H. Fowler had been apprenticed to the Lancashire and Yorkshire Railway and had become the Chief Engineer of the Midland Railway before becoming the Superintendent of the Royal Aircraft Factory; and S. T. A. 'Star' Richards had been apprenticed to the Great Western Railway before becoming Handley Page's personal assistant and, later, his Chief Designer in 1922. In these days, however, Mitchell's mathematical and draughting skills would have been his predominant qualification and applicants for Mitchell's post would only have offered aviation experience as an additional bonus.

Whatever had singled out the young man from the Midlands, first impressions were clearly not ill-founded, for Mitchell was promoted to Assistant Works Manager in 1918 and his improved financial position now enabled him to travel back to Stoke and to marry Florence Dayson, Headmistress of Dresden Infants' School at St Peters Church, Caverswall on 22 July. It is a striking fact that Florence was eleven years older than himself, which at least suggests that the young Mitchell's mentality was attracted to that of a mature professional woman and that he was displaying at an early age a single-mindedness and determination that overrode what, particularly at that time, must have seemed to his parents an unusual match.

One must imagine that some of Mitchell's early domestic evenings were spent in study as the same year saw the issue of *H.B. 806* by the Technical Department of the Air Board which contained a full account of the mathematical methods employed by the Department. And whilst a contemporary designer admitted that in such vital matters of weight/strength ratios and, therefore, safety margins, 'we did it by guess and by God', more theoretical information became available for Mitchell to study in 1919, with the publication of *Aeroplane Structures* by A. J. Pippard and J. L. Pritchard and *Applied Aerodynamics* by L. Bairstow. The recent appointment of a Professor of Aerodynamics at Cambridge also marked the development of something approaching a systematic and scientific approach to the new technology. But how far the Southampton company was to become a significant part of it was by no means certain: the Society of British Aircraft Constructors did not have a committee member from Supermarine until late in 1918 and the strong British presence at the 1919 Paris Air Show did not include any machines from Mitchell's firm either.

But, at least, the new Assistant Works Manager was busy early in 1919, being involved with another seaplane project, namely the conversion of some of the surplus Admiralty A.D. Boats into civilian passenger-carrying aircraft. Initially, ten of these two-seaters, which were in store, were purchased back from the Admiralty with a view to offering trips from Southampton to various seaside resorts on the Isle of Wight, and drawings were prepared for the installation of a more economical engine and four seats. The name Channel indicates the very modest transport ambitions of the company and, in fact, the first of the passenger services were only between Southampton and Bournemouth. Nevertheless, Supermarine had the distinction of obtaining the first British Certificate of Airworthiness for a passenger-carrying flying boat in the August of 1919. The flights, which cost four guineas single and seven guineas return, were claimed to be the 'First Flying Boat Passenger Service in the World'. (One of the pilots employed was Capt. Henri Biard who was to test all of Mitchell's designs in the next ten years.)

A Channel II 'on the step'.

Several trips were also made to Cowes but only for passengers who had missed the regular ferry; however, by the August of 1919, regular services to the Channel Isles as well as to the Isle of Wight were begun, weather permitting, in addition to joyrides at various venues on the south coast when opportunities arose. During that year, the steam-packets ceased operation in sympathy with striking British railwaymen and the Channels finally lived up to their name by operating a service to Le Havre for the duration of the dispute. The service, costing £25 return, began on the 28th of September and came to an end on October the 5th; the Bournemouth service gradually petered out with the onset of winter. (*See* Cozens p. 34.)

Four of the original A.D. Boats had gone to an embryo Norwegian air service and Mitchell was obviously also involved with the company's further purchase of another six A.D. Boats. Some were converted to a three-seat trainer version and some were bought by the Norwegian Government for use with their Naval

Air Services. There then followed a Mark II conversion with a more powerful engine and three such aircraft went to the West Indies in 1920; two more were modified for photographic reconnaissance and used for surveys of the Orinoco delta in Venezuela. Other Channels were also delivered to the New Zealand Flying School, the Royal Swedish Navy, Chilean Naval Air Service and the Imperial Japanese Navy. (The visiting Japanese officials were reported to have been particularly impressed by the handling of the Channel in strong winds and heavy seas by Henry Biard, who had joined the company as test pilot in 1919.) Meanwhile, a separate Air Ministry had been created, the RAF staff college had been established at Cranfield and a marine training section and aerial navigation school had been set up at Calshot, only a few miles from the Supermarine factory.

Completed Channel for Venezuela (note the anchor).

Thus, although Supermarine's order book was soon not always to be so healthy, the newly married Mitchell, with the optimism of youth, could see here signs of an expanding new transport system and of Government support, particularly for its military aspect, and he could look forward to his company taking a full part in these developments. Looking out over the River Itchen, down to Southampton Water and then, in his mind's eye, to the destinations of eighteen of the Channel flying boats – Europe, Japan, South America, or New Zealand – the industrial smog of Stoke-on-Trent must have seemed a long way away, particularly as, in 1919, he had just been appointed Chief Designer, at the age of twenty-four. Other quite young designers must have had similar dreams: Chadwick at A. V. Roe was twenty-six and Pierson at Vickers was twenty-seven, Sopwith, Fairey, Handley Page, Folland and Blackburn were all in their early thirties, leaving Oswald Short at thirty-six and de Havilland at thirty-seven, as the old men of the group.

The total dominance of landplane travel nowadays ought not to disguise the fact that, to Pemberton Billing, to Scott-Paine or to Mitchell, the marine side of aviation would have been seen as the most natural of developments, not the minority, specialist aspect of the business that it is today: with motor travel in its infancy and, with Britain being surrounded by sea, flying boats represented the most obvious next step in the development of long-distance travel. Boats had brought most of their ancestors to the British Isles in the first place, trade depended on them, and they had been essential to establishing and serving the far-flung outposts of the British Empire. Britain also had numerous reasonably sheltered stretches of water which offered generous level areas for machines which, as we have seen, might be erratic in behaviour and need a considerable area for take-off.

Although the original Pemberton Billing firm, whose experience Supermarine and Mitchell inherited, had diversified into landplanes as part of the war effort, the original intention of the enterprise was best signified by its telegraphic address 'Supermarine'. The owner's original goal had been to produce 'boats which fly and not aeroplanes which float' and, when the new company was formed, Scott-Paine, the new managing director, chose 'Supermarine' for the company name (literally, the opposite of the more familiar 'submarine') to indicate where his hopes for future aircraft also lay.

But despite the foresightedness of Jacques Schneider in 1912 and despite the fact that World War One had given a more than considerable boost to aircraft and engine design, the marine side of British aviation had not profited so well – with the large Felixstowe reconnaissance flying boat being the main exception (despite its inadequacies, in 1919, it made an untroubled 2,450 mile tour of Scandinavia). Unfortunately the Felixstowe had not been a Supermarine product and, with the Armistice, new orders, especially for marine aircraft, did not now seem to be promising such a very prosperous future for the company that Mitchell had joined. As we shall see in the next chapter, the third Schneider Trophy competition offered the chance of good publicity for marine products, including Supermarine's, but the entries showed what little progress had been made in this area of aero design.

* * * * *

Cozens' account of the early – and spartan – Supermarine passenger services:

In the summer of 1919 they purchased some three seater flying boats which they had built and sold to the Air Ministry and which were now surplus when the war was over, and hastily made them ready to take joyriders from beaches along the South Coast while the summer lasted. The conversion from wartime use to peacetime was fairly basic and consisted of taking out the engine and fitting a more economical Beardmore 160 hp which caused the machines already prone to porpoising, to be even worse. The rest of the preparation was to paint out the RAF roundels, but not the red, white, and blue colours on the twin rudders, and SUPERMARINE was painted on each side of the bow.

When the pilot and his mechanic got to Bournemouth, or Brighton, or Bognor or elsewhere they planned to operate they anchored and relied on the local boatmen to ferry out the passengers and help with the refuelling. This enterprise started in July and continued through August and September. When the Air Registration Board was set up the three machines were painted blue, which was to be Supermarine's distinctive colour, and the letters G-EAEE, G-EAED and G-EAEK made them our first commercial flying boats.

Towards the end of September they were modified to carry three passengers and began flying to Le Havre and so became the world's first international flying boat service.

The experience gained in September and October 1919 by flying passengers across 114 miles to Le Havre led Supermarine to plan further efforts for 1920 and they built more machines with a higher bow to give the passengers more protection against wind and sea. They were known as Channel II's but the passengers still had an uncomfortable ride on many occasions and I remember seeing a Channel come up to a buoy just below the Floating Bridge [at Woolston] and although the three people were wearing flying coats and helmets they looked wet and miserable as they got into a boat that was rowed out to meet them.

It would appear that ten machines were registered as civilian Channel I's, G-EAED to G-EAM, but only G-EAWC and G-EAWP seem to be registered here as Channel II's. However, a number of Channel types were sold, especially to countries having offshore islands, and the skilful handling by the Supermarine test pilot Captain Henri Biard both on the water and in the air resulted in these being shipped out to New Zealand, Fiji, Japan, Bermuda, and at least one was used by Instone Airlines in the Bristol Channel and English Channel area.

Captain Biard was yet another of the colour[ful] personalities that Supermarine seemed to attract and he was an extremely good pilot, indeed he flew every kind of aircraft as it was produced, be it single or twin engine, light or heavy, on land or sea. One of the little known but important things he did was to train pilots for flying boats, something that even experienced land pilots found difficult, and it was a common sight to see a Channel taxiing up and down Southampton Water trying to take off, or landing with a series of bounces.

Some of the men who became popular Imperial Airways pilots on the Empire flying boat routes with the 'C' class flying boats were trained on Supermarine Channels.

— *Chapter Two* —

1919 to 1922
Early Designs and
the Schneider Trophy

The early Pemberton Billing/Supermarine interests had ranged from the very large Nighthawk landplane to a small naval 'scout' for the Navy and Mitchell's early design experience was closely involved with this latter fast seaplane type and its various transformations.

The company's interest in the naval interceptor type had begun with an Air Ministry requirement, N.1B, for a fast manœuvrable single-seat seaplane or flying boat fighter with a speed of 95 kt (about 100 mph) at 10,000 ft and a ceiling of at least 20,000 feet. It was required particularly to combat the German Brandenburg fighter seaplanes which had been operating over the North Sea. The resultant Baby had been designed by F. J. Hargreaves, who was in charge of the drawing and technical offices at Pemberton Billing when Mitchell joined the company and who continued for a little while after the company became 'Supermarine' in 1917.

Hargreaves' close liaison with the Admiralty Air Department produced an aircraft with what appeared to be a dangerously small fin and rudder, typical of aircraft drawn up by this design team (presumably in response to accidents caused by over-ruddering which could lead to spinning in) but the Baby was, in other respects, a more 'in house' response to the ambitious N.1B specification.

N.1 Baby. Note the rather inadequate-looking fin and rudder.

This machine did not go into production because of the ending of World War One but, as Mitchell had joined the firm in 1916 and had then been involved at least with the Nighthawk, it is entirely likely that he had had some design input in the three N.1B airframes that were going through the works: by the time of the Armistice, N59 (*see* photo above) had been completed and was being evaluated by the Navy and N60 was largely complete. The third, N61, was under construction and was most probably (in view of its extensive departures from the N59 Baby design) the one bought back from the Air Ministry for entry in the 1919 Schneider Trophy competition – in the hope thereby to gain some very useful publicity from an event to be staged by Great Britain, the winner of the previous event in 1914. The modifications were such that it was renamed the Sea Lion I.

Sea Lion I and the Third Schneider Trophy Contest
The particular configuration of this aircraft suggests that the modifications to the original Baby design were largely those of Hargreaves. The fin and rudder were enlarged in a shaping not followed later by Mitchell; likewise, the base of the latter was used as a water rudder, the interplane struts were splayed outwards, and the balanced ailerons on the top wing had an inverse taper. Also, the hull was decked to keep down spray and so the front of the fuselage was far less sleek than Mitchell's later Sea Lion II and Sea King II:

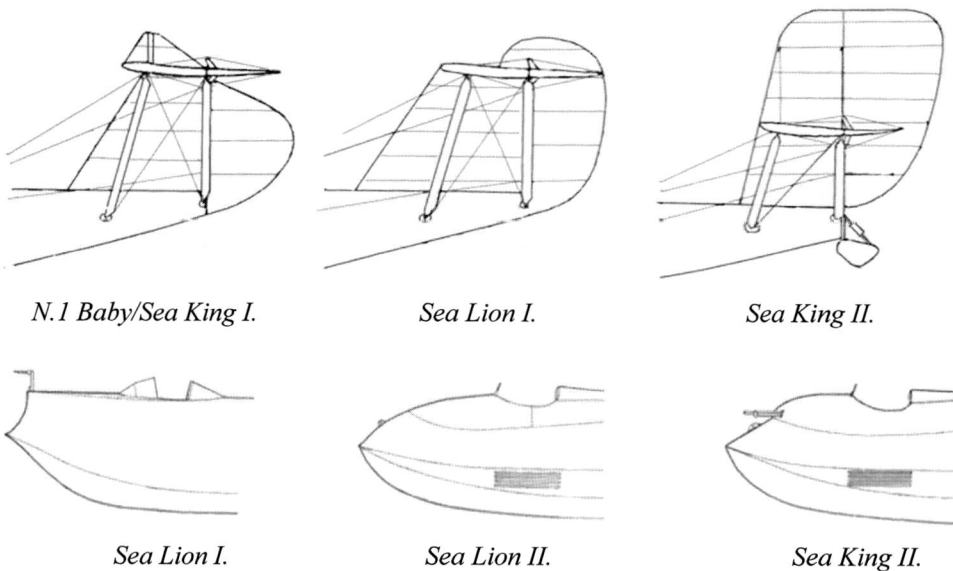

N.1 Baby/Sea King I.   *Sea Lion I.*   *Sea King II.*

Sea Lion I.   *Sea Lion II.*   *Sea King II.*

But Hargreave's Sea Lion was at least a dedicated seagoing design and so it represented the company's hopes of an Air Ministry order for a naval fighter. It was also powered by the water-cooled 450 hp Napier Lion, fast emerging as the outstanding power plant of the twenties, rather than the reliable and ubiquitous Rolls-Royce Eagle engine (which had recently powered the first crossing of the Atlantic by Alcock and Brown). The new engine, however, produced a 24% increase in the loaded weight, resulting in an increase of the Sea Lion's wingspan to 35 feet. Surprisingly, the challenge of the Schneider Contest seemed to have produced no special reduction in the Sea Lion's wing areas for the competition and the hull was far from sleek.

In fact, the design suggested that the man with overall responsibility for the aircraft seemed to have favoured rugged seaworthiness rather than speed through the air; as such, when it came to selecting the three aircraft to represent Britain in the Schneider Trophy competition, the Sea Lion was the Royal Aero Club's third choice over the slightly faster Avro 539A, possibly in order to hedge its bets because of the already proven seagoing qualities of Supermarine machines – the other two aircraft were floatplanes.

By the time of the Schneider contest, Hargreaves had left the company and it was Mitchell who would have assumed last-minute responsibility for this aircraft; he accompanied the new Managing Director of Supermarine in one of the company's current Channel flying boats to the event at Bournemouth.

Whilst Jacques Schneider had established his contest as a vehicle for the international development of marine aircraft in 1912, the main response to his challenge had come from French pilots, flying French aircraft, and powered by French engines. Even after World War One had given a considerable boost to British aircraft and engine design, the marine side of the industry had not profited so well and this situation can be seen when the entries for the 1919 Schneider Trophy competition appeared.

Pixton's previous win in 1914 meant that the Royal Aero Club would stage the next event in England and so a larger British entry than previously was forthcoming and might have provided a demonstration of bold, up-to-date, British aeronautical engineering. Unfortunately this was not to be. One entry was the Sopwith Schneider, an improved version of the 1914 winner but little different from the one five years earlier. It was still a converted landplane although it was now powered by a British engine, the 450 hp Cosmos Jupiter II air cooled radial. The Fairey IIIA entry, although specially prepared for the contest, and with its wingspan being reduced to a mere 28 feet, was also a converted landplane; it again featured a new British engine – the Napier Lion, mentioned earlier. Thus, despite the advances in engine design, the name 'hydro-aeroplanes' that had been used to describe these sort of machines was still apposite although, thanks to Winston Churchill (who had been appointed Secretary of State for War and Secretary of State for Air in 1918), they were now classified as 'seaplanes'.

The Sea Lion was in almost direct contrast to the Savoia S.13 flying boat, the only eventual overseas competitor for 1919 and the first Schneider entry of an Italian aircraft: its sleek lines and uncluttered design showed the extent to which Italy had chosen to develop this type of machine as a 'fighting scout' during the war, although its hollow-bottomed hull profile was only superseded by the more efficient vee-bottomed approach, which Mitchell had inherited from the Pemberton Billing days, in later Macchi aircraft – notably the M.33 of 1925.

On September 10, the day for the competition, no aircraft were able to profit from the potential publicity of the Schneider contest nor was it possible for dispassionate observers to judge whether the 450 hp British engines would pull their very traditional airframes through the air faster than rather similar French floatplanes or the more streamlined Savoia with the low-powered 250 hp Isotta-Fraschini power plant. The reason was a combination of incompetent organisation and the British weather. Two French entries eventually arrived, along with the Italian machine and the three British entries, and these six competing aircraft were to fly in to Bournemouth beach for the contest from their base provided by Saunders Ltd in the Isle of Wight but, as C. G. Grey commented: 'The idea of housing competing machines at Cowes and asking them to fly all the way to Bournemouth, with attendant risk of collision with yachts in the Solent, is as foolish as could well have been achieved'.

Also, the Royal Aero Club did not consider crowd control (or even providing food for the contestants) to be its responsibility. The result was that, at about 11.30 a.m., a Spad floatplane and the Savoia flying boat braved persistent fog and flew across, the latter nearly hitting a rowing-boat and then being surrounded by a large and enthusiastic crowd which threatened its safety. The Spad was towed in amidst bathers by the Supermarine motor launch and was later seen to have a leaking float, which was further impaired when hauled on to the beach by enthusiastic 'helpers'. By 2.30, the other French plane, a Nieuport, arrived and the contest was set to begin; but the fog increased and a start time of 6 p.m. was announced. This time was later brought forward, to the consternation of the

French who, it transpired, had repairs to do to the Nieuport also and who, in the end, did not start.

And so, at 4.50 p.m., the Fairey IIIA got away, followed by the Supermarine Sea Lion; the Sopwith went next, and then the Savoia. However the Fairey and the Sopwith aircraft soon returned, their pilots considering, rightly, that the poor visibility made flying too dangerous, particularly at the Swanage turn. It was here that the pilot of the Sea Lion, on his first lap, had to land in order to try to establish his position in relation to mist-shrouded cliffs but the aircraft hit something in the water. The pilot took off, nevertheless, managed to complete the first lap despite the poor visibility and made the first of the two mandatory navigability landings; unfortunately, the earlier damage to the hull caused the plane to upend and threw the pilot into the sea. Only Guido Janello in the Savoia 13 completed the required 10 laps but then it was found that he had flown a shorter distance, having mistakenly rounded a reserve marker boat anchored in a cove two miles short of the real marker. He was accordingly disqualified although the Royal Aero Club did the decent thing and, at a meeting on 22 September, recommended that Italy be awarded the trophy. However, the Fédération Aéronautique Internationale (which oversaw the contests) ruled that the flight was invalid and it rather pointedly awarded the next venue to Italy.

Publicity for the Sea Lion I had reflected Supermarine's hopes of military orders:

> This machine, which is said to be the fastest flying boat in the world, is a small, fast, single-seater, designed primarily for war purposes. With the Napier 450 hp engine, 2 hours' supply of fuel, and a load of 140 lbs of guns and ammunition, the speed is 147 miles per hour. The hull is guaranteed to stand up to practically any weather, and the machine itself may be looped, rolled, spun, or put through any of the manoeuvres demanded by aerial fighting.

Sea Lion I at Bournemouth, an engineer in cockpit. The pilot is Sqd. Ldr B. D. Hobbs, decorated for shooting down Zeppelin L.22 on May 14, 1917.

However, the 'non-event' of the 1919 Schneider Trophy contest was of no help to Supermarine's hopes for this type and no orders were forthcoming. The next two Schneider Trophy events took place in Italy but flyers from other countries apparently did not feel inclined to finance entries to what was not yet a major aeronautical event. Meanwhile, the company persisted with their fighter flying boat concept with their next two fast flying boats, the Sea Kings.

Sea King I
Little is known about Mitchell's involvement in N60, the second Baby version mentioned above, which had also been bought back from the Air Ministry; it seems likely that it became the Sea King I, which appeared at the 1920 Olympia Aero Show – that is, after Mitchell's appointment as Chief Designer. But how long it had been in existence in its new guise before this date is unknown; certainly a photograph from the Show reveals a direct repetition of the earlier, apparently inadequate tail configuration and so represents past practice rather than the future. Perhaps its original specification with a Beardmore 160 hp engine was not expected to present directional problems, given that the Baby had flown with a 200 hp Hispano-Suiza engine, but Supermarine's proposed fitting of a 240 hp Siddeley Puma engine might have produced some design responses from Mitchell, such as the fin and rudder of his later Mk.II version (*see* drawings p. 36). One speculates that, at this time, the profitable modifications to the A.D. Boats, also bought back from the Air Ministry, had so preoccupied Supermarine that N60, unmodified by Mitchell, except perhaps in respect of its altered and raised cockpit fairing, was sent to the Olympia Aero Show essentially as a marker for the company's continuing interest in the naval fighter scout concept.

C. G. Grey has left an interesting memory of this display:

> as one stepped onto the stand one had a feeling of confidence that here were people who really knew what they were at in the sea-flying game. Hubert [Scott-Paine] and his two brothers wore yachting caps and double-breasted reefer jackets and looked real sailor men. Their helpers wore jerseys … and in each corner of the stand was a large coil of tarred rope flemished-down in workmanlike seafaring fashion.

Despite the showmanship, there is little information about the aircraft having been flown, although the following publicity for this aircraft would seem to imply that control, even with the original less powerful engine, might not have been found to be quite adequate; it also reveals that the company was hoping to sell to the many private flyers that World War One had produced, if military orders could not be achieved:

> The 'Sea King' is a small fast single-seater which for general purposes follows the structural methods of the 'Channel Type' boat. With its 160 hp Beardmore engine it puts up a speed of 96 knots, so that it is either a thoroughly sporting little vehicle for the single or unhappily married man, or is a useful small fast patrol machine for Naval work along troublesome coasts. Its chief difference in design from the 'Channel Type' lies in the fact that it only has a monoplane tail of the depressing kind and so takes rather more flying on the part of the pilot than does the bigger machine.

Had there been any sales, perhaps Mitchell would have wished to modify the tail surfaces but, unfortunately, neither the military nor the 'single or unhappily married man' came along to buy one and it had to await the Mk.II, clearly modified by Mitchell two years later.

The COMMERCIAL AMPHIBIAN – Mitchell's First Design
New orders for the services were likely to be scarce with the ending of World War One – the flying boats that had been bought back by Supermarine for

conversion were among some five thousand serviceable airframes now surplus to RAF requirements, including over a hundred seaplanes. As Cozens reported:

> When the Great War ended so did the orders for aircraft, and many jobs as well. On the Marlands in the centre of Southampton there were hundreds of crates of surplus aeroplane engines and the common between Spring Road and Firgrove Road in Sholing was covered with aircraft frames, but the industry had to plan for peacetime or go out of business.
>
> May, Harden and May closed after making two or three large experimental types based on the now successful Felixstowe. Avro, easily the most wealthy company, continued its design of large twin engine and single engine machines and they now had Roy Chadwick on the design staff and Bert Hinkler was their test pilot. Wight Aircraft in West Cowes closed.

One hundred and eighty-eight RAF combat squadrons at the end of the war were reduced within months to thirty-three, with the result that there was also a large number of redundant pilots – although this reduction of service fliers did imply some slender hope of expansion in the field of private aviation. This potential was in addition to the embryo commercial sector represented by the Supermarine Channel operations of 1919 mentioned earlier and by the regular British commercial passenger and goods services between London and Paris, begun on August 25th of the same year by the Aircraft Transport and Travel Company. The formation of the Department of Civil Aviation at the Air Ministry could also be taken as a hopeful sign.

The first aircraft involved in commercial flying after the end of World War One were conversions of military machines – like the Channels, which we saw earlier were by no means well suited to their new roles. And so it was in March 1920 that the new Department of Civil Aviation at the Air Ministry announced two competitions for commercial designs 'of British Empire origin' to promote 'Safety, Comfort and Security' in air travel. With a view to developing international travel (bearing in mind the few airfields available, compared with large stretches of water worldwide) one of these competitions was specifically for amphibian seaplanes with a first prize of £10,000 and a second one of £4,000. It was not surprising that Mitchell was asked by Supermarine to design an entry for the seaplane competition, which was to commence on September the 1st of that year.

It might be true to say that the newly formed aviation ministry had no very clear idea what should be specified and there were few passengers to consult; so organising a competition between designers was a sensible approach. Mitchell, therefore, had quite a free hand at this early stage in his career in trying to satisfy the modest performance criteria that were laid down for the amphibian class. These requirements included: seating accommodation for a minimum of two persons exclusive of crew; a range of 350 nautical miles at 1,000 feet at a speed of not less than 70 kt; and a load of 500 lb to include passengers and life belts but not including crew. There was also a requirement of a flight of three minutes at 5,000 ft to check if the machine would fly itself and the machines were restricted to a four hundred yard take-off run to clear balloons at a height of twenty-five feet. From the experimental station at Felixstowe, the amphibian competitors were required to take off, pass as high as possible between marker boats 600 yards from the start buoy, and land at the experimental (land) station at Martlesham Heath. Taxiing on water had to include figures of eight, taking off and landing in rough weather, and mooring out for at least twenty-four hours in moderate weather. (These marine trials were not unlike those which Mitchell's Schneider Trophy racers had to complete before the actual flying contests and, of course, reflected the same concern to develop aircraft that had practical seagoing features).

As the response to the above requirements was the first major project which Supermarine's newly appointed designer was called upon to undertake, it surely deserves close attention. However, it is also very understandable that the end-product would be a conservative one. Even had Mitchell been an experienced aircraft engineer at this time, he would still, in all probability, have mainly followed previous best practice, in view of the little theoretical data that was available and as wind-tunnel experimentation or tank testing (for flying boat hulls) was not available to his small firm. Also, there were only about twenty weeks separating the announcement date of the competition and that of the trials, leaving little time for innovative thinking. In the event, *Flight* reported that 'for the Martlesham amphibian trials the Supermarine Company designed and completed a flying boat in all respects in 10 weeks from the time when the first drawing was commenced to the time the aircraft was in the air, the actual building time being 4½ weeks.'

Based on a Channel airframe, the Commercial Amphibian had a biplane layout in which similar dimensions of height and length were adopted and the sea rudder was similarly placed – vertically below the leading edge of the tailplane – but now converted to act also as a skid when taxiing over land. Between the Amphibian's struts there were canvas stabilising screens, full length between the inner pairs and quarter length between the middle ones. These screens were relatively uncommon by this time but survived on several later Supermarine designs as well as on the Channel and Sea Lion, perhaps mainly to protect the engine and propeller from spray on take-off or landing; but it was the present machine which was most extensively fitted with them and, in this respect, it did not look particularly like an advanced design. The wing tip floats were also of the Channel sort and the oval hull and the general arrangement of its built-on planing surfaces employed the Linton Hope/Channel principles of hull construction. Not surprisingly, therefore, Supermarine described the new design as 'practically a "Channel" type boat, with a wheeled undercarriage hinged on each side' although, in fact, side views show considerable changes in the Mitchell design:

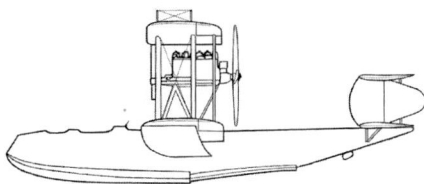

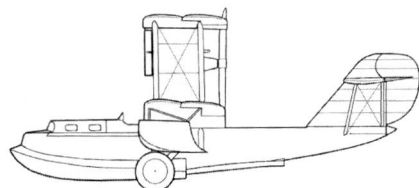

Channel.   Commercial Amphibian.

Mitchell also incorporated features of a much smaller aircraft, the Sea Lion I: the fin and rudder outlines were similar, although a proportional increase in surface area above the tailplane allowed the designer to provide a more symmetrical appearance to the fin. And the Sea Lion's outwardly raked inter-plane struts were repeated in the new, and larger, machine. (Thereafter, it would seem that Mitchell preferred the simplicity of equal span wings supported at right-angles by the inter-plane struts.)

As many of the features from both the Channel and the Sea Lion I were thereafter abandoned by Mitchell, the present design can be regarded as something of a 'time capsule', a summing-up of earlier practices rather than a statement of the way forward. But Mitchell also showed an early instance of boldness and originality by abandoning the biplane tailplane and twin rudders of the Channels in favour of a single fin and tailplane – and it is worth noting that the competition rival, the Vickers Viking III went through three more variants before the Mark VII, the Vanellus, appeared five years later with a more modern-looking single tailplane. Also, Mitchell's rudder was a departure from the minimalist approach of previous Admiralty inspired rudders – perhaps his work alongside Hargreaves on the Sea Lion I had had some influence in this respect. Likewise, Mitchell significantly remodelled the nose with a prominent boat-like entry to counter spray, a feature which was to prove successful in his future Sea Eagle, Scarab and Seagull designs.

A further novel feature was Mitchell's design for a retracting undercarriage, necessitated by the Air Ministry competition for amphibian aircraft. At this time, an American landplane, the Dayton-Wright RB-1 Racer, had an innovative fully retracting landing gear, designed specially for the Gordon Bennett race of 1920, whereas the present concern was merely to lift the wheels out of the water, in order to facilitate take-off and alighting. The first European design of this sort was the Sopwith Bat Boat of 1913 which, like the present rival Viking, had a mechanism which rotated the wheels upwards and forwards. Supermarine's concern for 'boats which fly' offered no previous experience of retractable undercarriages for Mitchell to call upon and so it is noteworthy that, for his specially designed mechanism, he chose a geometry which displaced the wheels outwards rather than forwards – thus avoiding any change of trim when the wheels were moved up or down. (Mitchell retained this sideways mode of retraction for all of his future amphibian undercarriages.)

Flight reported that:

> for the Martlesham amphibian trials the Supermarine Company designed and completed a flying boat in all respects in 10 weeks from the time when the first drawing was commenced to the time the aircraft was in the air, the actual building time being 4½ weeks. This hull is considerably bigger than the standard type of Supermarine construction.
>
> It is worthy to note that during the firm's trials at Southampton this aircraft landed in a ploughed field, with furrows of about 15 in., which not only pulled the amphibian up in a very short distance, but allowing for the heaviness of the ground, threw the boat forward on its nose, leaving the section of the hull on the earth for a distance of about 15 yards. The boat was dropped on its tail, which also speaks of the rigidity of tail construction, to put up with such treatment, and was flown off and landed at Eastleigh Aerodrome. No damage of any kind, with the exception of scratched varnish, was experienced, and later during the trials the aircraft was often tipped in this manner by the pilot for turning purposes. [*See* photo, opposite.]

And one other particular feature of the Commercial Amphibian must also be mentioned: the enclosed passenger cabin for the two passengers, who were seated in tandem. The competition's intention of ascertaining 'the best type of Float Seaplanes or Boat Seaplanes which will be safe, comfortable and economical' might have seemed to make an enclosure for passengers inevitable but it should be noted that the other two amphibian entries had open cockpits for their three passengers, one seated next to the pilot and the other two side-by-side behind. Open cockpits at this time were the norm and they saved weight, but they were far from ideal in northern climates – one remembers Cozens' previous description of Channel passengers looking 'wet and miserable as they got into a boat that was rowed out to meet them'.

The concern of Supermarine that passengers in the open would not be very comfortable, and possibly a safety hazard, might very well have resulted from the experience of Supermarine's pilot, Biard, on the Channel service to Le Havre

The Commercial Amphibian with ground handlers, at Martlesham.

A Commercial Amphibian 'arrival'.

Commercial Amphibian with cabin hatch open. Mitchell centre, Scott Paine right.

on the 30th September, 1919. He recorded that the weather on that day had developed into a gale with sleet and snow but, with a flask of rum donated by Scott-Paine for heating, a Belgian financier braved the open cockpit of the Channel flying boat. The cold was such that Biard could hardly feel his feet and hands and then he was nearly blinded when the passenger, a Capt. Alfred Loewenstein, tried to pass the flask to the rear cockpit but only succeeded in causing the rum to blow back into the pilot's eyes. Then the passenger tried to put up an umbrella, presumably hoping to keep the hail at bay. He obviously was unaware that the force of an aircraft slipstream would wreck the umbrella (even in those slow-flying days) and had clearly not considered the effect of even such a relatively fragile item being blown into the propeller – which was directly behind the two of them. As it was impossible to converse in the conditions, even if there had been time, Biard had to resort to hitting the Belgian about the head, whereupon he disappeared into the well of the cockpit. (Loewenstein was to become a 'mystery of flying', on July 4 1928, when he disappeared from a Fokker F.VII over the Channel.)

This glimpse into the pioneering days of aviation might seem amusing (although not to the pilot at the time!) as was the arrival of the Supermarine crew for this 1920 competition dressed in heavy jerseys and sea boots – to become known by their competitors as the Supermen. The Vickers people turned up in sailor's hats with 'Viking III' in gold on the hat bands and *The Aeroplane* at the time noted that 'all the competitors treat[ed] the affair as a very good joke'. Despite the apparently light-hearted or amateur approach to the event, the same correspondent did note that the amphibian entrants hedged their bets by reserving their maritime tests until last 'as they wanted to complete land tests before chancing damage to their machines by awkwardly handled launches or a sudden squall'.

No adjustments or replacements to the Mitchell aircraft were required, despite its one-off design and the short notice of the competition, and the Supermarine entry was the only one which completed all the tests that were stipulated and whose landing gear did not give trouble at any time. The judges also noticed with approval an effective tiller arrangement for steering whilst taxiing on water, the equipment for sea use, and the way in which the shape of the forward part of the hull kept spray off the passengers' compartment. It might be expected that the company's marine experience would produce such comments; equally, it might not be too surprising that the very novel retracting undercarriage gave rise to criticism for being none too clean, from the mechanical and the maintenance points of view. The lateral control of the Commercial Amphibian was also considered not immediately responsive enough. On the other hand, a comparison between the clumsy consuta-plywood sheeted approach of the Vickers entry and the boat-like hull that Mitchell inherited showed that he had had the good fortune to have joined a firm with an effective flying-boat hull, inherently more streamlined than most landplane designs of the time, that was to stand the firm and its chief designer in good stead for the rest of that decade.

Possibly, such criticisms would not have prevented Mitchell's aircraft taking first prize, in view of its performance in the competition as a whole and in view of the various features of the machine to receive favourable comment. But an increase of nearly 150 square feet of wing area compared with the Channel had been necessitated by the fitting of a more powerful but heavier Rolls-Royce Eagle engine, in order to lift the additional weight of the amphibian landing gear that was called for and to address the various other specifications of the competition. Thus the performance of the Commercial Amphibian did not match that of the Viking and its passenger carrying capacity was only two – the minimum allowed by the competition rules. So the low power/weight ratio resulting from the engine of the Commercial Amphibian, available cheaply to the company from war surplus supplies, produced a significant loss of certain competition points and resulted in its coming second to the Viking.

The final official report on October 11th stated that 'the results achieved for amphibians show that considerable advance has been attained ... and the competing firms deserve congratulations on their enterprises.' They also recommended an increase in the second prize money as 'the proportion of the monetary awards does not adequately represent the relative merits of the first two machines'. The Company's own assessment of the Amphibian was as follows:

> The Supermarine Aviation Works ... machine followed previous types of Supermarine in the general characteristics of structure and put up an extraordinarily good show in that competition. It completed all the tests satisfactorily, and was only beaten by competitors with engines of considerably greater power in the matter of speed and climb. [The competition entrants also included a Fairey III floatplane with wheeled attachment and a Napier Lion 450 hp engine which came third and was awarded £2,000.] It was awarded the second prize, and in view of the general excellence of the design and construction, the amount of this prize was increased from £4,000 to £8,000 by the Air Ministry.

The Commercial Amphibian being readied for launching from the Supermarine slipway, Woolston. Mitchell is standing nearest to the hull.

One can imagine how Mrs Mitchell must have felt at this promising start to her husband's career but, from a technical point of view, it was a modest beginning to be sure. Nevertheless, as we shall see later, many of its features and its overall performance gave rise to a call from the Air Ministry for a development of this machine which led to the Sea Eagles and the Seagulls between 1923 and 1926.

* * * * *

Early Military Designs – Seal II and Sea King II
Meanwhile, Mitchell's success with the Commercial Amphibian, which was wrecked shortly afterwards by a bad landing, had brought an order from the Air

Supermarine COMMERCIAL AMPHIBIAN (1920)

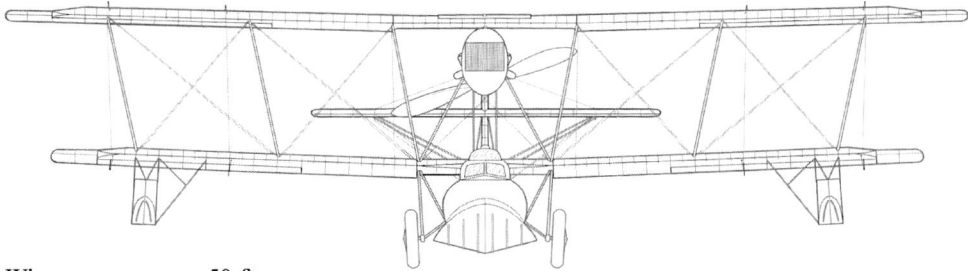

Wingspan	50 ft
Wing area	600 sq ft
Loaded weight	5,700 lb
Maximum speed	94.4 mph

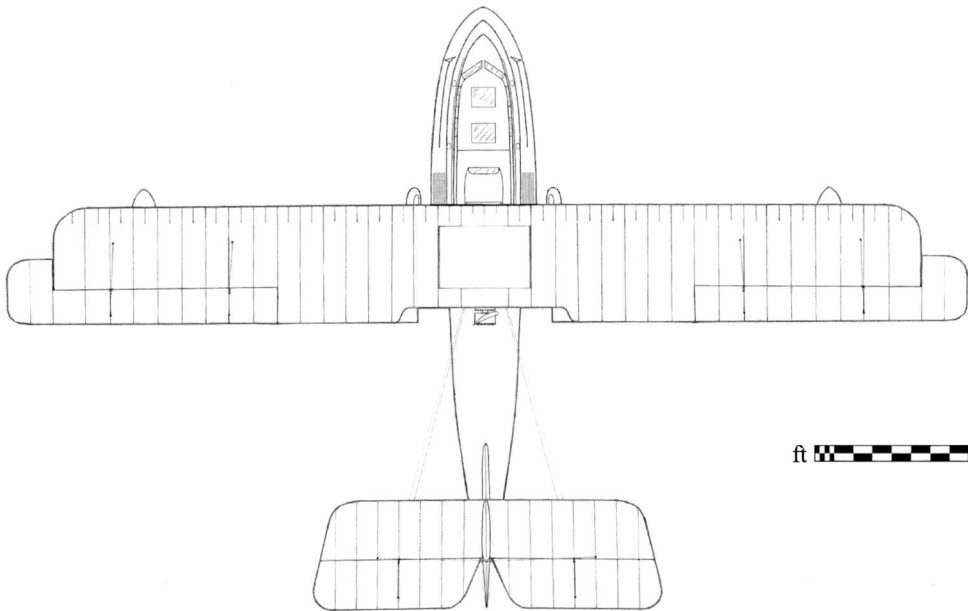

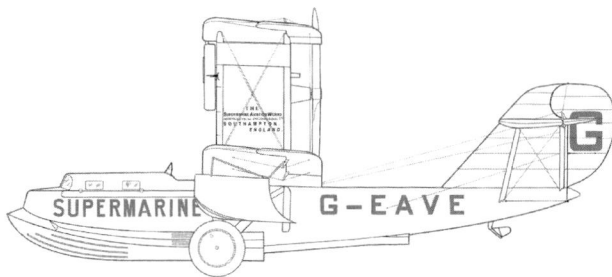

ft

SUPERMARINE G-EAVE

Ministry for a military development of this larger and more utilitarian type as part of its policy to assist the struggling aviation companies to stay in business. This requirement represented a recognition that British air power needed the support of a healthy aviation industry especially as, by 1920, the RAF had been in action again with the new military tactic of 'control without occupation' – against the Bolsheviks in Russia, against the Afghans on the Indian frontier and against tribesmen in Mesopotamia, Transjordania, the Sudan, and Somaliland. The Air Estimates of that year accordingly allocated £1,389,950 (but compared with £54,282,064 in 1919) for the purchase of aeroplanes, engines and spares, and recognised that contracts for new, experimental types would have to be spread around the various aviation firms in order to maintain the technical staffs which had been built up. (Sopwith had already turned to making motorcycles and Fairey, Gloster, Blackburn, Shorts, and Bristol were manufacturing bus or car components.

In response to the new Air Ministry initiative, and Specification 7/20 in particular, Mitchell's next design was to be a three-seat amphibian for use as a fleet spotter, to be extremely seaworthy and to have the lowest possible landing speed with good control – in order to land on to aircraft carriers. Mitchell's response was known as the Seal II, presumably with the Commercial Amphibian being regarded as the Mark I predecessor – which had also been a three seater, as well as being fitted with a retracting undercarriage. The new aircraft is important in the Mitchell design history as it, basically, establishes the pattern for all his later military, medium-size, amphibians, up to and including the well known Walrus.

SEAL II
As the new design sprang from his work on the Commercial Amphibian, the Seal II can be regarded as the second design for which Mitchell had full responsibility. However, many design changes show Supermarine's young designer eager to improve upon his previous effort, in particular the outwardly retracting landing wheel geometry which shows that something had been learned from the criticism of the earlier plane's mechanism: the earlier machine had an undercarriage consisting of two steel tubes, hinged below the lower wing centre-section join with the lower main planes and the wheels were raised or lowered by sideways movements of a tube in the hull to the wheel axles; the new undercarriage (*see* drawing overleaf) now had a single strut, suspended from the lower wing and braced by two tubes hinged to positions on the hull. For retraction, the top of this main strut was moved inwards by means of a worm-and-bevel gear, thus reversing the previous method and siting the retracting mechanism further from the water; it was utilised on all future Supermarine amphibians up to and including the Sheldrake of 1927.

Mitchell also improved on the previous aircraft by siting the tailskid/water rudder at the sternpost and this had the effect of increasing the wing incidence during taxiing and so improved the take-off performance (on land at least) which had not been very impressive in the earlier machine. A *Flight* commentator noted what had been a further design consideration: 'it is much easier to provide the necessary strength and water tightness than it is with a rudder working in the trunk of the hull'.

Also, the wing shape was new (*see* G.A. over page) and this planform was retained by Mitchell for all his subsequent single-seat naval aircraft up to the Scarab. The wing-tip floats were less clumsy than before and, because of their decreased side-area, were carried on struts to the waterline, so decreasing frontal area.

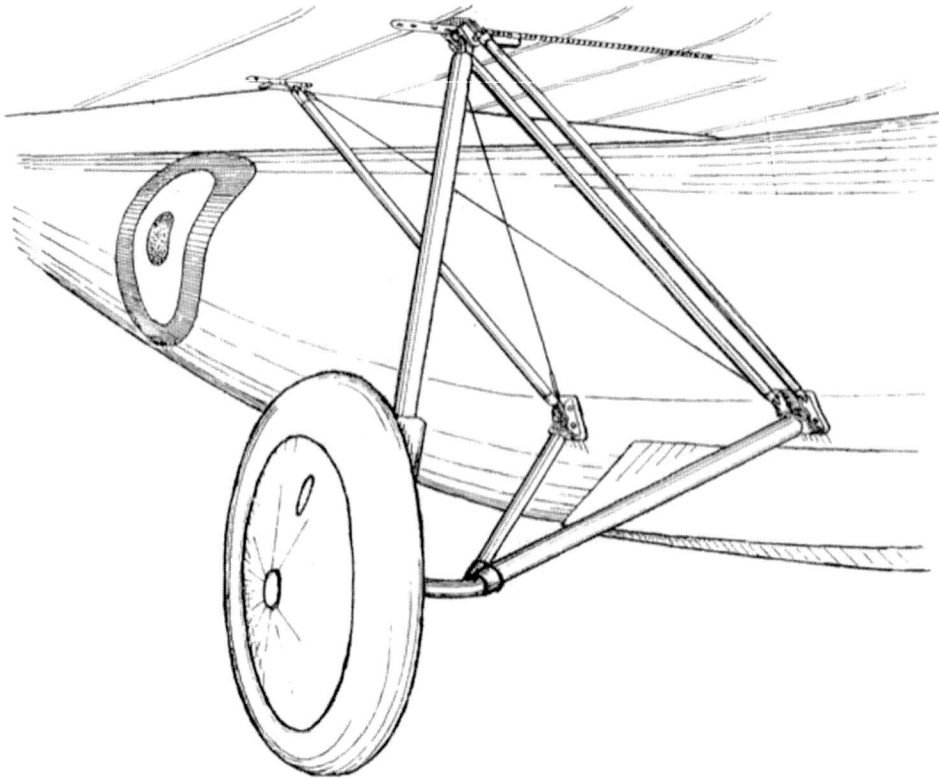

Sketch of Mitchell's improved retraction arrangement –
used from the Seal to the Sheldrake.

The rearward folding wing requirement for a shipboard aircraft had not been tackled by the Supermarine company since the Baby of 1918 and Mitchell adopted a similar approach – one which he, again, persisted with in military aircraft until the Scarab. The forward wing strut at the joint between the wing centre section and the folding main plane was doubled so that one member carried the weight of the leading section of the wing when folded back. And in order to keep storage space to a minimum, large cut-outs were made in the trailing edges of the wings so that they could fold close to the plane's centre-line and the wings were placed further forward than in the Commercial Amphibian in order not to project behind the trailing edges of the tail assembly when folded.

The pilot was placed well forward and was now supplied with a machine gun which could be retracted and shielded during take-off and landing; the wireless operator was just aft of the wings and the rear gunner behind him, but with the fuel tanks separating the pilot and these other two crew members. Because two of the crew members were placed behind the wings, a tractor layout had to be chosen for the engine to prevent the centre of gravity moving too far back. *Flight* believed this to be 'the first British flying boat to be designed as a tractor' and in the Supermarine publicity for the Seal (*see* p. 51), attention is drawn to this placement because of its relative novelty (at least in single-engined machines) – after all, the pusher configuration kept the propeller as far back as possible, out of the spray generated at take-off or landing. (One notices that Supermarine made the offer of the more conventional pusher layout – presumably in the hope of civilian versions that would not, one might reasonably assume, need provision for a gunner behind the wings.)

The company publicity also draws attention to the company's faith in the suitability of the Linton Hope type of hull construction and its seaworthiness and to the manœuvrability of the design, all features which had been paraded before

Supermarine SEAL II (1921)

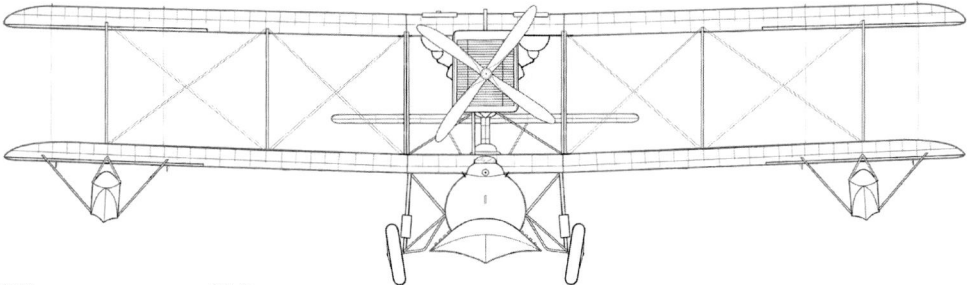

Wingspan	46 ft
Wing area	620 sq ft
Loaded weight	5,600 lb
Maximum speed	112 mph

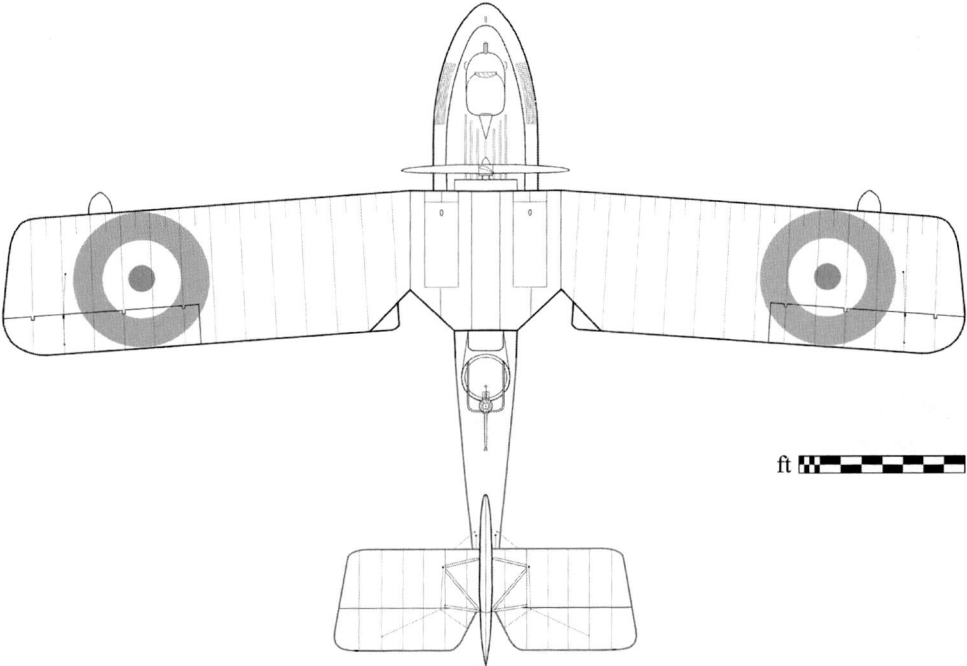

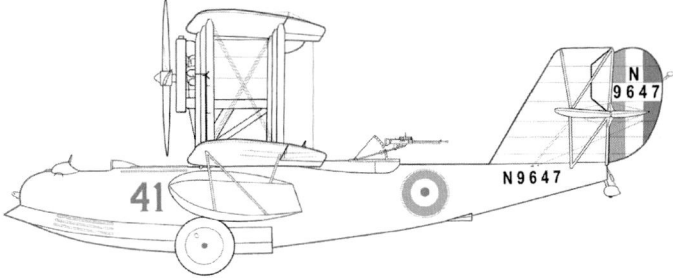

ft

41

N9647

N
9647

Seal II with machine gun in nose.

Seal II at the Supermarine slipway showing retracted undercarriage.

by the company. Mitchell's own improved method of undercarriage retraction is amply described and attention is also drawn to another aspect of the new designer's typical concern with the practicality of his machines – in this case, the ease of access to the engine.

The Seal II first flew in May 1921 and, in the following year, one machine was sold to Japan, a nation keen to be kept abreast of Western technology.

Supermarine publicity:

THE SUPERMARINE AMPHIBIAN 'SEAL MARK II'

The machine depicted in the accompanying illustrations is one of the most recent products of the Supermarine Aviation Works, Ltd, and has been built for the Royal Air Force as a deck landing amphibian for Fleet 'spotting' purposes.

The special requirements for a machine of this type are the lowest possible landing speed combined with a high degree of manœuvrability at that speed, in order to permit landing on the deck of a seaplane carrier, combined with a very high degree of sea-worthiness when used as a seaplane. The results of many tests with this particular machine demonstrate that these requirements and many others have been amply met by the 'Seal'.

Generally, the design of the machine follows standard Supermarine practice. The hull is of the well-known Supermarine circular construction, combining in a high degree the requisites of strength, lightness and resiliency. This hull is fitted with one cockpit right forward for the pilot, and a second, well aft the wings, accommodating in tandem observer and rear gunner, together with such equipment as is required for their use.

The upper wing is flat from end to end, the lower wing has an appreciable dihedral, and carries at its outer extremities a pair of wing-tip floats. Ailerons are fitted to both upper and lower wings.

At the rear end of the hull is mounted the tail unit. This consists of a monoplane tail, of the inverted wing section type, which is braced to the hull by steel tube stays running out to about one-third of the half span on each side. The tail is therefore largely overhung and there is little bracing to restrict the field of fire from the aft gun.

In addition there is a large fin and a balanced rudder, together with a small water-rudder beneath the hull.

The wheeled undercarriage, which is based on the design used in the very successful Supermarine amphibian which did so well in the Air Ministry Competition, is arranged in such a manner that no landing loads are imposed on the hull structure when the wheels are in use. The loads on the wheels are transmitted through a nearly vertical telescopic strut furnished with shock-absorbers, to the lower wing centre section front spar, immediately under the centre section and engine mounting struts. The wheel axle is carried by a projecting triangular structure of steel tubes hinged to two points on each side of the hull. The nearly vertical strut is movable. When in its nearly vertical position, it maintains the wheels in position for landing. When the top end is withdrawn inboard, the wheels fold up round the hinges on the hull side and take up their position close under the wings and well clear of the water.

The engine is the Napier of 450 h.p. The engine mounting is unusual in that it is of the tractor type. This has been rendered possible by the fact that in this case the greater part of the useful load carried is aft of the wings in the tandem cockpits, and the success of the tractor mounting will allow this type of boat to be arranged either as a tractor in such a case as this or as a pusher in cases where the greatest useful load is concentrated forward.

Very great attention has been paid in designing this engine installation to securing accessibility for inspection and adjustment of the engine and its accessories.

SEA KING II

Whilst it might be simpler, from a narrative point of view, to consider next the development of the Seal, culminating in the well-known Walrus, and to have dealt with all the early commercial amphibians in another, separate, section, it is a more faithful reflection of the way in which various design requirements were placed on Mitchell to consider next a *third type* of machine, the Sea King II. Responding to Supermarine's perception of the needs of the emerging aviation scene, the luxury of an orderly development of a single type was not available to Mitchell and, instead, the new Chief Designer had found himself having to produce three distinct types, each with novel requirements, all conceived and built within the short space of one year.

But at least Mitchell was on more familiar ground with this third type as it represented the company's continuing faith in the naval 'scout' concept which had been first evidenced in the Pemberton Billing N.1B Baby. Mitchell now produced an amphibian version, designed to meet an Air Ministry interest in a fighter design for shipboard use – as company publicity proclaimed:

> a high performance fighting scout, specially adapted for getting off gun-turret platforms of capital ships, or getting off and landing on the decks of aircraft carriers. The strength and design of the hull are such that it can operate on and from the water under any weather conditions in which it would be possible to operate any other sea craft [boats] of equal size.

Sea King II, without registration letters, perhaps on taxiing trial.

The Sea King was produced at the end of 1921, after the Seal, but was much more of a direct modification of earlier Pemberton Billing/Supermarine practices. Nevertheless, its modifications can be attributed entirely to Mitchell and, indeed it bore distinct evidence of his taking over the design department at Supermarine.

The most obvious revision of the earlier design was the more generous fin and rudder area than that handed down to Hargreaves by the Admiralty team (*see* sketches on p. 36) – particularly as the 160 hp Beardmore engine of the Sea King I was now replaced by the much more powerful 300 hp Hispano-Suiza engine – and it would appear from the Supermarine publicity quoted below that this redesign had a noticeably beneficial effect.

As with his Seal, the tailplane was now placed almost midway up the fin, the retracting gear of the Seal was again utilised, and a Seal type combined tailskid

Supermarine SEA KING II (1921)

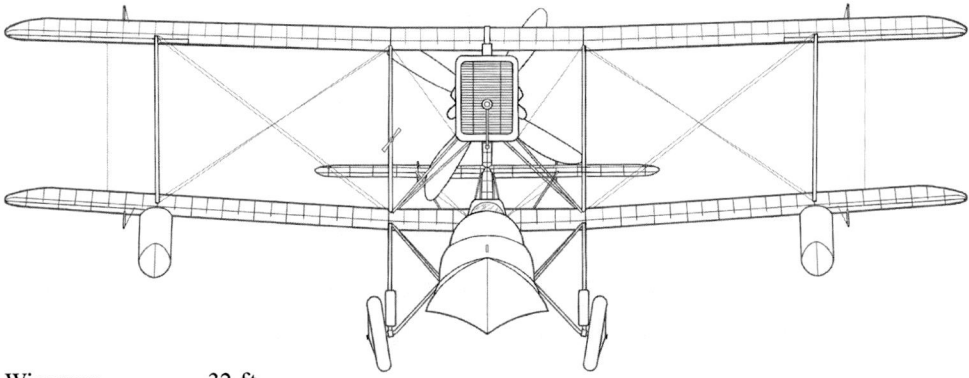

Wingspan	32 ft
Wing area	352 sq ft
Loaded weight	2,850 lb
Maximum speed	125 mph

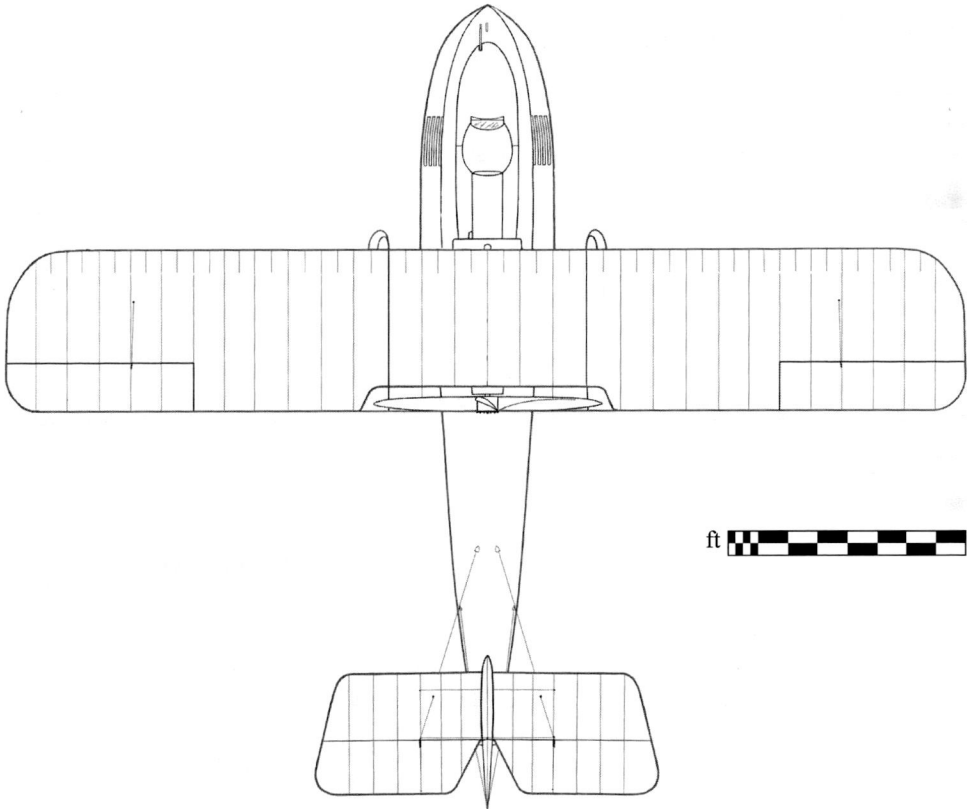

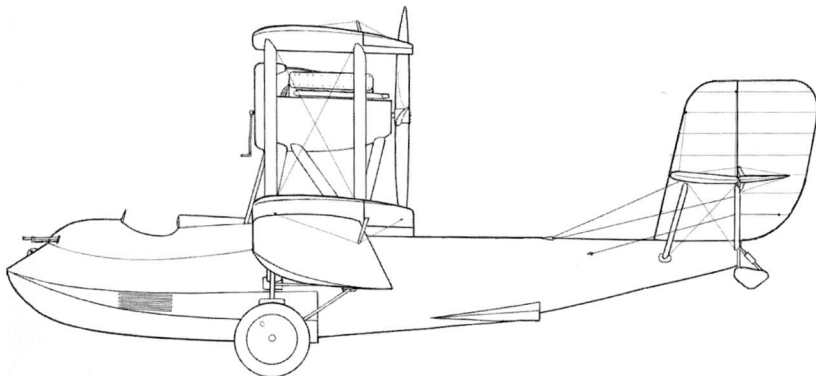

ft

Supermarine publicity:

THE SUPERMARINE 'SEA KING' MARK II

This single-seater amphibian flying boat has been designed as a high performance fighting scout, specially adapted for getting off gun-turret platforms of capital ships, or getting off and landing on the decks of aircraft carriers.

The strength and design of the hull are such that it can operate on and from the water under any weather conditions in which it would be possible to operate any other sea craft [boat] of equal size.

The manœuvrability of the 'Sea King' Mark II is one of its most important features. It can be looped, rolled, spun, and stunted in every possible way.

Longitudinally, the machine is neutral, and flying at any speed throughout its entire range either with engine on, gliding, or climbing, no load is felt on the control stick. This balance has been obtained entirely on the stabilising surfaces, and no mechanical adjustment by the pilot is required.

The hull is of circular construction with built-on steps, which can be replaced in case of damage. The steps are divided into watertight compartments, the top side being of single-skin planking, covered with fabric, treated with a tropical doping scheme.

The engine, a 300 hp Hispano Suiza, is mounted in a streamlined nacelle, which contains oil tank, radiator and shutters, piping, controls, etc. The whole unit is very accessible and the engine can be replaced very easily.

Interchangeability and ease of upkeep and repair have been carefully studied. The complete wing structure, including power unit, can be removed from the hull by withdrawing eight bolts. The wing structure consists of top and bottom centre sections, and top and bottom planes of equal span. One set of struts are carried on either side of the centre section. The top planes have a dihedral angle of 1° and the bottom planes one of 3°. The engine unit is carried on two sets of inwardly inclined N struts, and can be removed and replaced without interfering with any wing structure member.

The petrol supply is by pressure, and every effort has been made to reduce the length of piping and eliminate as much as possible the carrying of piping into the hull.

The amphibian undercarriage, which can be removed by the undoing of ten bolts in all, folds up under the wings, and when folded is well clear of the water. It is raised and lowered by a worm and bevel gear.

The pilot's cockpit is in the nose of the boat and gives an almost unobstructed view in every direction. The equipment consists of a complete set of instruments, anchor and cable, bilge pump, towing fairleads and make-fast cleats, boathook, engine and cockpit covers, towing bridle and lifting slings, 'Pyrene' fire-extinguisher, Lewis gun and six double trays of ammunition.

The tail-unit consists of fin and rudder of ample dimensions and a monoplane tail plane with reversed camber.

The steerable tail skid is carried under the extreme end of the hull.

and sea rudder was also employed. The aerodynamically balanced ailerons and rudder of Hargreaves' Sea King I were again abandoned and, at the same time, Mitchell also devised a very simple method for the removal of the undercarriage system which enabled the company to offer the choice of flying boat or amphibian with minimum extra production costs.

In other ways, the Sea King II reverted to previous Supermarine designs, particularly the Linton Hope hull construction: 'with built-on steps, which can be replaced in case of damage … divided into watertight compartments, the top side being of single-skin planking, covered with fabric, treated with a tropical doping scheme'. The wing-tip floats were the same full depth type as employed on the Baby, Sea Lion I and Sea King I and the tailplane outline was similar to that of the Sea Lion I or the Seal II but with the lower position of the latter; and its reversed camber ('of the depressing kind') continued the Baby method of counteracting nose-down tendencies at higher engine revs).

Once more, no orders for this type of aircraft were forthcoming from the Air Ministry but it was to feature prominently in 1922 as the Sea Lion II.

Mitchell's First Schneider Trophy Success
The two Schneider contests, in 1920 and 1921, took place in Italy and did not involve Supermarine or, indeed, any other country other than the sponsoring nation. The drawings overleaf of the winners of these two meetings, nevertheless show the continuing Italian tradition of sleek hulls, compared with those that Mitchell had to work with, as will become obvious later. In view of the importance of the Schneider contests to Mitchell's future career and Supermarine's flying boat approach to their next two entries, it is worth recording these intervening 'events' (*see* 'Chronicles' overleaf) as it shows the still precarious nature of the Trophy Competition and of flying boat development.

And when, in 1922, the Air Ministry issued Specification N6/22 for a floatplane single-seat fighter capable of operations from aircraft carriers, the Sea King, being a flying boat, was not in the reckoning – in spite of a potentially competitive performance. The winner of the profitable Air Ministry contract was the Fairey Flycatcher (196 machines built) which, in 1923, had a top straight-line speed of 133 mph; it was just as aerobatic and its short span allowed it to be struck down to the carrier hangars without folding wing arrangements; its extensive aileron-cum-flap arrangements produced very low minimum take-off and landing speeds.

With hindsight, it might thus be seen that the days of the flying boat fighter were numbered but, despite the success of the Fairey floatplane and there being no orders for the Sea King, Scott-Paine, the managing director of Supermarine, was still determined to continue with the fighter flying boat approach and sought publicity for it by deciding to enter a more powerful version of the aircraft in the 1922 Schneider Trophy contest. Apart from the precarious state of Supermarine finances, another, more patriotic, reason might have been a determination to prevent the Italians from winning the Trophy outright which, according to the rules of the contest they could do, in the forthcoming competition, given the two flyovers in the previous two years. As it turned out, the Supermarine entry was to become the only machine to try to stop Italy because the expected two French entries did not eventually present themselves for the competition.

Meanwhile, one of the fortuitous events which influenced the progress of Mitchell's career was the French winning of the Gordon Bennett Cup for land-planes for the third consecutive time. As with the Schneider contest rules, this third win gave permanent possession to the winning country and, more importantly for the present narrative, brought about the end of the other international prestige air race. Thus, despite the poor attendances and even worse performances in the fourth and fifth events, the Schneider Trophy Contest ended up as the only international speed competition left. This would obviously suit Supermarine very well because of their concentration on seaplanes and, in particular, on flying boats because the recent contests had set the trend for entering this type of machine rather than the improvised landplane.

In addition, the rules had been changed in 1920 to encourage a more practical type of aircraft rather than an out-and-out racer: 300 kg of ballast had had to be carried and this favoured flying boat designs – in fact, the winner that year had coped very well with appalling weather during the preliminary navigability trials. Although the ballast requirement was dropped in the following year, it was replaced by a watertightness test in which the aircraft had to remain afloat fully loaded for six hours after the navigability tests. Again, this rule tended to suggest the flying boat's suitability for the Trophy contest and it had no doubt been noted that the recent Italian designs to meet these requirements were, indeed, flying boats and that those which had failed had done so because of overambitious power uprating.

Schneider Trophy Walkover for Italy. Venice, September 21, 1920

Following the chaos of last year's competition, the choice of venue for the fourth Trophy Race was awarded to the Italian Aero Club. The Italians' bad feelings about last year's disqualification of Guido Janello were put behind them and they chose the Lido for the event.

The rules of the competition allowed the Aero Club of each country to nominate three aircraft but only Italy showed up with any machines this year. Their choice was to be made from four aircraft but their new types, the Macchi M.19 and the Savoia S.19, suffered from teething troubles and were withdrawn. Labour troubles then made for transportation problems, which resulted in the Macchi M.12 not showing.

And so it remained for the Italian Navy Savoia S.12, to achieve a flyover with Lt Luigi Bologna, covering the 230.68 mile course at an unhurried average speed of 105.97 mph.

Italy Again Unopposed in Schneider Trophy Contest Venice, August 7, 1921

This year's contest produced a considerable number of Italian entrants but, again, the new designs by Savoia did not fare well, the S.22 crashing before the eliminating trials and the fast, tiny S.21 (20' top wingspan) was withdrawn when the pilot entrusted with it, Guido Janello, fell sick. The Savoia S.13s were pretty much the same as the one flown by Janello at Bournemouth and were eliminated and the two Macchi M.18 flying boats were also withdrawn because they proved not fast enough (having been designed as passenger-carrying aircraft). So only two of the earlier Macchi M.7s were selected along with a third and favourite machine, the Macchi M.19, similar to the one that had been intended for the 1920 contest. At least there was some foreign competition this year but the event once again became an Italian flyover when the floats of the solitary French entry were damaged during the navigability tests.

The apparently inevitable Italian win almost failed as only one Italian machine completed the course. Arturo Zanetti's M.19 (the only two-seater ever to compete in a Schneider competition) achieved a very creditable 141 mph before the crankshaft failed early on and the aircraft caught fire – the pilot and his engineer both escaped unhurt. Then Piero Corgnolino's M.7 ran out of fuel with the finish line in sight. This chapter of accidents left Giovanni de Briganti the winner flying a Macchi M.7 at an average speed of 117.85 mph.

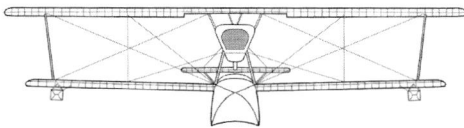
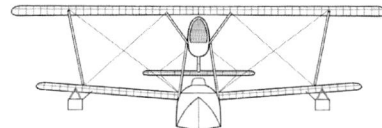
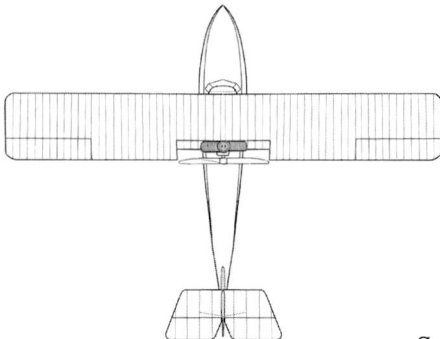
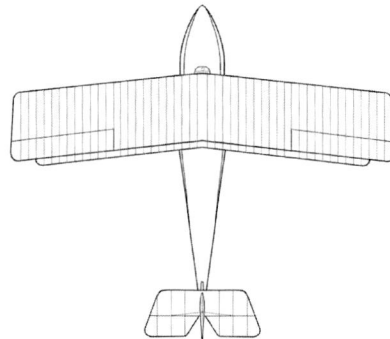

Savoia S.12.   *Macchi M.7.*

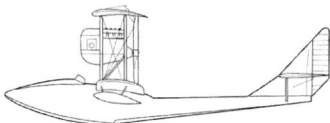
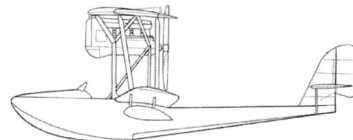

SEA LION II

Despite the omens favouring flying boats, any such Supermarine entry was unlikely to have any financial backing from its government, unlike the firms from Italy. And the uncertain financial outlook of the Supermarine company at this time was such that, when its Managing Director entered an aircraft, he had obtained the loan of a Lion engine from the manufacturers, a high speed propeller, petrol and oil from other companies, and having negotiated a reduction in insurance rates. Scott-Paine also negotiated with the Royal Aero Club for the payment of the competition entry fee and shortly afterwards a payment of £100 towards the Company's costs. Nor was the company intending to incur the cost of designing and building an entirely new airframe. The fuselage of the salvaged Sea Lion I was with the Science Museum at South Kensington and so the fuselage of the Sea King II (which, in any case, was aerodynamically cleaner) was utilised.

Sea Lion II fuselage and Napier Lion engine (Mitchell standing to right of Scott-Paine). For further photograph of Sea Lion see p. 17.

Mitchell, no doubt having taken note of the sleek Savoia machine at Bournemouth, aimed for increased speed by making the entry of the fuselage (which was originally shaped to house a gun – *see* drawing earlier) somewhat smoother and his simplified method for the removal of the undercarriage system also enabled the hull to be easily stripped of the extra weight and drag of this item. Also, as the Napier replacement developed 150 hp more than the Hispano-Suiza engine originally fitted, he was able to decrease the area, and therefore the drag, of the wings by reducing their width.

Another modification, probably mentioned to Cozens by his neighbour, Biard, was necessitated by the pilot's refusing to test fly the aircraft until the rear fuselage had been stiffened up (*see* quotation on p. 62). Again, in response to the extra power of the engine, an additional increase in fin area was called for. Mitchell achieved this with least expenditure of time and money by merely modifying the vertical surfaces above the tailplane – the leading edge of the fin was given a pronounced forward curvature which proved to be effective but certainly won no prizes for elegance.

Although the finished machine was, in essence, Mitchell's Sea King, it was now registered as G-EBAH and named the Sea Lion, thus drawing attention to the name of the loaned engine. It was also designated a Mark II, to distinguish it from Hargreaves' earlier design but, as a result, misleadingly suggested that it was a direct development of it – contrary to its actual pedigree, as described earlier. Another indication of the unadventurous state of British flying boat technology was that it was more cluttered with struts and wires than the Macchi M.7 winner which had not been particularly impressive in the previous competition.

Because of bad weather and the date of the competition being put forward by fourteen days, there was little time to test out the result of these modifications. Indeed, there was a distinct possibility that the Sea Lion would not survive to compete at all for Biard, Supermarine's test pilot, barely avoided crashing into a litter of ships when his engine stopped two hundred feet above Southampton Docks on the take-off of the very first test flight. Then Supermarine ran into difficulties over the delivery of the aircraft to the Naples venue; most fortunately, the General Steam Navigation Co., with whom Scott-Paine had business contacts, came to the rescue with the assistance of the SS *Philomel* which diverted specially to Naples with the aircraft and personnel, free of charge. But persistence with the challenge was justified as the test flying had shown that the machine was not only very manœuvrable but was capable of coming close to the 160 mph target thought by Scott-Paine to be necessary for a win – given the indifferent showing of the Italians in the previous two competitions.

On arrival, Biard once more nearly came to grief when he flew the Sea Lion over Vesuvius where the 'thermal' from the crater swept him upwards and left him several miles away before he could return. He was a notorious practical joker but his undoubted professional skills were well exemplified in Naples. Putting sightseeing behind him, Biard gave the Italian opposition no clue as to the real capability of his aircraft by taking the corners of the competition circuit widely in practice whenever he thought he was being observed. He also kept his practice speeds close to that expected of the machine in Italian aviation circles – the estimated speed of the Sea King I with a Napier Lion engine had been 141 mph and that achieved by the Sea Lion I had been 147 mph.

The sixth Schneider Trophy Contest, 12 August, 1922 at Naples
The rules for 1922 were similar to those of the previous year with the navigability tests involving taxiing over the start/finish line, ascending and alighting, taxiing for half a mile twice at no less than 12 mph, flying round the circuit, and taxiing again over the line. These requirements were followed by the

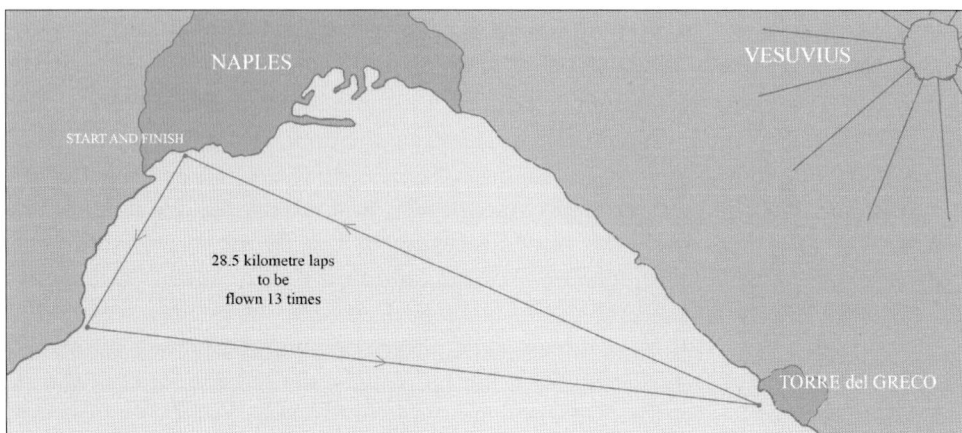

Schneider Trophy Contest Course, 1922.

Supermarine SEA LION II (1922)

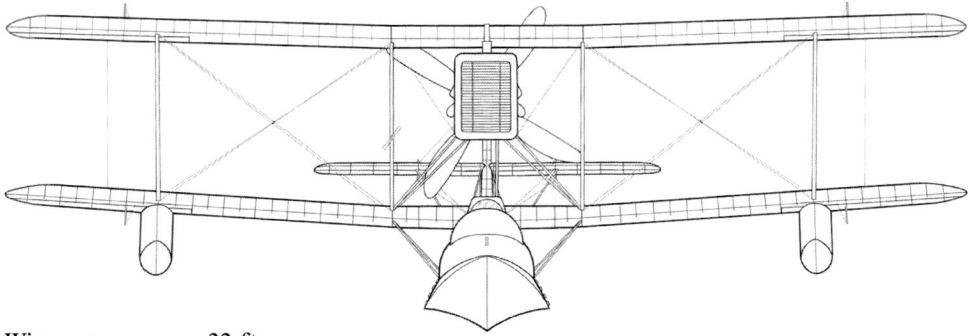

Wingspan	32 ft
Wing area	384 sq ft
Loaded weight	2,850 lb
Maximum speed	160 mph

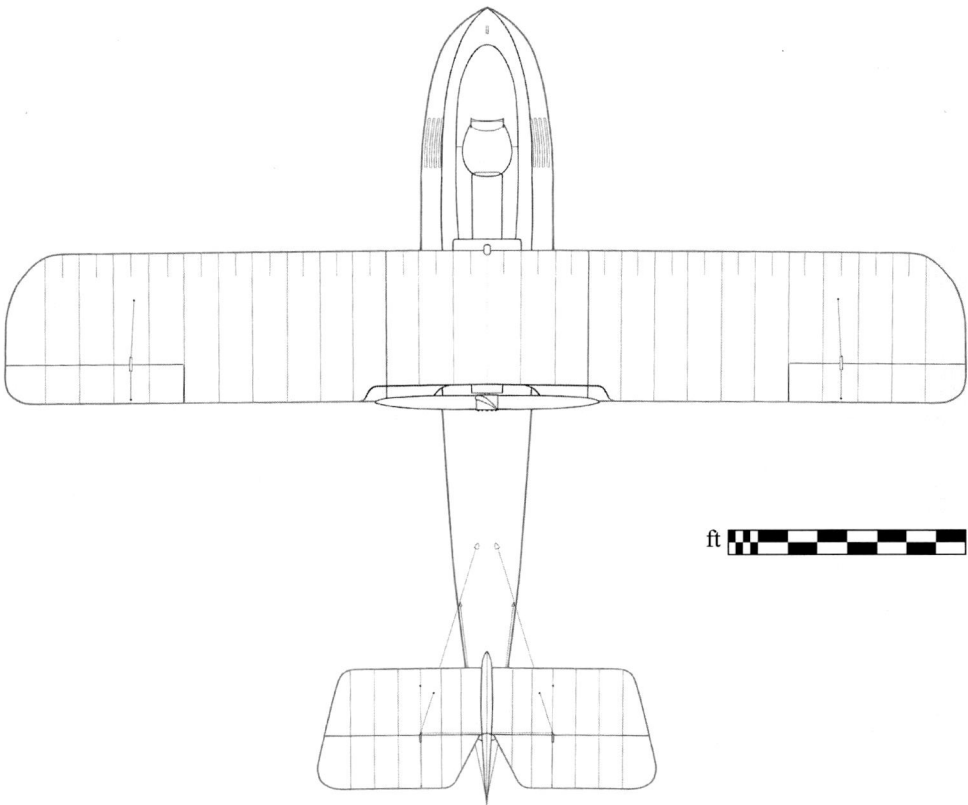

ft

G-EBAH

six-hour flotation test, which had been introduced in 1921, and the course was new, having thirteen laps instead of the previous sixteen. As the distance overall was to be 200 nautical miles, the longer straights, with only three corners, would obviously be good news for the faster machines – and the Sea Lion's Napier engine was 150 hp more powerful than the rival Italian Savoia S.51 and 190 hp more than the Macchi M.7 and M.17.

After their previous experiences, the Italians had not opted for ambitious engine upratings but had put their faith in airframe innovations. Unknown to Scott-Paine, all three Italian aircraft embodied the tried and tested approach of previous models but the hulls were sleeker and the engine nacelles more streamlined. In addition, the S.51 also had a much smaller lower wing. This 'sesquiplane' arrangement was designed to reduce 'biplane inter-ference' caused by lower pressure above the bottom wing affecting the higher pressure below the upper wing. A lower wing was still felt ne-cessary to give light structural integ-rity to the upper one but at least the sesquiplane configuration allowed the removal of complex drag-inducing wire bracing.

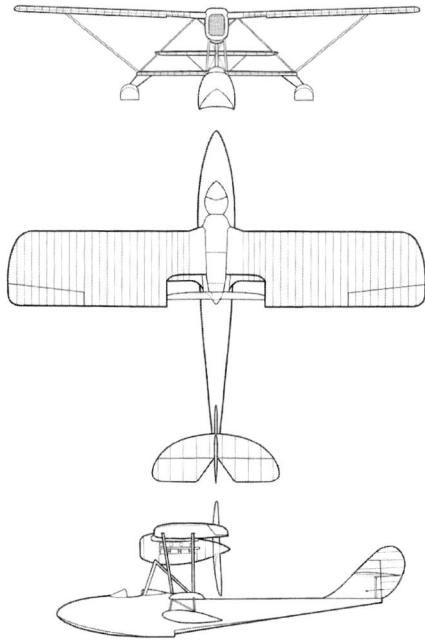

Unfortunately, the short span of the lower wing, allied with the high mounted engine and the small volume tip floats, were insufficient to prevent the S.51 from capsizing in a squall during the six hour mooring-out test, whereupon Scott-Paine sportingly raised no objection to its being salvaged and returned

Savoia S.51.

to the contest. Because of the heat, the contest was delayed until 4 p.m. when Biard, who had been drawn the first to start, took off. His shirtsleeves and flannel trousers gave no hint of the speed that he had kept from observers but, in his eagerness to surprise, he came so close to the first marker balloon that he decided after that to be a little more circumspect.

Despite the time required by his take-off, Biard's first lap was calculated to have been flown at over 150 mph thanks to the Lion engine having just been retuned to suit the local conditions. According to various accounts, including Biard's 'ghosted' autobiography, the Italians suddenly realised that the Macchis could not compete if Biard kept going at that rate and so, when their pilots joined the circuit, they tried to prevent the Englishman from getting round them by flying close together in order to make overtaking difficult. The consistent claims by Supermarine for the exceptional manœuvrability of their 'scout' flying boat type were then demonstrated when, as Biard narrated, he was able to close up, keep control in the prop wash, pull upwards and dive round and ahead of them at full throttle. Unfortunately, Edward Eves' more recent book records lap details which suggest that his overtaking in this way was confined to Zanetti's M.17; Pegram, on the other hand, notes that Biard passed Zanetti once and Corgnolino twice. In neither account did he encounter the Italians bunched together. It is at least certain that Biard was able to ease back to avoid overtaxing his engine and, as a result, his sixth lap was only 6 seconds better than that of Passaleva's S.51, then on its fourth lap. But its previous ducking had affected the bonding of the propellor's wooden laminations and so the Italian machine had to be flown with the engine throttled back to avoid excessive vibrations.

Sea Lion II rounding a course marker balloon.

It would appear that Mitchell's upratings had produced a straight-line speed of about 160 mph – nearly 30% faster than his Sea King II – and that he was thus able to taste the sweet pleasure of success with his design winning the Schneider Trophy race for Britain at an average speed of 145.7 mph. Biard flew an additional two laps, gaining in all the following first FAI World Records for seaplanes:

(i.)   Duration – 1 hr 34 min. 51.6 sec.
(ii.) Distance flown – 230 miles
(iii.) Fastest time for 100 km closed circuit – 28 min. 41.4 sec. (130 mph)
(iv.) Fastest time for 200 km closed circuit – 57 min. 37.4 sec. (129.4 mph).

* * * * *

On Mitchell's return to Southampton, all the city dignitaries turned out in full ceremonial dress in recognition of Supermarine's international success. In their subsequent publicity, Supermarine tried to capitalise on their achievement by pointing out (rather disingenuously) that the aircraft was, basically, a standard machine:

> The machine is in general essentials of the same design as the 'Sea King' type already described, but is fitted with the Napier 'Lion' engine of 450 hp.
>
>   The fact that the machine was in all essentials a standard type suitable for ordinary service duties, but was, nevertheless, able to defeat the specially designed racing machines with which it had to contend, is a very great tribute to the excellence of Supermarine design.

Subsequent lack of military interest in the Sea Lion type must have been a disappointment and its designer would have been only too aware that the Savoia S.51 which was beaten in this current Schneider competition should really have been the winner had it not been for the handicap of a damaged propeller – and this with an engine only two thirds as powerful. One also wonders if Scott-Paine, bearing in mind his indebtedness to sponsors, would have been so sporting as to allow it to compete if he had known what the S.51 was really capable of: on December 22 of that year, it captured the world speed record for seaplanes at 174.08 mph. Nevertheless, a win was a win and Supermarine and its designer profited from the result. The next two Schneider contests were to show how transient was this success.

Cozens' account of the 1922 event:

The price for such an engine was beyond the means of a such a small firm as Supermarine but on the other hand they offered an ideal opportunity to test its capabilities, so Montague Napier loaned it, knowing that if the Sea Lion won the publicity would repay him. It was a water-cooled 12 cylinder with 3 banks of 4 cylinders in arrow formation, short and sturdy for its power output. Each cylinder barrel was a separate casting but the cylinder heads were in a single casting for each block so that the whole block bolted together into a well designed and efficient unit, sturdy and reliable. Each cylinder had its own copper water jacket. The tall Mr Pickett was responsible for the engine fitting and his skill and experience served Supermarine well for many years.

When the new racer was taken out on to the quay and the powerful Napier was run up Captain Biard took one look at the vibrating tail and said he would never fly that aeroplane, but, with typical press-on spirit the riggers wrapped doped fabric round the fuselage and made the whole unit stiffer. The next crisis was when the Italians advanced the date of the race by fourteen days and bad weather cut down Captain Biard's chances of getting used to the Sea Lion, and this was further jeopardised by a forced landing which began with the engine cutting out over the Dock. However, when he had had a few more flights he was satisfied and the speed and handling proved very good, indeed it was faster than any flying boat or seaplane of that time. Then, with the limited time available, it was doubtful if they could get the Sea Lion to Naples in time but the General Steam Navigation Co. agreed to take it and it was hurriedly dismantled, put into a crate and on to a lighter, and one of Ray's tugs took it down to the Solent and the freighter *Philomel* lifted it on board and took it to Naples.

Pre-race spying and counter-spying on both sides was all part of the event. This atmosphere continued throughout the whole series and it was the policy of each competitor to arrive at the start of a competition with a machine that was ahead of its rivals by virtue of some secret and outstanding advantage which was not revealed until it was too late for anyone to copy. In the case of the Sea Lion this meant that the wingspan was cut down to an absolute minimum and as the trials at Woolston had been curtailed even Captain Biard was not too well practised as to the machine's behaviour.

He kept his speed down in the practice flights but was quietly getting used to the course and conditions, and his engine fitter, Mr Pickett tuned up the engine to the higher temperature of the Bay of Naples, so that when the race started Biard was reasonably prepared ...

When Captain Biard and his victorious team came to the Floating Bridge with the great prize held above their heads no one bothered whether it was a Cup or a Trophy – everyone called it a cup, certainly Scott-Paine. I had parked my bicycle outside the Woolston Picture House and I saw the Supermarine workers run down to meet them. They had taken the two swivel chairs from the office and fixed them to poles and they lifted Captain Biard and Scott-Paine in the chairs shoulder high and carried them round the works ... Fireworks were let off and there was some horn blowing ... [*One suspects that Mitchell's relatively recent arrival in the firm at the time as well as his temperamental self-effacingness account for there being no mention of him in the celebrations.*]

— *Chapter Three* —

1922 to 1925
Military and Commercial Successes; Competition Failures

Despite the 1922 Schneider Trophy success, no military orders had resulted for the Sea Lion or for a development of the type, although all was not lost as the aircraft was purchased by the Air Ministry and used for research into high speed seaplanes. Nor were there any orders forthcoming for Mitchell's Seal II. After the hopes at the beginning of the decade, the company had now to come to terms with the post-war Anti-Waste League and the resultant work of the Geddes Committee Report which led to a slashing of all Government expenditure. Whilst, by the armistice, the RAF had been the largest air force in the world, with 22,000 aircraft and 188 operational squadrons, the new Secretary of State for Air, Sir Samuel Hoare, reported that, in 1923, only 371 front line aircraft remained, either in the British Isles or abroad, and assessed the current situation thus: 'Orders for military planes had almost come to an end and a demand for civil planes did not yet exist ... Only two thousand five hundred men and women were left in the industry and the few firms engaged on machines and engines were on the verge of closing down.'

On the other hand, the actions mentioned earlier of 'control without occupation' by the RAF had been adopted as a very economical and swift-acting alternative to the employment of large army land forces in the policing of colonial and League of Nations mandated territories and this new approach would need to be backed up by support for the ailing aircraft industry if there were to be an adequate response from the currently depleted provision abroad. And at home, it had been accepted that, over the next five years, thirty-four new squadrons should be formed, bringing the air defence of Great Britain up to fifty-two squadrons by 1928. In the event, the home squadron numbers only rose to thirty-four by the date proposed but at least the 1923 Air Estimates of £10,783,000 had risen incrementally to £16,042,000 by this time.

SEAGULL II
The first positive result of the new situation was seen when Commander James Bird, who had joined the Board at Supermarine in 1919, approached the Air Ministry and subsequently received a letter which suggested that it 'might be inexpedient' to close down the works entirely as Supply and Research were considering an order, 'the exact amount of which cannot yet be stated, but which might approach 18 machines, spread over the period ending March 31st, 1924'.

This Air Ministry lifeline first came in the form of a cautious, initial order for two flying boats of the Seal II type. By the time that this order was received in February 1922, the original fin had been further modified and later again increased in area, in view of the more powerful Napier Lion II engine fitted – again in a tractor layout. The two new aircraft, N158 and N159, were completed by March 1922 and, by July of that year, the type had been named Seagull II – in view of the modifications just mentioned. It was then displayed in the same year at the third annual RAF Pageant, Hendon.

A Seagull II with Mitchell (left) and Biard.

One particular improvement of the aircraft ought to be mentioned: the fuel tanks had now been moved from the fuselage to positions under the centre-section of the top wing, so supplying petrol to the engine by gravity feed. (In 1917, the Felixstowe flying boats, which had fuel tanks conventionally placed in the fuselage, had suffered so many forced landings from blocked pipes and fuel pump failures that a contemporary report stated that 'our real enemy is our own petrol pipes'. A Supermarine comment on the 1921 Sea King II design had been that 'the petrol supply is by pressure, and every effort has been made to reduce the length of piping and eliminate as much as possible the carrying of piping into the hull'.)

Had the First World War continued after 1918, such petrol feed problems where high-positioned engines are concerned would, no doubt, have been more speedily overcome by adopting gravity feed, even though stability considerations associated with placing large (and heavy) amounts of fuel under the top wing centre-section were not trivial. (The Fairey Fawn which had large tanks placed on the top wings was described at this time by its Martlesham test pilot, H. F. V. Battle, as 'a shocker', being 'unstable fore and aft as well as laterally'.) It is thus a reflection of the very slow pace of aircraft development after the Armistice that Supermarine drew particular attention to their adoption of gravity feed as late as 1923, with the Seagull II. In fact, one of the significant differences between the publicity for the Seal II (*see* earlier) and that for its direct descendant relates to the latter's fuel storage and delivery as well as to its resultant improvement in crew accommodation: 'The petrol tanks are carried under the top centre section plane, and thus direct gravity feed is obtainable'.

Subsequent experience of the high position of the Sea Eagle's two fuel tanks was to confirm that the basic Supermarine amphibian configuration, with its suspended boat-like hull, made possible such new fuel arrangements without stability problems. Despite the inconvenience of filling tanks in the higher position – especially when an aircraft was not on a slipway, there was also an important additional bonus: as a consequence of moving the fuel tanks from the hull, Supermarine was able to announce that 'inter-communication between

Supermarine SEAGULL II (1922)

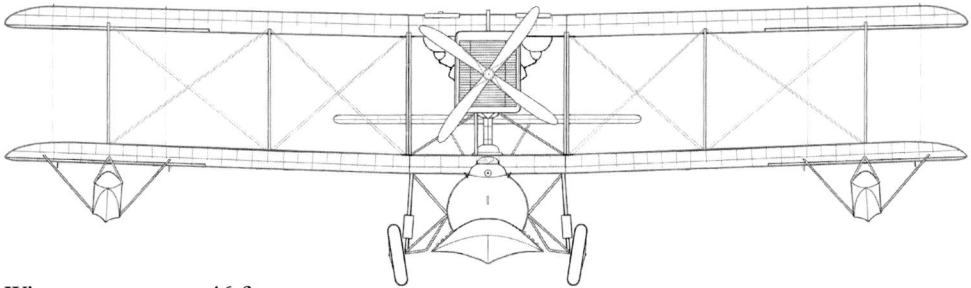

Wingspan	46 ft
Wing area	593 sq ft
Loaded weight	5,690 lb
Maximum speed	98 mph

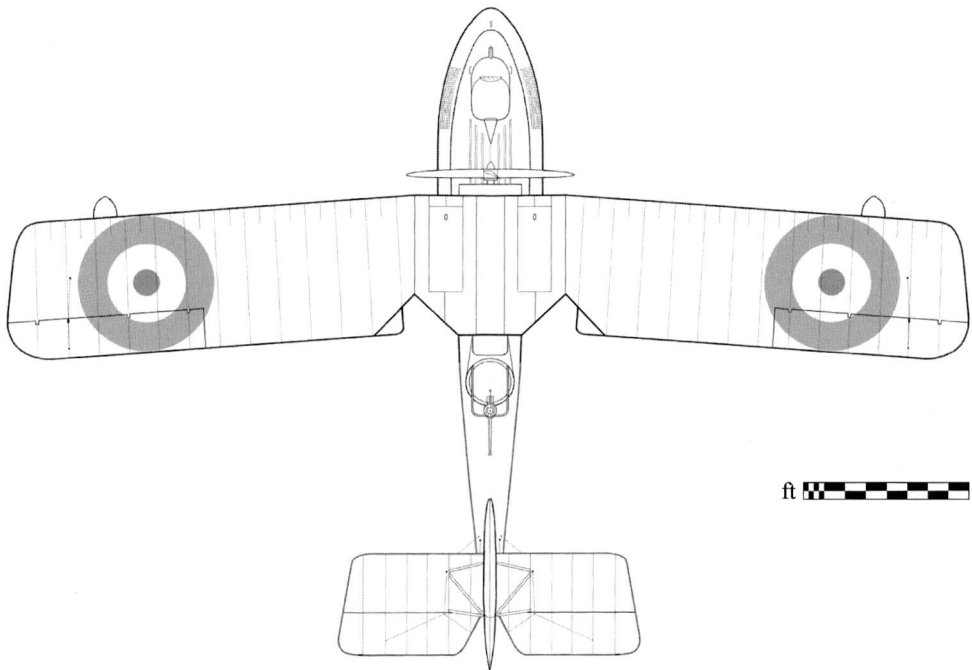

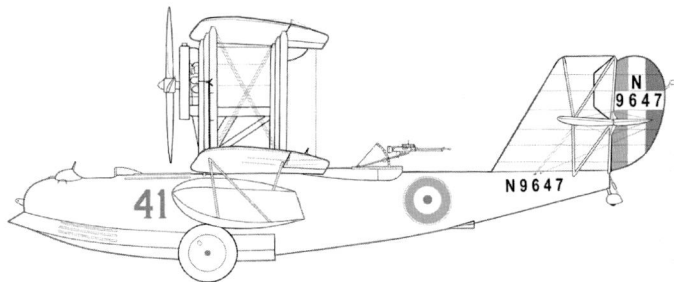

ft

N9647

N
9647

41

N9647

The first Seagull II of the original order.

Seagull II of 440 Flight, off south coast of Malta. Note rungs on rear centre-section strut, for access to fuel tanks.

crew has been considered fully, and a through passage is arranged for this purpose'. Thus Mitchell was not only solving possible supply problems from petrol in lower positions but was also making an important step forward in the matter of military crew communication – something that was particularly appreciated when the Southampton flying boat came into service two years later. In passing, it should be noted that the constructional methods of the Linton Hope hull here conferred another advantage as there were no structural bulkheads to be weakened or bracing wires to be forfeited by the cutting through of a passageway between the pilot and the other crew members.

A competitive test on HMS *Argus* between the Seagull and the Mark VII version of the Viking (whose predecessor had competed successfully with Mitchell's Commercial Amphibian in 1920) had found in favour of Mitchell's machine and, thereafter, Vickers concentrated upon landplanes. An RAF order for five Seagulls, N9562 to N9566, was received in February 1923 and the Under Secretary of State for Air, the Air Vice-Marshall, and the Director of Research visited the Supermarine works on February 23, 1923, to view the progress of the order.

The Ministry was sufficiently pleased with the aircraft that a further order for five additional Seagulls (N9603 to 9607) was received, and this was followed by a requirement for another thirteen (N9642 to N9654). These aircraft equipped No. 440 (RAF) Fleet Reconnaissance Flight and some were placed aboard the aircraft carrier HMS *Eagle*. An additional machine was again sold to Japan. By this time, the wing tip floats had been redesigned, the wings given a slight sweep back, the ailerons redesigned, and the fin area further enlarged.

The SEA EAGLES
Whilst the Seal II was being developed into the Seagull II, Supermarine began to see possibilities of again producing and selling a civil aircraft as the Air Estimates now also included allocations for civil aviation, although they represented something less than 2.5% of the total. Thus it was that, in 1922, the Air Ministry gave approval for an air service between Southampton, Cherbourg and Le Havre. The route, with a subsequent extension to the Channel Isles, was to be operated by an air service named the British Marine Air Navigation Company and Hubert Scott-Paine and James Bird of Supermarine were to be its directors. Not surprisingly, the first Supermarine aircraft for this service was already being built when the Air Ministry granted the company a subsidy of £10,000 and agreed to pay £21,000 for aircraft and spares (later revised substantially downwards as the air miles generated were less than the Company had undertaken to fly).

The aircraft which Mitchell was required to design was to be powered by the Rolls-Royce Eagle IX engine and thus Supermarine were able to continue the Company's practice of finding maritime animal names beginning with 's' which, if possible, incorporated the name of the engine to be used. For the new Sea Eagle, Mitchell reverted to the customary pusher configuration for single-engined flying boats and went back to the more boat-like hull shape of the larger Channel and Commercial Amphibian designs. In fact, the fore section of the Sea Eagle resembled a cabin cruiser of the time, with its high, pointed prow, enclosed accommodation for passengers, large windows and grab-rails running the length of the passenger compartment above the cabin and along the top of the hull. As the two planing steps were also joined by a continuous hard chine which ran three-quarters of the hull length, it embodied more than any other Mitchell design the original Pemberton Billing concept of 'boats which fly'; indeed, the Sea Eagle was somewhat reminiscent of his very early P.B.7 (*see* sketch p. 80).

As with the earlier, intermittent, service using Channel flying boats, the Air Navigation Company had to consider local hangarage for the aircraft and so wing folding was adopted. It would seem that width rather than length was the important consideration because a forward folding arrangement was again adopted (which reduced the width of the Sea Eagle by 54% whilst *increasing* the length by 15%). There was also the structural advantage of folding at the main spar, with no possibility of a wing folding backwards in flight, although this arrangement necessitated a cut-out in the leading edge of the wings, which did nothing for aerodynamic efficiency.

Mitchell continued the practice of gravity feed for the engine of this latest flying boat with apparently little qualms about stability problems, for the fuel tank (and subsequently a second tank) was now attached to the top of the centre section. On 28 June, a *Flight* correspondent wrote that 'this machine represents a great step forward in the development of the seaworthy [commercial] amphibian' having appreciated the 'most important innovation' that, in place of the usual tank in the hull, 'the main petrol tank has been mounted on top of the top plane, so that direct gravity feed, with its attendant simplicity and freedom from breakdown, can be used'. The writer also added, 'the fact that the engine is mounted high above and some distance aft of the cabin has resulted in reducing the noise audible in the cabin to a minimum, and as a matter of fact, in the "Sea Eagle" it is possible for the passengers to converse in ordinary tone of voice, without having to shout to one another'. (One remembers Biard's earlier 'communication problem' with Loewenstein in the Channel flying boat.)

One departure from all previous (and future) practice was the use of a pronounced stagger of the two wings, as the weight of the forward passenger cabin and its six passengers necessitated bringing the centre of lift of the top wing well forward of the engine. Again, one suspects that Mitchell's usual preference was for the simplicity of directly opposed biplane wings.

Sea Eagle with Biard aboard (Note sea anchor below cockpit and original, single, fuel tank on top wing).

The first of the completed Sea Eagles had made its maiden flight in June 1923, and received its Certificate of Airworthiness on 11 July. Two days later, Supermarine entered the new aircraft in the King's Cup Air Race that had been initiated the year before, also to encourage aviation development. As it was a handicapped event, the entry of a commercial flying boat might not seem too strange but the carrying of four passengers must have had much to do with the Company being mindful of publicity generated by air races. Unfortunately, circumstances involving a burst tyre and its replacement led to the aircraft being disqualified.

On the 5th of the next month, the Director of Civil Aviation at the Air Ministry, Sir Sefton Brancker, came to Southampton and was given a display of the machine's ability to negotiate the (usually) crowded seaway as well as a demonstration flight. He announced himself to be well satisfied with the Sea Eagle's potential contribution to the development of civil aviation, both in terms of performance and comfort and, along with other senior members of his department, had another flight nine days later. (As we shall see when considering the Southampton flying boat, such good impressions made on these members of the Air Ministry were very valuable not long afterwards.)

Particular comment was made on the very sensible placing of the passengers and, in the following publicity, the Company makes reference to the advantages of this arrangement:

This machine was specially designed as a commercial amphibian or flying boat for passenger carrying work. It carries six passengers and pilot, with fuel for a distance of 230 miles. Extra tankage is fitted so that the range can be increased by reducing the number of passengers.

The passengers are accommodated in a roomy cabin in the fore part of the hull. This cabin is very comfortably fitted out. Its position in front of the engine makes it very quiet and free from engine exhaust, gases, oil, etc.

It is very efficiently heated and ventilated, and is fitted with sliding triplex windows along the two sides for use in the warm weather. The machine is very strongly built and very seaworthy, and has proved itself quite safe in the roughest of seas usually experienced in the Channel. It is fitted with either a Rolls-Royce 'Eagle IX' engine of 360 hp or a Napier 'Lion' of 450 hp.

Supermarine's experiences with the Channel service and with the Commercial Amphibian had obviously influenced Mitchell to give considerable thought to an enclosed cabin arrangement, such that one passenger recorded descending into the Sea Eagle and finding 'a delightful little room' that the Company had fitted with 'reposeful armchairs'.

A Sea Eagle on the Supermarine slipway prior to a demonstration flight for Sir Sefton Brancker. Note passenger hatch at front.

Intermittent proving services began in August and regular daily services between Southampton and Guernsey began on 25 September, 1923, so constituting the very first British scheduled flying boat service; it was advertised to leave Woolston at 11.15 a.m. and return from St Peter Port at 3.30 p.m. (The French section of the service did not materialise.) The service, often with breaks due to bad weather, continued with the Sea Eagles for the next five years, even though the single fare to the Channel Isles was not cheap for the 1920s at over £3 single and £7 return. Compared with boat transport, however, the normal flight time of one and a half hours was very attractive although, in adverse wind conditions, it might be almost an hour more. Nearly four years later, the fleet of three Sea Eagles was down to one: G-EBFK having crashed on May 21 (due to a bird strike, according to Cozens) and G-EBGS was rammed and sunk when moored at St Peter Port on the 10th of January, 1927 – a reward of £10 for the identity of the culprit was never claimed.

In the event, Mitchell's machines had not only operated the first scheduled flying boat service in Britain but they also had the distinction of forming part of the basic fleet of the organisation which eventually became British Airways.

Supermarine SEA EAGLE (1923)

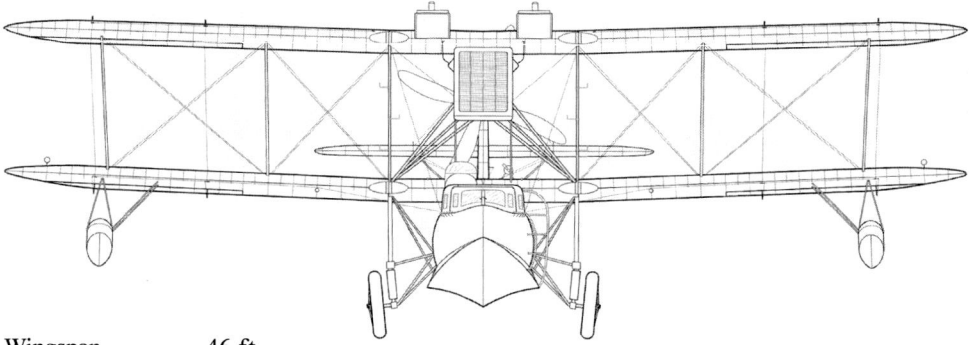

Wingspan	46 ft
Wing area	620 sq ft
Loaded weight	6,050 lb
Maximum speed	93 mph

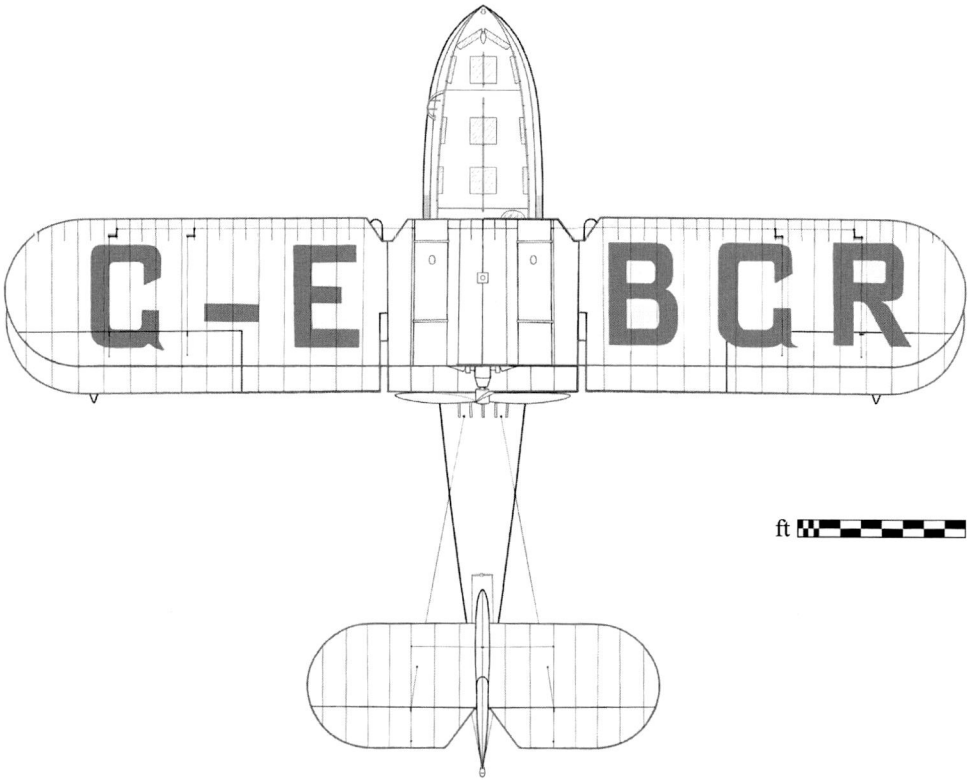

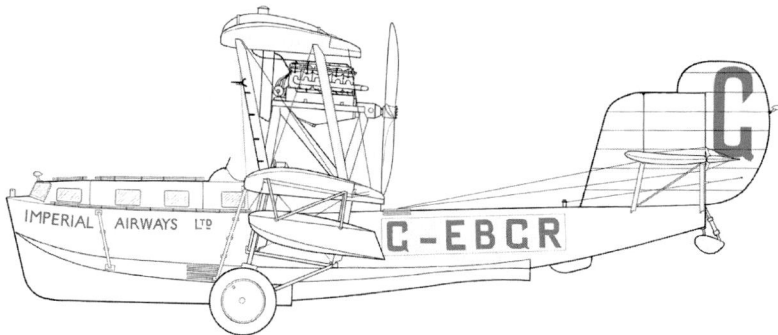

G-E BCR

G-EBCR

IMPERIAL AIRWAYS LTD

G-EBCR

ft

The third Sea Eagle being christened 'Sarnia' after its maiden flight to Guernsey, 13 October, 1923.

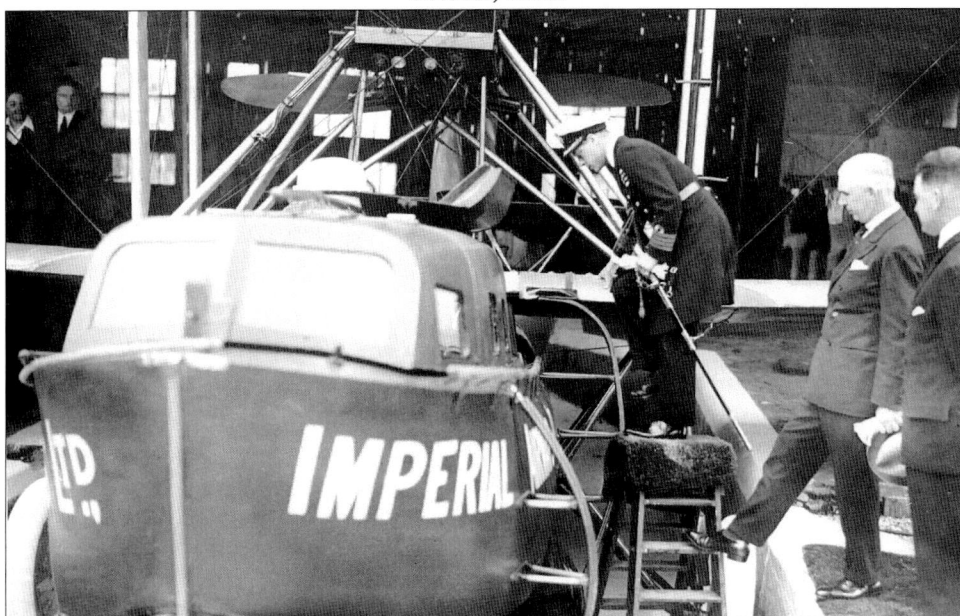

The visit of HRH Edward, Prince of Wales, 27 June 1924. The Prince's cockpit inspection accompanied by Imperial Airways directors, Hubert Scott-Paine and Col. Frank Searle (anxious that the royal visitor should not fall off the stepladder).

On 31 March, 1924, Imperial Airways Ltd was incorporated as the 'chosen instrument' of the British Government for developing national commercial air transport on an economic basis and the British Marine Air Navigation Co. was one of the four companies taken over for this purpose. Scott-Paine was a member of the Imperial board of directors and so was able to use his influence to keep the Solent area, and Southampton in particular, to the fore in British commercial flying boat operations. The two Sea Eagles which were remaining by that time now had their fuselages painted with prominent 'Imperial Airways' lettering and they continued their accustomed service to the Channel Isles under the control of the new national company from May 1, 1924.

In the following year it was reported in *The Aeroplane* that the Sea Eagles, 'during their hibernation have grown another 100 hp' and are 'now equipped with Napier "Lions"'. The last of the three Sea Eagles, G-EBGR, was finally retired in 1928, thus justifying Supermarine's claims that this type was 'very strongly built and very seaworthy'. A photograph from a correspondent to *The Aeroplane* showed a Sea Eagle hull at Heston Airport in 1954. At this time, the Imperial Airways marking had been painted out and the extant letters G-EBGS were now of a different character from those seen in photographs of the machines when in service. Andrews and Morgan state that the hull in question was that of G-EBGR, the last survivor of the Eagle fleet, rather than that of G-EBGS retrieved from Guernsey. Whatever the truth of the matter, a hull was presented to BOAC in September 1949 and intended for restoration and display at the new London Airport; but nothing came of this proposal and this piece of industrial archaeology was burnt on 13 February 1954.

Sea Eagle in Imperial Airways livery at the Woolston terminal. Note 'Airport' on roof of terminal (first use of word?) and leading edge cut-outs for forward wing folding; also note fixed ladders for passengers and for crew.

Cozens' recollections of the Sea Eagle service:

The land planes flying from Croydon to the Continent carried wireless transmitters and receivers and had the benefit of a simple direction finding service, by which stations at Pulham in Norfolk and Lympne in Kent and Croydon could take bearings on an aircraft when its generator was running. Then Pulham and Lympne passed their bearings to Croydon, who would plot them, together with its own bearing, and so reach a 'fix' which would be passed to the aircraft pilot. The Sea Eagles … operated outside the sector where the system could be applied, but they could speak to their bases at Woolston and Guernsey.

I lived a mile from the Supermarine works and could hear an engine start, in those days a rare sound and quite distinctive from the riveters at Thornycroft or the rumble from the coaling wharf in the Docks … From the front gate I could watch the machine fly across the Dock Head and then go indoors and pick up the headphones of my crystal set. Soon I would hear 'leaving the coast at Beaulieu', and 'Passing the Needles', and about 90 minutes later and hoping the cat's whisker [fine wire detector] had not moved I would hear 'Passing the French coast at Ushant' and sometimes very faintly 'Approaching Guernsey and winding in'.

When the time came for the return trip I would hear the call in the reverse order … Then I would cycle to the Floating Bridge in time to see the flying boat taxi up to its mooring … 'Winding in' referred to the trailing aerial, a long wire with a weight on the end, which had to be wound on to a drum before the Sea Eagle could touch down … By this time the Airline's operations were carried out from buildings along the side of the works and the staff were glad to have proper offices instead of the exposed Jopling's Wharf and along the side of the boatshop the words 'Woolston to Guernsey in 90 minutes' were painted in blue and white. The Sea Eagles were finished in copal varnish and white, and the windsock was an orange colour and the motor boat varnished, making the whole operation smart and attractive …

Sea Lion III – An Unhappy Return to the Schneider Trophy Competition

Whilst the Sea Eagle was being developed for the Southampton–Guernsey service, Mitchell was asked to prepare an aircraft for the 1923 Schneider Trophy competition. As the next Schneider Trophy contest was to be held in England, it was to be expected that Supermarine would be only too happy to capitalise on their 1922 publicity by competing, successfully it was hoped, without the cost of overseas travel and accommodation. Indeed, they could hardly have been better sited as the competition was to be based at the Saunders Roe factory on the Isle of Wight, less than 20 miles from the Supermarine works at Woolston.

But, whilst the second batch of five Seagulls in February 1923 was a welcome addition to the order book, it was still part of the Government's initial lifeline to the ailing aircraft industry; only with the large-scale production of the Southampton, which began two years later, might Supermarine have felt justified in the cost of designing and building a one-off specialist racer; and, as the top speed of the Sea Lion II was significantly less than the record breaking Savoia S.51, Scott-Paine did not immediately respond to the challenge.

When Scott-Paine was persuaded to submit an entry, he confined himself to asking his Chief Designer to do his best with the previous year's Sea Lion II airframe which had presumably not gone back to the Air Force but was still with the company. Mitchell designed new wing-tip floats, which offered less frontal area, mounted them on streamlined struts and added fairings around the main strut attachment points. Because an extra 75 hp was available from the new Lion engine, the rudder and fin were increased in area, with the resultant redesign looking somewhat like an extension of his early Sea King II outline and certainly less improvised than that of the Sea Lion II.

Mitchell fitted the uprated Napier Lion III engine and radiator into a more streamlined nacelle and the more powerful engine also allowed for a reduction in the wingspan by four feet; and he also had fairings made behind the two hull steps – which might have inhibited the ability of the aircraft to take off, had it not been for the extra power now available.

Nevertheless, Mitchell's changes could hardly prevent the Supermarine entry from showing its, by now, venerable pedigree – as did the second British entry, the Blackburn Pellet, which was, like the Sea Lion, based on a hull built for the 1918 N.1B contract; a third entry, from Hawker, was to be a further uprated version of the Sopwith Schneider machine of 1919. This impoverished showing prompted requests in Parliament and in the press for government assistance with the building of new machines but no help was forthcoming – even though it was known that America, at least, was sending three government-sponsored aircraft which might show up the British entries in front of a British crowd, expected to match the size of that which had turned out at Naples on the previous occasion.

It meanwhile transpired that Italy would have had to rely solely on last year's S.51 because of economic and political problems but, as C. G. Grey commented: 'there is no Italian entry due to the volcanic state of the Italian people'. Also, the Hawker aircraft was written off, following a dead-stick forced landing and the Pellet sank in the tricky tidal waters of the Humber. It was salvaged and crated to Cowes but arrived without time for air-testing until the day before the navigation tests were to start; it was then found to be excessively nose-heavy and prone to engine overheating. Whilst the trim problem was corrected without too much difficulty, the solution to the overheating required the bypassing of their innovative wing radiators and the substitution of large draggy Lamblin radiators.

The seventh Schneider Trophy Contest, 28 September 1923 at Cowes
Stiffer opposition appeared to be coming from the French whose two state-supported Latham L.1 flying boats were each powered by two 400 hp engines mounted in tandem. They got as far as flying over to England for the contest but one experienced loss of power in gale conditions and suffered such damage as to be eliminated; the second sheared a magneto drive at the outset of the contest and was also out of the competition There were also two CAMS flying boats of

traditional design – that is, very similar to the Italian Savoias S.12 and S.51 or the Macchi M.7 – but their 360 hp engines were much less powerful than the now uprated 550 hp of the Sea Lion's Napier power plant.

As it turned out, one of the French machines which had survived to compete collided with a yacht and the damage to its hull put it out of the contest; also, the Blackburn Pellet sank after serious porpoising on take-off. Perhaps therefore, the Supermarine gamble with an obsolescent aircraft might pay off, after all – although one feels that the rather whimsical sea lion motif painted on the nose and floats of the Supermarine entry (*see* photograph below) was almost a self-deprecating gesture in the face of the expected serious opposition from America.

Supermarine Sea Lion III at Cowes, 1923. (Biard in cockpit and fuselage 'all painted up to represent its name.')

Prior to the 1923 contest, the Americans had not competed in the Schneider Trophy competitions. Their design and production of combat planes in World War One had been negligible and the surplus post-war stock of their Liberty engine had inhibited significant development of aero engines afterwards. But in 1921, the US Navy contracted with the Curtiss Aeroplane and Motor Co. for the development of a pursuit (fighter) plane, the CR-1, which was soon to be tested in the newly established Pulitzer Trophy Race for landplanes and which won at an average speed of 176.6 mph.

The US Army then took over the development of the Curtiss racers which, in 1922, were fitted with of one of the great aero engines in aviation history, the Curtiss D-12, whose frontal area was about 50% less than the rival Napier Lion. It was also delivering 465 hp with a power ratio of 1 lb per horsepower. The winning of the 1922 Pulitzer race was also due to the incorporation of radiators flush-mounted on the wings and to the use of metal propellers (the newer engines were producing airscrew tip speeds approaching the speed of sound, for which thicker, wooden blades were beginning to prove inadequate.) Racing was once again bringing about significant technical advances in aeronautical engineering and design – it was thus fortunate for Mitchell and Supermarine that the Pulitzer Trophy, which came to be dominated by competition between the aviation divisions of the American Army and Navy, never established the same international following that the Schneider Trophy did for seaplanes.

Supermarine SEA LION III (1923)

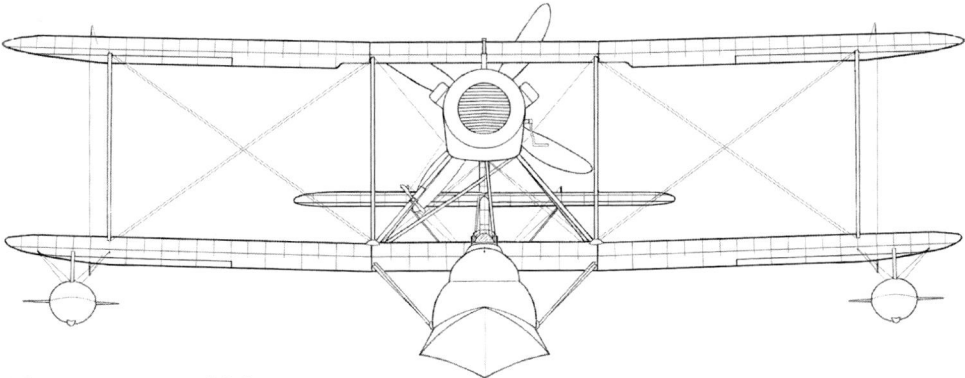

Wingspan	26 ft
Wing area	360 sq ft
Loaded weight	3,275 lb
Maximum speed	175 mph

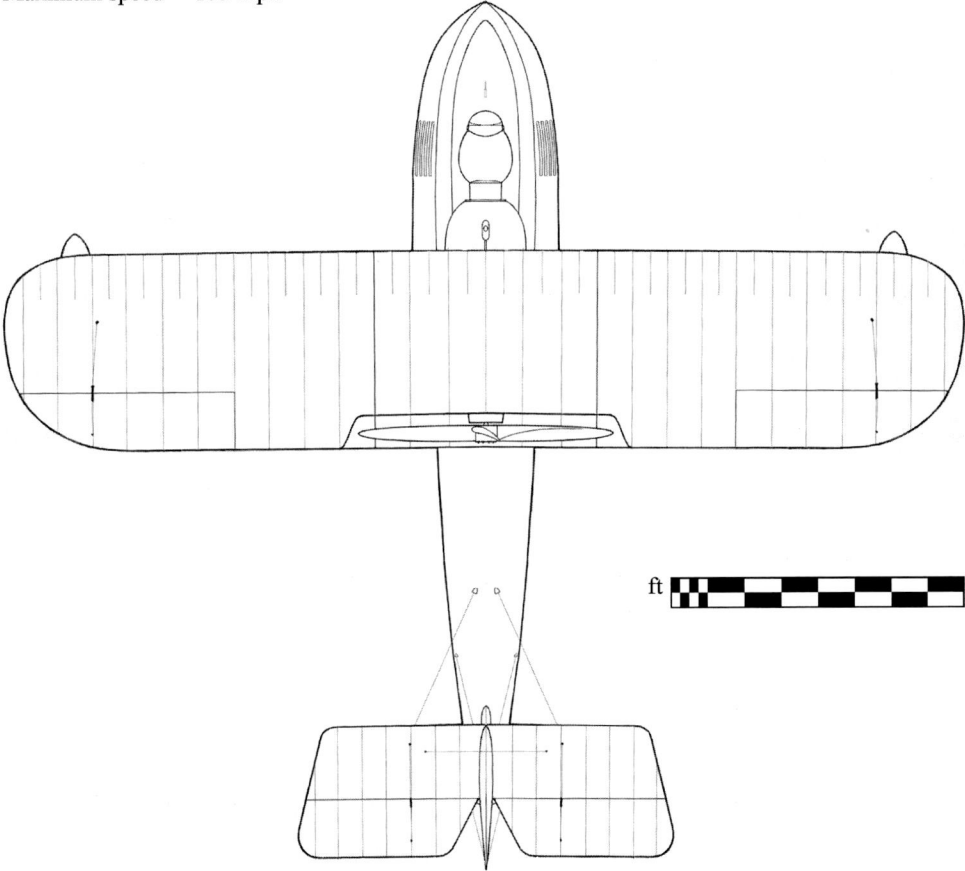

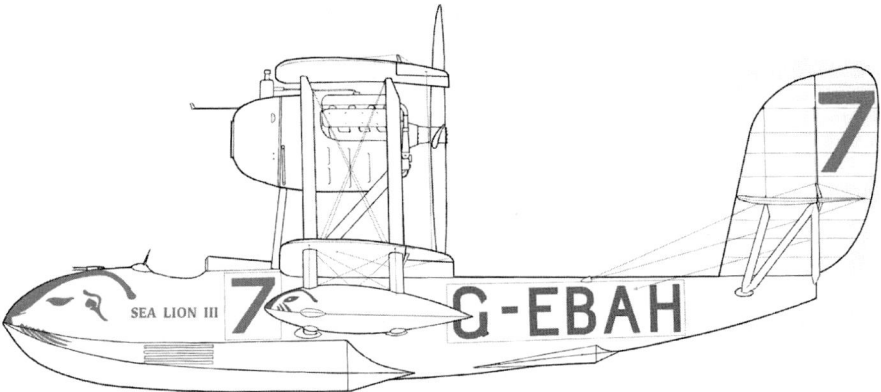

SEA LION III

7

G-EBAH

Nevertheless, one effect of the Pulitzer races was immediately apparent: the American Schneider entries also brought floatplane design back into the competition again after the dominance of European flying boat machines from 1920 onwards. At the time, this reversion might not, in itself, have seemed a clear aerodynamic choice but rather the result of America capitalising on their land-plane racing designs which meant that landing wheels had, of necessity, to be replaced with floats.

Perhaps the British Government might have been persuaded to follow the French example and give some backing to Supermarine and other British aircraft firms had they known how the American challenge was to be conducted. Instead of leaving the competition to the various firms supplying the aircraft, the Americans gave the responsibility for the event to the US Navy who were soon seen to approach the contest as a military operation, bringing with them an impressive array of equipment and neatly uniformed support crews, several weeks in advance of the competition date; and a huge warship, the USS *Pittsburgh*, anchored off Cowes. At their disposal were four aircraft: a naval Aircraft Factory TR-3A (a carrier based fighter), two of the specially designed Curtiss CR-3s, and a formidable Navy/Wright NW-2 racing design, with an engine reported to deliver 700 hp – which would have made it the highest powered single-engined floatplane in the world.

The American pilots were all navy fliers and had been practising since early July, including familiarisation time with their race aircraft. Nevertheless, not all went smoothly for them. Four days before the contest, the powerful engine of the NW-2 blew up on a full-throttle test flight and the aircraft crashed into the sea, though fortunately its pilot was uninjured. Then, on 27 September, the day of the navigability tests, the American reserve machine, the TR-3A had to be withdrawn due to a failure of its starting gear.

The eventual contestants completed the navigability tests successfully and, on the following day, prepared for the flying contest. As the remaining two American aircraft had performed perfectly well in the calm conditions of the 27th, there was always the hope that poorer sea conditions on the next day might not be so kind to their new, possibly more fragile, floatplanes. Unfortunately for Supermarine, there were good conditions on the next day and so the contest between the remaining two American and the single British and French aircraft would be decided by speed and reliability, not by the weather.

The Royal Aero Club had originally announced that, by lot, the flying order would be Italy, America, Britain and France, with each team taking off as a group at 15 minute intervals. As the Italians had not shown, the first American took off at 11 a.m.; at 11.15 Biard, who was then manœuvring for take off, was

Left to right, Curtiss CR-3, Sea Lion III, and CAMS 38 at the Saunders Roe base, Cowes.

Sea Lion III at the Saunders Roe base, Cowes.

astonished to see the Americans completing their first laps and this indicated something like a 170 mph lap – including crossing the start line on the water. Biard was thought to have contravened this requirement and was disqualified for having become airborne too soon. The surviving CAMS machine began the race correctly but its engine failed on the second lap.

Of the three finishing machines, that of Lieutenant David Rittenhouse came first, with an average speed of 177.3 mph, and the aircraft of Lieutenant Rutledge Irvine was second with an average of 173.46 mph. James Bird, the newly arrived Managing Director of Supermarine had forcefully protested about his aircraft's

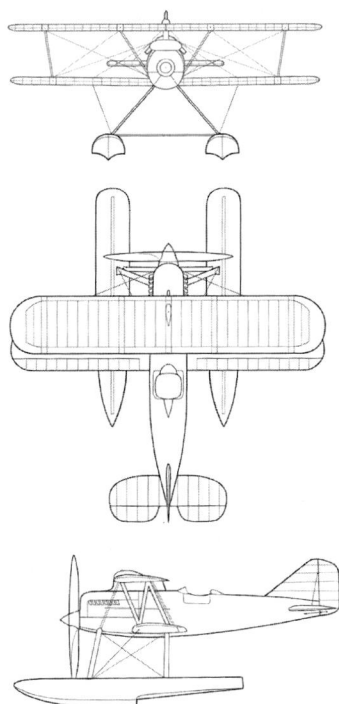

Curtiss CR-3.

disqualification and the Aero Club eventually ruled that Biard had only been making several bounces prior to the actual take-off. Nevertheless, the reinstatement was somewhat academic for, despite 75 more horsepower, Biard could only manage an average of 157.17 mph and so his Sea Lion had been well beaten into third place.

It must now have been very clear to Supermarine that an aircraft with a formidable in-line engine and propeller combination and cooled by flush-fitted wing radiators was going to be hard to beat; it must also have been clear that the European flying boat approach with an engine mounted above the hull was no longer likely to be the best approach: the American pattern, despite the drag of floats, allowed for an engine to be neatly cowled so as to merge into the streamlines of the fuselage. No doubt Mitchell had also noted that the Curtiss CR-3 configuration limited the number of certain drag inducing items to 16 struts, with 20 wires, whereas the Sea Lion tradition he had inherited had required 33 struts and 42 wires, even without the Americans' penalty of float struts and bracings.

At the end of the race, Biard made a consolatory gesture by pulling up to a considerable height and then descending in a series of tight spirals before alighting in front of the British crowd. In a speech afterwards, Scott-Paine praised the Napier engine 'that would have gone on for ever' and said that he felt the need 'to apologise to Capt. Biard because we did not give him a good enough machine'. Afterwards, Supermarine publicity also did its best in the circumstances, by repeating its Sea King II attributes – drawing attention once again to the aircraft's strength, seaworthiness, manœuvrability, and its essential military specification (the racing aircraft being still

Sketch of typical Supermarine empennage from Sea Lion I (1919) to Scarab (1924).

referred to as a 'fighting scout'). It was also described as an amphibian which was not the case at the time of the Schneider Trophy contests:

> This single-seater amphibian flying boat has been designed as a high performance fighting scout, specially adapted for getting off gun platforms of capital ships, or getting off and landing on decks of aircraft carriers ... The fact that the [Mark II] machine was in all essentials a standard type suitable for ordinary service duties but was, nevertheless, able to defeat the specially designed racing machines with which it had to contend, is a very great tribute to the excellence of Supermarine design.
>
> The 'Sea Lion' Mark III was third in the Schneider Cup Race, 1923, and completed the triangular course at an average speed of 157 mph. For this race the machine was generally cleaned up in design, but was still a service type machine and by no means a specially designed racing machine. The maximum speed over a straight course is 175 mph.

The Sea Lion, which had retained the Mark II registration G-EBAH, was returned as N170 to the Marine Aircraft Experimental Establishment at Felixstowe with its undercarriage now restored but its career was short lived, owing to its extremely lively take-off performance. When he first saw the aircraft, Biard had predicted that 'she was going to be a bit playful to get off the water' and so '"gave it the gun" from the start, took a phenomenally short run on the water and

Sea Lion taking off and 'on the step'.

went straight into the air with speed enough to climb straight away.' As he reported, 'It was an interesting sensation; you switched on the engine, and before you could count 1, 2, 3, 4 fast – she was flying.'

Unfortunately, on 25 June the next year, an Air Force test pilot, Flg Off. E. E. Paull-Smith, did not appear to heed the warning that the machine actually wanted to rise before sufficient flying speed had been attained. As a result, he took off and fell back onto the water, rose to about forty feet, stalled again, and dived into the sea. The pilot was killed and the machine was too extensively damaged to be considered worth restoring. The Air Ministry placed no orders for a seaplane scout with Supermarine and this incident marked the end of Supermarine's attempts to interest the Air Ministry in such a concept. It also marked the beginning of Mitchell's search for a worthy and dedicated Schneider Trophy competition machine.

* * * * *

The defeat of the Sea Lion III also marked a change in ownership of Supermarine and in Mitchell's terms of employment. In November of 1923, James Bird resolved his increasing disagreements with Scott-Paine by buying him out for £192,000, a very large sum at that time. And Gordon Mitchell has recorded his father's new contract, safeguarded by solicitors in Stoke, his home town. Acknowledgement of his contributions to Supermarine's developing military and commercial activities and, thus, his increasingly important position in the company is evidenced by the negotiation of a higher salary and other benefits, and an undertaking to be awarded a technical directorship at the end of 1927.

* * * * *

Scylla and Swan – Mitchell's Ugly Ducklings
Whilst developing single-engined flying boats, either medium-sized military or civil machines or the much smaller racing aircraft, Mitchell had also been proving his worth by having to consider more ambitious designs. The first of these, N174, to be known as the Scylla, was to have been a five-seat military seaplane. It had been ordered to Specification 14/21 in 1921 as part of the Air Ministry attempt to improve upon the Felixstowe F series flying boats which, as we saw earlier, had had both porpoising and fuel supply problems. The F.2A seaplanes of 1917 were followed by the larger F.3; this machine had a wingspan of 101 feet but did not handle as well and there were problems with leaking hulls. Nevertheless, such machines had been operated successfully in coastal-reconnaissance duties and so were replaced after the armistice by the F.5 from the same makers. The strengthened hull of the new machine made it considerably heavier than previously and D'Arcy Greig (*see* Chapter Six) found it no better a performer on water than its predecessors:

> they were grossly underpowered by two Rolls-Royce Eagle VIII engines, and if there happened to be a flat calm at time of take-off, they frequently refused to unstick. On such occasions the pilot had to taxi frenziedly up and down the Solent and around in circles in order to disturb the surface of the water before trying again, but even then they sometimes failed to get airborne.

The designer of the F series, John Porte, died in 1919. Had he lived, he would most probably have had much to do with improved flying boats, although the Air Ministry were also concerned to see if the Linton-Hope type of hull could be adopted on aircraft of the Felixstowe-Porte size. Thus, in 1917, Specification N.4 had been issued for this purpose but could not be contracted out to specialist manufacturers of seaplanes as these companies were fully committed to the wartime production of standard service machines. The slow process of fulfilling this requirement by various manufacturers around the country lasted well after the war, by which time Supermarine had come into contention.

SCYLLA

The new machine was to be a triplane with two main engines and a much smaller, auxiliary one, sited in the hull to drive a water propeller for slow-speed taxiing. This last feature seems particularly old-fashioned, looking back to the earlier Pemberton Billing days of the company and, in particular, the P.B.7 (which had also shared other features with Mitchell's Commercial Amphibian of about the same time). Additionally, the biplane stabilisers and the squarish fins also looked backwards – in this case, to the Nighthawk, to the A.D. Boat/Channel of those early days, and to a proposed Torpedo Carrier triplane of 1921.

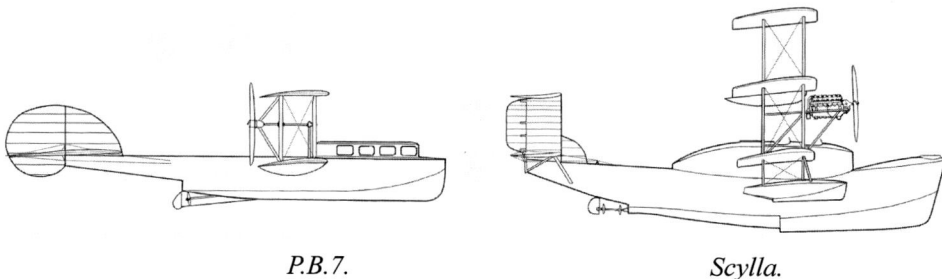

P.B.7.　　　　　　　　　　　　　*Scylla.*

　　Why only the hull of the Scylla was completed is unknown, as was its final fate. At this distance in time, one must accept the view of the authors of *Supermarine Aircraft* that, after taxiing trials with the hull in February or March, 1924, 'the fate of the Scylla design is wrapped in mystery'. Perhaps Mitchell's rapidly developing confidence as an innovative designer, which is the main subject of the next chapter, might well have been an important factor. As a new contract to Specification 21/22 was received soon afterwards for a large commercial amphibian, his thoughts could turn from the traditional thinking represented by the Scylla to a more forward-looking aircraft, given the confidence that must have followed from his successful Commercial Amphibian design and the small fleet of passenger-carrying Sea Eagles.

　　This new requirement, to be known as the Swan, was ordered by the Air Ministry as a commercial, twin-engined amphibian and, as such, had the further appeal of becoming the first of its type in the world. As the new machine was to carry twelve passengers, it would come out at about the same size as the military Scylla and one wonders if the first machine was soon relegated to merely providing information for the new design – which, a biplane-to-be, Mitchell must have predicted would be a more efficient aircraft. It could also be easily retrofitted for the military purposes specified for the Scylla. Certainly the former machine seems to have been used only for water taxiing trials and a photograph in *Supermarine Aircraft* shows a very basic framework erected on the hull to accommodate (perhaps temporarily) the two engines for this purpose.

SWAN

But at least it can be more certainly recorded that the more forward-looking Swan first appeared in 1924 as an equal-span biplane and with forward-folding wing arrangements like the Sea Eagle of the previous year. In some ways, the Swan might best be regarded as a scaled-up Sea Eagle, although its original conception as a twelve-passenger aeroplane resulted in the necessity of accommodating the passengers in the main body of the hull rather than in the advantageous fore position provided in the earlier aircraft. Again, Sea Eagle practice was followed with fuel tanks situated high enough to provide gravity feed to the inter-plane engines, as well as to make the accommodation of the passengers roomy as well as fume-free. The fin and rudder outlines also resembled the earlier aircraft.

On the other hand, Mitchell's less complex use of dihedral only on the outer sections of the lower mainplane was an innovation and the three vertical tail surfaces looked forward to his larger designs of the next decade. The single plane stabiliser was also new to larger Supermarine aircraft and was kept well clear of the water by the upward slope of the rear section of the hull – very unusual for the time although not quite an innovation in flying boat design (*see* the smaller French Tellier T3 or the Latham HB3); nor was it as graceful as the upward sweep of the future Southampton rear fuselage but it did represent a bold new step in Mitchell hull design, without previous experimentation with smaller hulls.

The need to mount three fins had also led to a reversal of his earlier practice, from the Commercial Amphibian onwards, whereby the tailplane was supported by the fin and numerous struts. The new feature also anticipated most of Mitchell's later seaplanes and was a more elegant configuration than the traditional biplane-tail approach which had been proposed for the Scylla. On the other hand, the similar upswept fuselage and the upturned prow, flared outwards at the top to counteract spray, were features in common which had probably been proven to be effective by the taxiing trials of the earlier machine.

The raised cockpit superstructure was also very reminiscent of the unfinished Scylla and it contributed significantly to the clumsiness of the hull profile. Cozens' comments are complementary, if not always complimentary:

> The Swan had several features which showed improvements on previous designs, and no doubt these led to its success. The keel had an upward curve towards the tail that enabled it to take off more readily and this feature was noticeable in all later flying boat hulls built throughout the flying boat era, even to the Saunders-Roe Princess of the nineteen fifties, and it is very apparent if one compares the pictures of the Swan with that of the Channel I. The struts of the Swan's centre section formed large W's which made for great strength and the large fins and rudders and the considerable spacing between the wings made this aircraft a success from a handling point of view. At any rate, Captain Biard was pleased and so was the Air Ministry, but no one could say that the Swan was a handsome machine with its rounded bow and strange-looking cabin and the pilot's cockpit at the top. There was a hatch on the foredeck for a crewman to use when picking up a mooring, and Mr Trotter [assistant to 'Digger' Pickett – *see* p. 62] told me that he was invited to occupy this on the maiden flight, but he declined.

The 'strange looking cabin', which housed a crew of two, sat on the top of the main fuselage so as to maximise the hull space for passengers; and, as the proposed passenger windows had yet not been fitted, the offending side-view was unrelieved. The same had been true of the Scylla and, whilst Supermarine had no doubt chosen the latter's name to suit its proposed military role, it might seem to others that the name reflected its stark appearance – according to Ovid, the beautiful Scylla was turned into a thing of terror and, in Homer, Odysseus manages to sail past the monster but not before she catches and devours six of his men. As the new design was to have a more pacific role, the new aircraft was named Swan although, despite its size, 'ugly duckling' comes more to mind.

The new aircraft was first flown by Biard on 25 March, 1924 and, at this time, had triangular cut-outs in the leading edges of the wings to enable them to fold forwards. The Swan also had the retracting undercarriage arrangement of the sort Mitchell had designed for his single-engined amphibians but the vastly

Supermarine SWAN (1924)

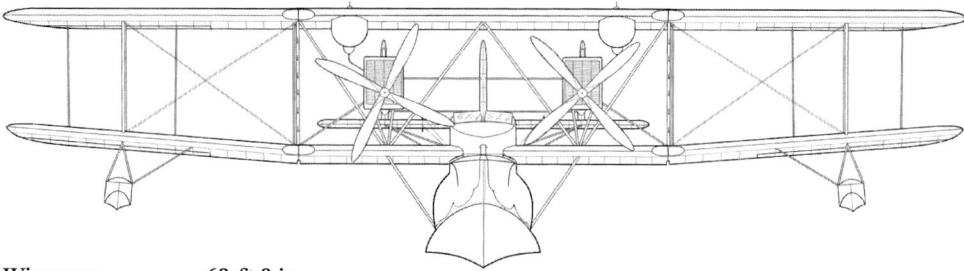

Wingspan	68 ft 8 in
Wing area	1,265 sq ft
Loaded weight	12,832 lb
Maximum speed	105 mph

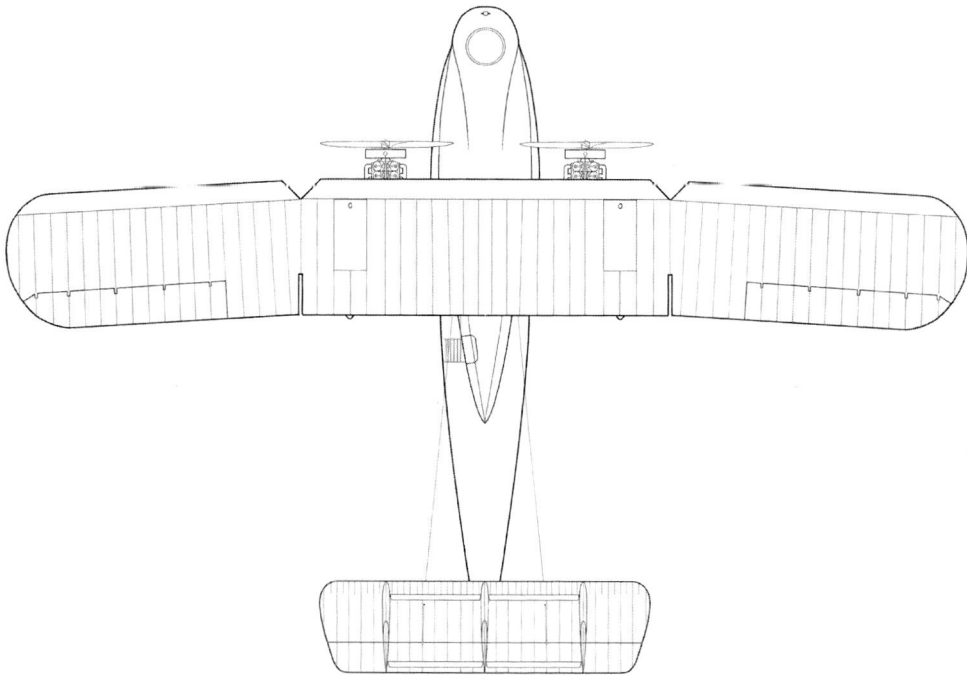

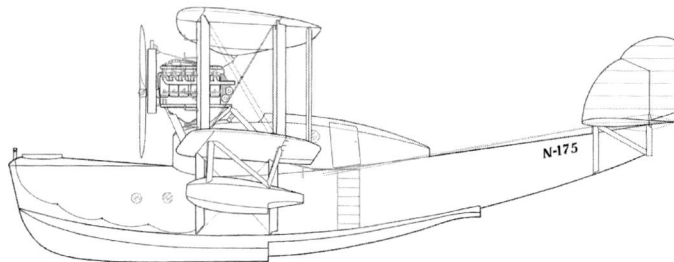

N-175

The Swan nearing completion.

increased size of this particular machine necessitated the novelty of some form of servo assistance. Biard described Mitchell's solution as follows:

> it would have been quite impossible to wind down the six-foot wheels and powerful landing-carriage, which had to stand the weight of several tons of aircraft and passengers! So a neat device was fitted to the machine to do the work quickly and efficiently for us. This consisted of a small propeller, which, when not in use, was set sideways to the direction in which we were flying. When we wanted to lower the landing-gear, this propeller was swung round to face the direction of our course, and the whirling propeller was connected by cogs to a handle which wound very rapidly round and lowered the wheels into place; by turning the propeller rearward the wheels were wound up out of our way under the wings, and the machine was then able to descend on water. This gear, after one or two adjustments following minor troubles during tests, when the Swan behaved neither like fish, flesh, nor fowl, proved remarkably efficient, and wound the heavy landing-gear into place in about half a minute or less.

Biard also describes the visit of the Prince of Wales to Supermarine, and to the Swan in particular, on the 27th of June in the same year. (On being invited to climb the ten-foot steel ladder to the cabins, he declined because of his dress sword.) His reported conversation was about England's need to possess efficient commercial aircraft and fighting air forces to safeguard her Colonial trade routes. As we shall see later, Supermarine and Mitchell had an important part to play in both of these developments.

By the time of the Prince's visit, the Swan had been re-engined and had been converted from amphibian to flying boat by the removal of the undercarriage (*see* photos overleaf).The folding of the wings and the leading edge cut-outs had also been dispensed with: whilst the Sea Eagle folding system had reduced the aircraft width by 54% for an increase of 15% in length, the similar Swan arrangement only achieved a 30% reduction at the expense of a 28% increase in length.

Plans for an RAF version of the Swan were also being actively pursued at this time – which may also have had some influence on the change to the fixed-wing layout and possibly throws more light upon the decision to terminate the

An informal shot of the Royal visit, showing the impressive size of the Swan. Note the substantial Samson post at the prow. Mitchell to right of Prince.

Swan after removal of wing leading edge cut-outs and undercarriage.

Swan after conversion into the world's 'first twin-engined amphibian flying boat'.

development of the military Scylla. The Swan's Eagle engines were replaced by two Napier Lion IIBs, each developing 90 hp more than the Rolls-Royce units, thereby increasing the Swan's top speed by 13 mph.

Meanwhile, the first of the N.4 specification aircraft, the Fairey designed Atalanta, had made its maiden flight and, whilst being regarded as an excellent machine in its own right, its 139 foot wingspan and four Rolls-Royce 650 hp Condor engines, now made it a less attractive proposition for the drastically reduced post-war Services. In comparison, the Swan had only two 450 hp engines, had only half the wingspan, but nevertheless was only 10 mph slower and, of course, also embodied a Linton-Hope type of hull in which the Air Ministry was interested. Successful trials at the Marine Aircraft Experimental Establishment, Felixstowe, followed and these were to have important results for the fortunes of Supermarine, as will be described in the next chapter.

When the Swan was returned to Supermarine, its fitting out as a passenger-carrying aircraft was completed and the company was particularly concerned to stress that its standard of accommodation was extremely competitive as well as being the first twin-engined commercial amphibian:

> This is a passenger-carrying flying boat or amphibian flying boat, fitted with either two Rolls-Royce 'Eagle IX' engines or two Napier 'Lion' engines. It carries twelve passengers, with luggage, together with a crew of two, complete navigation equipment and fuel for 300 miles. This is the first twin-engined amphibian flying boat to be built in the world and it may also be fairly claimed to be the first twin-engined commercial flying boat.
>
> An important feature of this machine is that the whole of the hull is devoted to passenger accommodation. There are no internal obstructions of any kind, and the amount of room in the saloon far exceeds that of any commercial landplane. The internal accommodation consists of one large passenger saloon, elaborately furnished and upholstered and with every comfort. Forward of the saloon is the luggage compartment, fitted with racks for the stowage of passenger baggage. Aft of the saloon is the buffet, with all necessary fittings to supply light refreshment during the journey. Still further aft are the lavatories, which are efficiently and fully equipped.
>
> The pilot and the navigator are accommodated in a cabin specially built on top of the hull. In this position they are given an extremely good view in all directions, and are in a fine position for starting the engines, handling the machine on the water, and for operating the craft under all conditions.

The above publicity mentions the originally planned 12 passenger-carrying capability although, in fact, the machine was finally fitted out with only ten seats. In this form it was registered as G-EBJY and first flew on June the 9th, 1926, with a representative of Imperial Airways and eight excited female employees of Supermarine as passengers.

Whilst the placing of the built-on crew compartment above the main hull did nothing for the lines of the Swan's hull, its position and the slight reduction in passenger seating did allow Supermarine to set new standards in passenger accommodation which its test pilot, Henri Biard, fully confirmed: 'the Swan, then the world's largest amphibian … was a real cabin liner of the air, with comfortable armchairs, big porthole windows, a commodious passage along the centre of the living accommodation, and all sorts of luxuries and refinements which were very new to aircraft at that time.' A more neutral observer from *The Aeroplane* agreed, saying that 'the appointments are exquisite' with 'a commodious passenger saloon padded luxuriously and in which there are ten cosy armchairs. An ample porthole is provided for each chair.'

Mitchell (right) in the Swan. Note the Linton-Hope fuselage structure of close-spaced hoops on longitudinal stringers.

Cozens describes a later, and more mundane, use for the passenger compartment in the course of his recollections of this aircraft:

> Both Rolls-Royce Eagles and Napier Lions were tried but the Lions were finally chosen. The writer recalls this machine when it was used in the colours of Imperial Airways as a freighter to bring bags of early potatoes and boxes of daffodils from the Channel Islands. The large mid-section made this possible and highly suitable for bulky cargoes. I remember seeing the horse drawn market carts down on the slipway being loaded from the Swan.

The aircraft had been loaned by the Air Ministry to Imperial Airways in order to supplement the service of its remaining two smaller sisters, the Sea Eagles, on their Channel Isles service. It operated during 1926 and 1927 but, as the *Guernsey Evening Press* reported, 'during the normal rigorous inspection prior to leaving Southampton on April 12, a structural defect was discovered which necessitated the stripping of the whole machine'. As a result, the Swan was scrapped; Imperial Airways' next long-distance seaplane was not to be a replacement Swan, however, and so its main significance remained that of providing the prototype for the Royal Air Force's next standard maritime reconnaissance aircraft, the Supermarine Southampton of 1925 (*see* Chapter Four).

Sheldrake and Scarab – an Enigma and An Amphibian Bomber
Whilst, later, the Mk.III Seagull went to a dominion of the British Empire, one particular design by Mitchell was sold exclusively to a foreign power. This machine was the Scarab, which first flew on 21 May 1924, and was a powerful fighting machine for its period. And at this time, there was also a basically similar amphibian, designed for the Admiralty; it was named Sheldrake, and appeared at an air pageant in 1927 but in most respects little is known about it.

SHELDRAKE
The result of the criticisms of the British Seagull II was that, as early as 1923, an Air Ministry order was placed for an improved version of the Seagull II. The resultant aircraft was the Sheldrake whose flying surfaces were virtually identical to those of the Seagull but which had a boat-like hull very similar to that of the Sea Eagle, if the latter's passenger cabin were discounted. As the new hull shape

Supermarine SHELDRAKE (circa 1924)

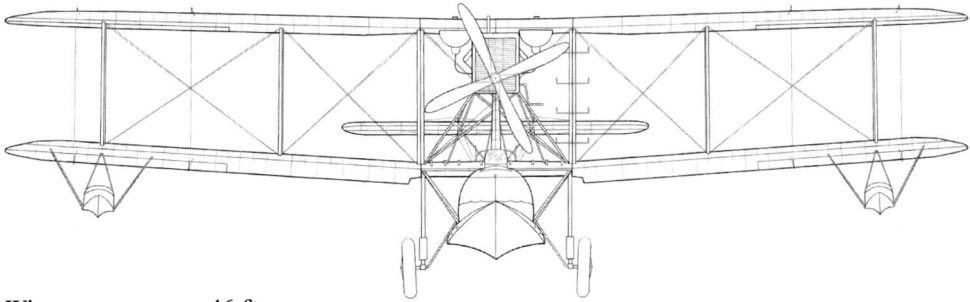

Wingspan	46 ft
Wing area	593 sq ft
Loaded weight	6,100 lb
Maximum speed	103 mph

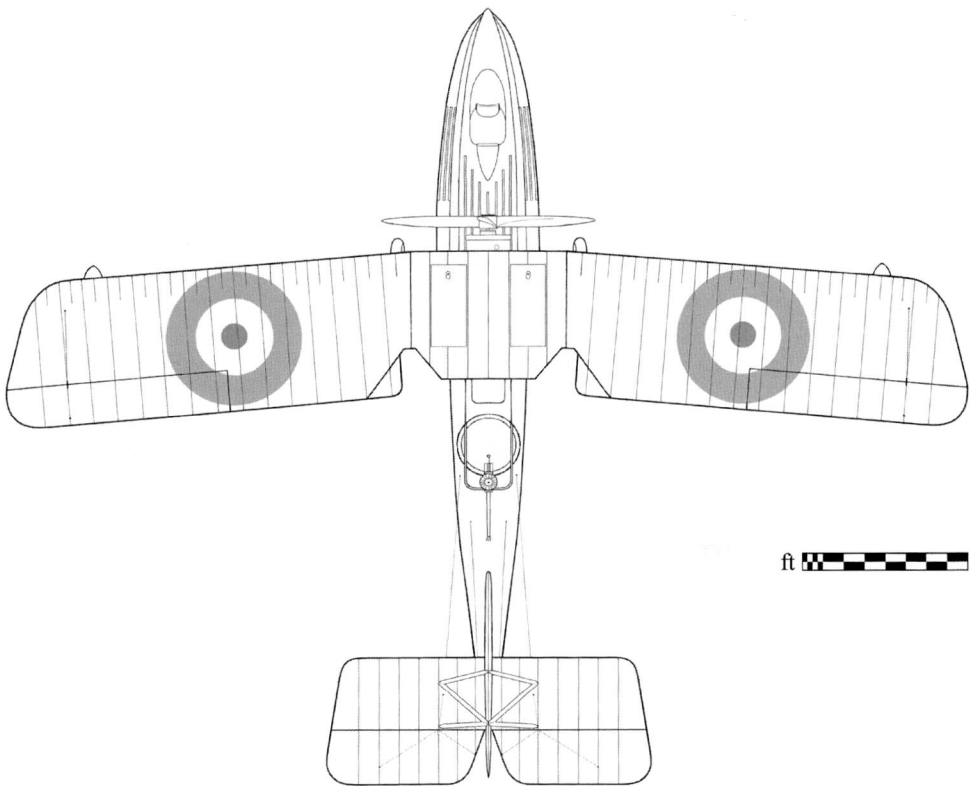

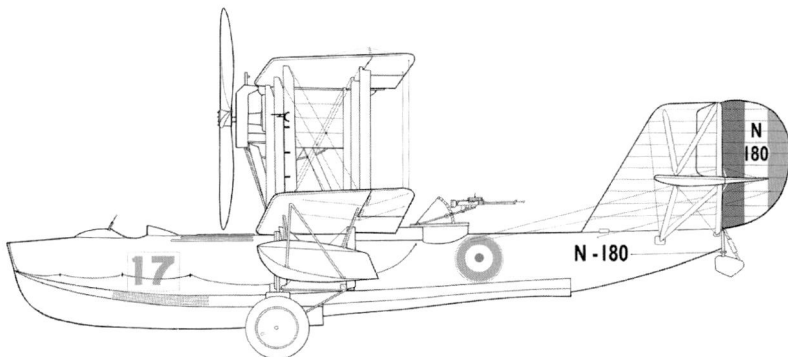

ft

was selected in response to one of the RAF's criticisms of the Seagull, it is surprising that the stabilising screens were still retained, as was the separation of the pilot from the rest of the crew. Even more surprising was the apparent inactivity around the Sheldrake, for it only appeared in public in 1927.

The Sheldrake at the Hamble Air Pageant, 1927.

Cozens' retrospective view of this aeroplane's antecedents contains an interesting sideline on its Hamble appearance:

> All the Seagulls were anachronisms in the [late] nineteen-twenties, and this was shown most clearly when they appeared at the land based RAF Air Pageants and the writer especially remembers seeing a Sheldrake at the end of the line of aircraft on show at Hamble. This machine, N180, looked very much like an 'old gaffer' at Cowes in modern times, but it did have a certain dignity when it took off and was 'attacked' by three Gloster Gamecocks in their silver paint, black and white squares [No. 43 Squadron markings] and RAF roundels.

No doubt contributory factors to the tardy appearance of the Sheldrake were the other aircraft being developed at the time or on the drawing-board between 1923 and 1927: the first flights of the Sea Eagle and of the Sea Lion III, the first production batch of the Seagull II, the larger flying boats, the Scylla and the Swan being considered, the important activity around the twin-engined Southampton and the new breed of Schneider trophy racers. Certainly, by 1927, N180 was an obsolescent type and no production orders were placed for this enigmatic machine. It was, however, important in respect of a Spanish order for twelve aircraft, named Scarab, which must also have affected the Sheldrake development.

SCARAB
A year after the Sheldrake prototype was ordered, the second of our aircraft made its maiden flight. There was no prototype for this aircraft and, as it was in a great many respects similar to the Sheldrake, it seems obvious that the Air Ministry order for the earlier aircraft provided most of the design work for the Spanish order – at little extra cost to Supermarine.

King Alfonso of Spain was a regular visitor to the Hendon RAF Pageants and must have had an early appreciation of the British 'control without occupation' tactics, in view of the war which was being fought in the early 1920s between Spain and the Berbers in Morocco. Accordingly, the Spanish Royal Naval Air Service asked Supermarine for an amphibian to be produced, capable of carrying a bomb load of 454 kilos, as the load carrying capacity of the Sea Eagle

Supermarine SCARAB (1924)

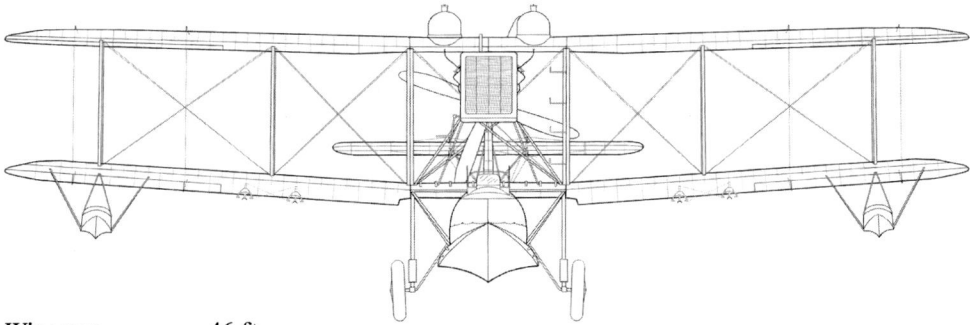

Wingspan	46 ft
Wing area	610 sq ft
Loaded weight	5,750 lb
Maximum speed	93 mph

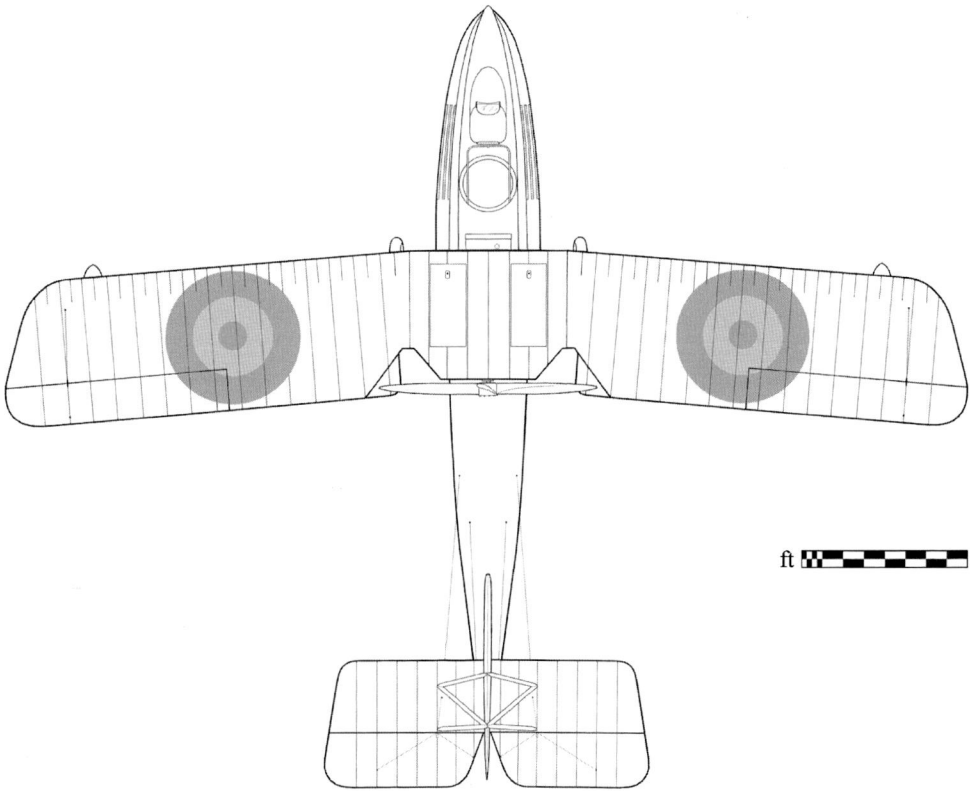

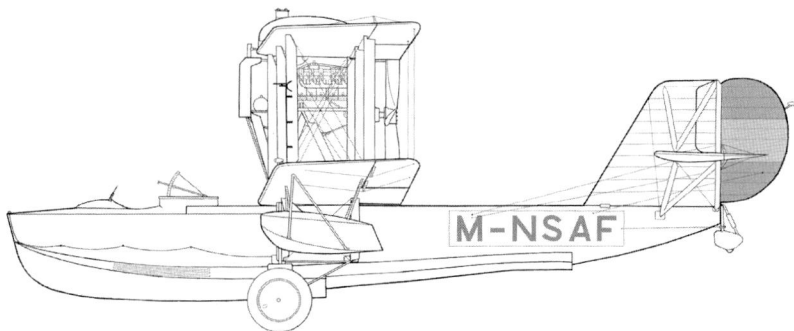

ft

promised a suitable basis for a design. Supermarine suggested that the plans being drawn for their new Sheldrake would be a more suitable model for the project and, doubtless, Mitchell was already envisaging improvements in respect of the Sheldrake.

When the new machine appeared, it could be seen that the Sheldrake upper wing fuel tanks, which had been similar to those of the Seagull II, were now replaced by larger ones situated on top of the centre section, following Sea Eagle practice. The engine was returned to the more familiar pusher configuration as the crew were all now, more conveniently, grouped together in front of the wings, with the navigator also having a cabin in the hull immediately behind and below his cockpit position. Dual controls could be fitted in the second cockpit and the space not now required for fuselage petrol tanks was used for twelve 50 lb bombs which could be dropped via a sealable aperture in the bottom of the hull. Four 100 lb bombs were also carried under the wings. The total weight of bombs carried amounted to the equivalent of six men; it would also have a crew of three and would carry a machine gun, ammunition and a considerable amount of fuel, thus making it an attractive single-engined proposition to its buyers.

Two views of the first Scarab for the Spanish Royal Naval Air Service.

The first Scarab made its maiden flight on 21 May 1924, bearing the letters M-NSAA but whether all twelve actually saw service is unclear. One of the Scarabs was damaged on acceptance trials when its Spanish pilot hit the side of a Union Castle liner when taking off; the ship sent to collect them was reported to have a cargo lift four inches too small in one dimension, and the machines were subject to a severe Bay of Biscay storm stowed under tarpaulins as deck cargo. Nevertheless, Scarabs were seen above Barcelona at the 1925 Royal Review of the Spanish forces by King Alfonso and they equipped a seaplane carrier, the *Dédalo*. This vessel was a converted merchant vessel and lacked a landing deck; the aircraft were lowered into the water or raised from it by crane, like their Seagull III predecessors.

Based at Carageno and commanded by the king's nephew, the unit took part in actions against Riff and Jibala insurgents in the Spanish Moroccan campaign, including bombing raids in support of an amphibious landing at Al Hoceima. The Moroccan conflict ended soon afterwards in 1926 and Supermarine publicity, no doubt mindful of the new RAF tactics of control without occupation, ran as follows:

THE SUPERMARINE 'SCARAB' BOMBER
AMPHIBIAN FLYING BOAT

A recent development of the single-engined flying boat is the Supermarine 'Scarab', which has been specially designed for naval bombing and reconnaissance work. A large number of these machines have been bought by the Spanish Government, and these have been in operation for the past year in Morocco *with the most satisfactory results* [my italics].

The hull is constructed on standard Supermarine lines and is very seaworthy. It is practically impossible to ship water over the cockpits. The machine handles very easily and effectively on the water by means of its water rudder and can be turned on a radius of one span. It tows very nicely under all conditions of wind and tide.

The whole of the crew are placed in front of the main planes, and thus intercommunication is excellent. The pilot is seated in the very bows of the machine, behind him is the gunner, and aft of the gunner is the navigator and wireless operator, who has both a cockpit in front of the main planes and a spacious wireless cabin in the hull immediately behind his cockpit.

The undercarriage is similar to that used on the Supermarine 'Seagull'.

The engine, to which is attached a pusher airscrew, is either a Rolls-Royce 'Eagle IX' or a Napier 'Lion'.

The fuel supply is by direct gravity feed from the petrol tanks situated on the top of the top centre section plane.

Provision is made for carrying a maximum bomb load of 1,000 lbs.

Only the Rolls-Royce engine was fitted to the Spanish Scarabs (when the Sheldrake made its belated appearance, it was powered by a Lion engine) and no other foreign orders resulted with either engine being used, even though the

Mitchell explaining features of the Scarab to HRH Edward, Prince of Wales, on his 1924 visit.

Greek government had sent observers to the Scarab acceptance trials. The Super-marine company did not receive any British orders for the Scarab and the sole Sheldrake was not heard of again. It might, however, be noted in passing that the more successful hull planing configuration developed for the Sheldrake, and also employed on the Scarab, was an important influence upon the eventual Seagull V/Walrus design. All these machines continued Mitchell's development of the medium-sized amphibian biplane (of about 46 feet in wingspan) and so, thanks to Mitchell's original Seal which first flew in May, 1921, the financial position of Supermarine began to improve as the total of 33 Seal/Seagull orders were in addition to the small fleet of four Sea Eagle passenger amphibians and the twelve Spanish amphibian bombers. It is clear that its Chief Designer, R. J. Mitchell was becoming a valuable asset to the company.

Mitchell's First Landplane
Early in 1924 and whilst the Swan and Scarab were being prepared for their first flights, the Air Ministry announced its competition for a two-seat, light, all-British aircraft to take place at Lympne in the September of that year. The previous year, a similar competition, had been sponsored by the Duke of Sutherland, the Under-Secretary of State for Air, and later by the *Daily Mail*, but it had not been of great interest to most of the larger aircraft companies as the various prizes had encouraged designs specifically for height or speed or for distance performances: the prize for the longest distance flown on one gallon of petrol, for example, had produced a succession of, essentially, gliders powered by adapted motorcycle engines.

As a result, the revised 1924 competition rules called for more all-round attainment whereby marks would be awarded for high-speed and low-speed performance, control, shortest take-off and landing runs, and dismantling and re-erecting (so the aircraft could be housed like a family car). The engine size was raised from 750 cc to what was thought to be a more practicable 1,100 cc and the total prize money had also been raised from £2,600 to £3,900, of which £2,000 would go to the winner. Later in the year, the enticing prospect was held out to prospective entrants that the Air Ministry would assist the development of ten light aeroplane clubs by providing an approved design selected from the competition entrants.

As specially designed small aero engines were now being developed and as the aircraft had to be dual control two-seaters, major British aircraft producers became interested in this more practical proposition, bearing in mind the general paucity of orders at the time, the possibility of potential buyers among the large number of ex-service pilots in the country, and the fact that the main prize money was now more substantial. Thus Mitchel now became engaged upon overseeing another 'first' – a very light aeroplane and, exceptionally, a landplane as well.

SPARROW I
Supermarine's publicity for their entry, the Sparrow, mentions a choice of two engines, but the only type used by Supermarine for the competition was the 35 hp Blackburne Thrush, whose name was perhaps its only endearing feature. Biard reported nineteen engine failures in the fortnight before the start of the competition and matters were no better at Lympne. As a result, the Sparrow was not airborne enough to complete the preliminary flying tests and was eliminated.

This result could hardly have pleased the company as Supermarine had been joined by most of the other major manufacturers in producing an entry. Four of these were conventional single bay biplanes like the Swallow: the Vickers Vagabond, the Avro Avis, the Westland Woodpigeon, and the Hawker Cygnet.

Supermarine Sparrow I (1924)

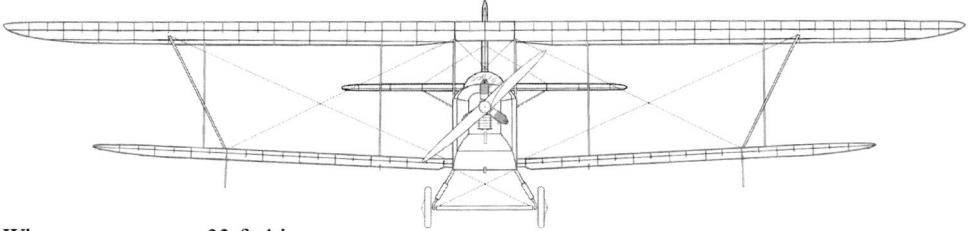

Wingspan	33 ft 4 in
Wing area	256 sq ft
Loaded weight	860 lb
Maximum speed	72 mph

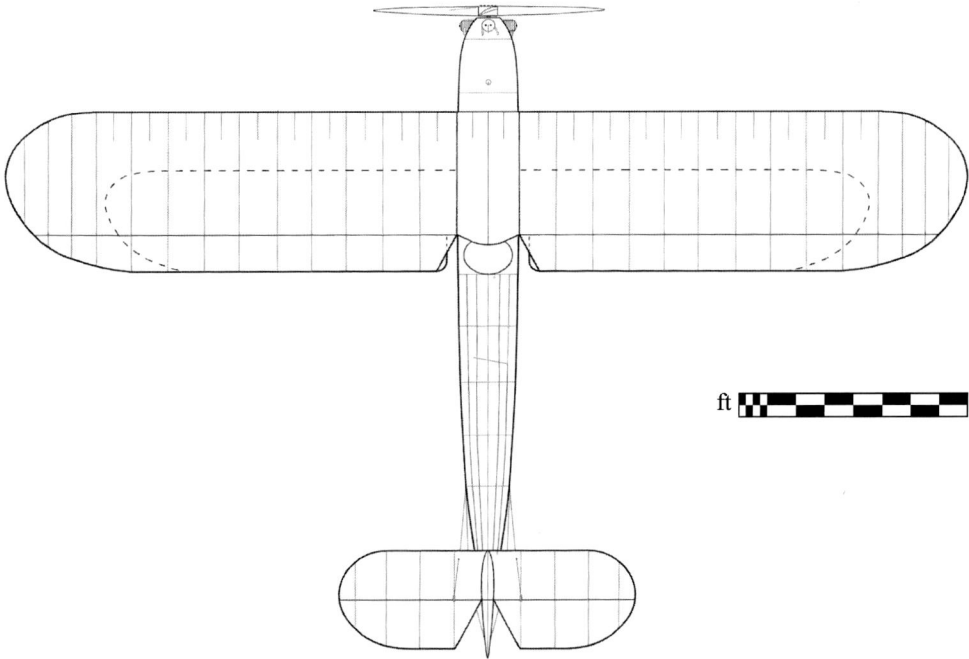

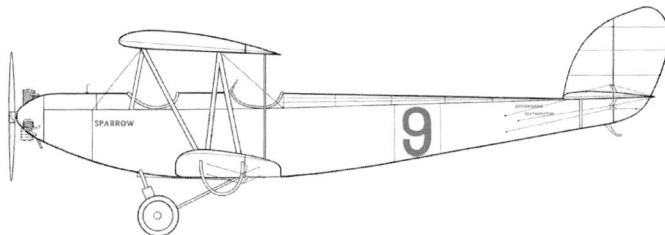

ft

Of these, the last three had full span combined ailerons and flaps like the Mitchell machine – all designed, as the Supermarine publicity said, 'for reducing the landing speed and the runs to land and get off'. Hawkers had arrived at an even more similar formula to that of the Sparrow in that it was also a sesqui-plane, 'with a top wing of much greater chord, and appreciably larger span than those of the bottom'. In contrast to these entries, the Bristol Brownie, the Beard-more Wee Bee, and the Short Satellite represented the monoplane approach to the specification.

With so many companies addressing themselves to the same problems, with the same top limit for power, seating, and other features, the prospect of seeing which designer came up with the best machine at the very same point in time was an intriguing one. This was particularly so because it would also have provided a comparison of the competing claims of the biplane and the mono-plane approach to aircraft design which was intensifying now that structural advances were making the latter type a much more practicable proposition than had been the case earlier. To those able to look back later from the perspective of the Supermarine Spitfire and the Hawker Hurricane days, it would have addition-ally been very interesting to have compared the design of Mitchell, who had now been a Chief Designer for just over four years, and that of Sydney Camm whose Cygnet was one of his first designs for Hawker.

The Sparrow I at Lympne.

An informed comparison was, unfortunately, not possible for there had not been enough time to match propellers to the various engines and continual engine failures made matters worse. Of the nineteen entrants, only eight survived the eliminating trials and only six finally competed for the prizes; all but seven of the competitors had failed because of some sort of engine problem. Even a comparison between the last six was not possible: the speed-range competition had to be preceded by ten trouble-free laps and, by the last day of competition, only the Beardmore and Bristol entries had done so – with the engine throttled well back. The fact that the Beardmore and Bristol machines took first and second prizes, respectively, was not so much a vindication of the monoplane philosophy as a result of their particular engines holding out the best.

Official Air Ministry reports were made of the entries and, despite having the lowest landing speed and making a generally favourable impression, the Sparrow was criticised for having the cockpits too far apart for good communication and for less than satisfactory vision from the front position; its exterior was also con-sidered to be too cluttered with external control rods and cables (not an unfamil-iar feature of Supermarine's single-engined flying boats of this time). With a redesigned undercarriage and a strengthened and widened fuselage, Sidney Camm's Cygnet might very well have been ordered by the Air Ministry to equip the light plane clubs; it probably had the best turn of speed although the Sparrow had the advantage of Supermarine's concern to produce a safe, slow landing aircraft.

But, in the end, no orders were forthcoming for any of the machines: it had become only too obvious that it took much longer to produce a new and reliable power unit than to design and build an airframe. H. F. V. Battle, test-flying for the Air Ministry, reported that 'these light planes caused us to have many forced landings' and it was also perceived that it would be necessary to think in terms of an engine which could develop more power than what was currently possible from the 1,100 cc limit imposed by the competition rules.

Supermarine were free of Air Ministry guidelines for naming their landplane and the name 'Sparrow' which was chosen was appropriate enough, in view of its non-marine associations and size. Supermarine described it as follows:

THE SUPERMARINE 'SPARROW'

A two-seater light aeroplane built for the Air Ministry Competition.

This – the first land machine built by the Supermarine works since the early days of the war – is a very neat biplane with a top wing of much greater chord, and appreciably larger span than those of the bottom. In addition, the wings are so staggered as to bring the trailing edge of the lower wing vertically below that of the upper wing. The two wings are also of different section, the upper wing being an American Sloane section, the lower known as A.D.1. Both these sections are of the low-resistance, high-speed type.

The fuselage is rectangular built with spruce longerons, spruce verticals and spruce diagonal members, and covered with three-ply.

The crew are seated one below the top centre-section and one behind the rear spar. The trailing edges of both wings are cut back from the fuselage, to give the passenger in the rear seat a view downwards.

The wings are built on spruce spars of channel section, with ribs of spruce and three-ply. Drag struts are box-form ribs and the leading edge nose is covered with three-ply. Outwardly raking interplane struts of 'N' type are fitted, one on each side. These are of steel tube with fairing. The usual streamlined wire flying and landing bracings are fitted.

The wings fold in the usual way about hinges on the inner ends of the rear spar, the wings flaps being folded right down when the wing is stowed.

The undercarriage is of the Vee type, with telescopic front legs, sprung on rubber rings. The tail unit is of normal form, comprising fixed tail plane, divided elevators, fin and rudder. The tail plane setting is adjustable on the ground.

The ailerons, which extend over the whole span of both planes, are also utilised as flaps for reducing the landing speed and the runs to land and get off. These features in performance have been given very careful consideration throughout the design, and as a result the landing speed of this machine has been reduced to a figure lower than that of any other two-seater aircraft.

Sensitivity and lightness of control are combined with marked efficiency in manoeuvrability, which render the machine not only highly suitable from the purely instructional point of view, but also remarkably safe.

The machine can be fitted either with a Blackburne 3-cylinder radial engine or a Bristol 'Cherub'.

The aircraft is equally pleasant to fly from either seat, ballast being unnecessary during solo flights.

Supermarine Southampton Mk.I.

— *Chapter Four* —

1925
A
Turning Point

It is surely no exaggeration to identify 1925 as the year when two of Mitchell's aircraft stood out dramatically from what had preceded them. Whilst he had had early successes, incrementally improving on conventional machines, this year marked his full emergence as a designer who had transcended the design conventions that he had inherited and who was now striking out boldly into the future. After joining his firm at the age of 21 and assisting with the designs of others, he now produced – nine years later – the first standard naval reconnaissance aircraft since the end of World War One and the racing floatplane which established the basic design configuration for all subsequent Schneider Trophy machines.

Whilst the sixty-nine foot Swan and the diminutive Sparrow were being tested and flown, the company was fully engaged upon finishing the third production order for Seagulls and the Spanish Scarab contract. On the other hand, there was no great Supermarine interest in an entry for the proposed 1924 Schneider Trophy contest, given the healthier state of the Supermarine production line and the lack of Air Ministry interest in the small fighter seaplane concept. Of much more potential interest was the fact that the improved performance of the modified Swan had been noted at the Marine Aircraft Experimental Establishment at Felixstowe, to which it had been sent in the August of 1924.

By this time, the question of a suitable replacement for the Felixstowe F. series of military flying boats was becoming critical and the aim of developing air links with the outposts of Empire seemed a long way off – as Hoare recalled: 'In 1922, there were no aeroplanes capable of maintaining a long distance service. The existing heavier-than-air machines were low-powered, very noisy and uncomfortable. Flying boats had almost ceased to exist and there was no plan for an Empire airline of any kind.' As mentioned earlier, the first N.4 specification replacement flying boat, with its large wingspan and four engines, was not such an attractive proposition in the post-war 'anti-waste' climate; also, the English Electric P.5 Kingston which flew in 1924 was, essentially, a continuation of the old N.3B specification of 1917 and the Short S.2 of the same year, whilst being important for future flying boat development because of its experimental metal hull, utilised the old Felixstowe flying surfaces.

As an alternative approach to the situation, the Air Ministry authorised the building of six airships, the first two to inaugurate a service to Egypt in 1924. However, the many technical and logistical problems raised by bulky lighter-than-air machines brought about many embarrassing delays. Thus the successful trials, at Felixstowe, of the more compact, twin-engined Swan had not gone unnoticed at the Air Ministry, whose officials had been very impressed by the standards set by the Sea Eagle in 1923. As a result of this appreciation of the new standards in flying-boat performance which Mitchell had now established, the Air Ministry took the unusual step of ordering (no doubt with considerable private relief) straight off the drawing-board, a number of reconnaissance flying boats on the basis of the Swan amphibian passenger carrier. It was fortunate all round that the Estimates of January had, at the expense of the Army and the Navy, provided for an increase of £2,500,000 for the RAF.

SOUTHAMPTON I – the new RAF standard reconnaissance flying boat

By the time that Supermarine received Specification R.18/24, in the August of 1924, for a modified and slightly enlarged Swan-type flying boat, Mitchell was already having this aircraft's hull lines redrawn to improve the streamlining. The eventual modifications were such that a 'Swan Mark II' designation was less appropriate than a completely new name. Since the Swan was built as a commercial aircraft, it was not subject to the 'Aircraft Nomenclature Committee' [!] which required the names of waterfowl for small multi-seat amphibians (e.g. Seagull, Sheldrake and Seamew) and names of 'seaboard British towns' for larger seaplanes. Thus 'Southampton' was chosen for the new Air Force machine – albeit an estuary port not a seaboard town, such as, say, the nearby Southsea. But, in negotiating this name, the company was now signifying Supermarine's increasing status where its factory was sited and whose dignitaries had welcomed home the successful Sea Lion II in 1922.

The first Southampton of the initial RAF order (with the originally designed floats).

The Cozens extract opposite contains the interesting information that a silver shield with the Southampton coat of arms was fixed to the bow of the first production aircraft (which is, indeed, visible in the photograph above). The company could, with some confidence, thus mark their increasing importance in the manufacturing community of the area as the Air Ministry order was substantial by the criteria of the day: it had called for six standard military aircraft (N9896–N9901) and for an experimental one, to be fitted with a metal hull. And these aircraft were also the largest aircraft yet to come from the Supermarine production line – the Seagulls and Scarabs had a span of 46 feet, whereas the Southampton spanned 75 feet (6 ft 4 in. more than the 'prototype' Swan).

The new machine continued the planing configuration that Mitchell had been developing since 1923 with the Sea Eagle but it was now a part of one of the most elegant hulls that Mitchell had ever been responsible for; indeed, the transformation of the lines of the Swan, was dramatic. Taking advantage of the new, more utilitarian military requirements, he removed the ad hoc looking high-drag

Cozens gives some interesting information about the construction of the Southampton hull:

Many considered the Southampton's wooden hull to be the ultimate in design and craftsmanship, in its shape and purpose Mitchell combined his experience in building with Captain Biard's reports on flying* and this was recognised by the people of Southampton who subscribed and had a silver shield with the Southampton coat of arms fixed to the bow of N9896.

It was about this time that Scott-Paine and his co-director Commander Bird had a tremendous quarrel which ended with Scott-Paine leaving Supermarine with a fortune which he used to begin the British Power Boat Company at the old boat sheds at Hythe, where the May, Harden and May company built the Felixstowe flying boats of the Great War period.

The late Mr Conrad Mann, who worked on the wooden Southamptons, told me they were built bottom side up so that the two steps and the curved keel, which had been a feature of the Swan, could be built. He said there were six men and two apprentices on each hull and the contract price agreed by them for each hull was £483 19s. 4d. so that the money worked out as follows:-

Contract price for 6 men to build 1 hull   £483 19s. 4d.
Wages for 6 men to build 1 hull   £357 19s. 4d.
                                   Balance £126   0s. 0d.
This was shared among the men giving each   £21   0s. 0d.
The two apprentices were paid by the company.
This arrangement seemed to work very well, giving both the management and the men every encouragement to build the machines as quickly as possible, and most of the men bought their bicycles with the lump sum bonus ...

Extracts from letters about the Supermarine Southampton which appeared in the *Southern Evening Echo*:
   'Southampton people had good cause to remember and honour that machine because it brought a good deal of prestige and a steady flow of work to the factory where it was built'.
   'This was the time of the Depression and the General Strike, when a machine was finished there was an order for another one. With a steady wage and the prospect of another bonus, the workforce was fortunate and happy. This air of well-being, stimulated by the success of the Schneider Trophy Races, made the firm, and Woolston generally, a vigorous and active area.'
   'About 20 [24 in fact] of the wooden flying boats were made and, as there was not enough floor space at Woolston, the main components were put on barges and ferried across to the boatsheds at Hythe for assembly. There were often three or four machines moored on the Hythe buoys, their varnished hulls and white wings making a picture that those who saw them would never forget.'
   This was brought to mind when a Group Captain who saw the picture of the wooden Southampton in the Echo remembered his early days on Flight 48[0], an RAF conversion flight stationed at Calshot to train pilots to fly Southamptons, looked in his logbook and found that he had flown that same machine in about 1928. He said that one day a second pilot wanted to change into the first pilot's seat and left his own seat and went to the one behind him, passing between the two propellers ... At the nearest point the two propellers were only nine inches apart!'.

[**Cozens mentions several times the importance of Biard's advice to Mitchell during these early years and this would have been especially important to Mitchell when there was little theory to guide designers; as Cozens was a neighbour of the test pilot, it would seem very likely that he heard of particular instances where advice was given — and no doubt listened to willingly (see Chapter One).*]

Supermarine SOUTHAMPTON (1925)

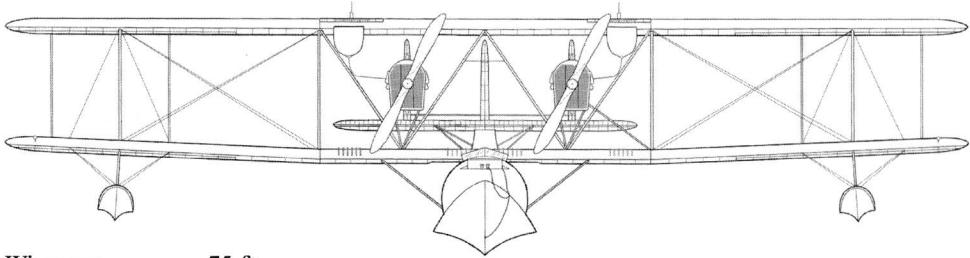

Wingspan	75 ft
Wing area	1,448 sq ft
Loaded weight	15,200 lb
Maximum speed	95 mph

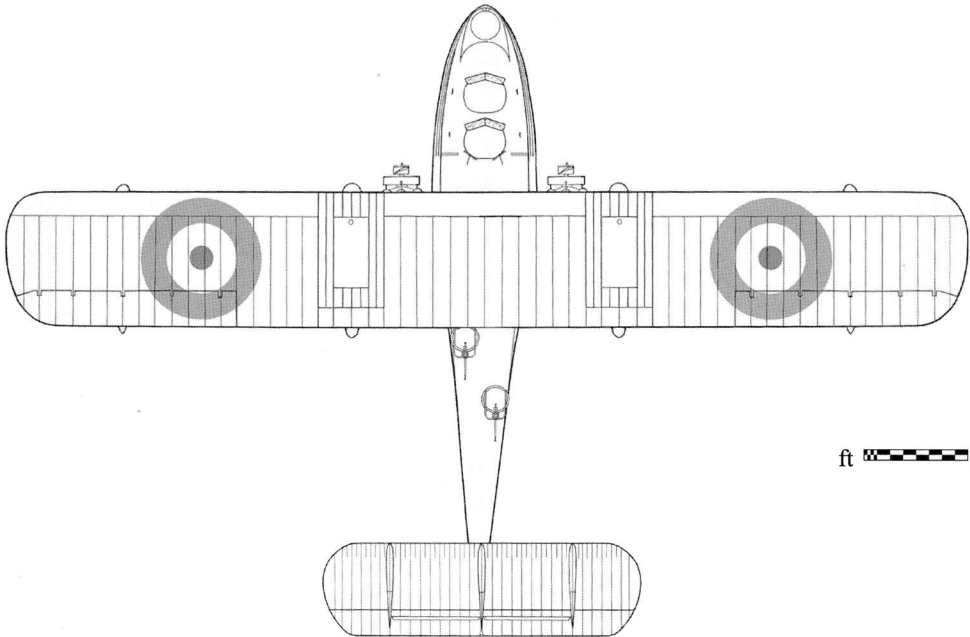

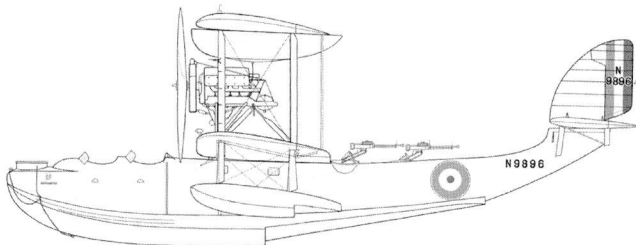

ft

crew compartment above the Swan's lower wing and utilised the passenger bag-gage compartment area for the pilot and navigator, sitting in tandem in open cockpits. He also streamlined the Swan nose and dramatically swept the rear of the hull upwards, terminating in an integrated stub pylon to keep the empennage well clear of the water.

Whilst this last feature had been seen earlier, on both the small FBA and the Latham L-1 Schneider aircraft as well as on the World War One Grigoravich machines, its incorporation in the large Southampton hull was a novel and bold move (*see* Supermarine Publicity, p.104) – of which Mitchell must have been aware, as a 1924 patent on behalf of himself and Supermarine draws explicit attention to the fact that 'the hull is curved upwardly and rearwardly'. Elsewhere, in larger hull designs, Curtiss and Sikorsky moved from the previous Felixstowe unswept approach to the employment of 'canoe' type hulls with the empennage attached by booms subtended from the wings and supported by girders from the hull:

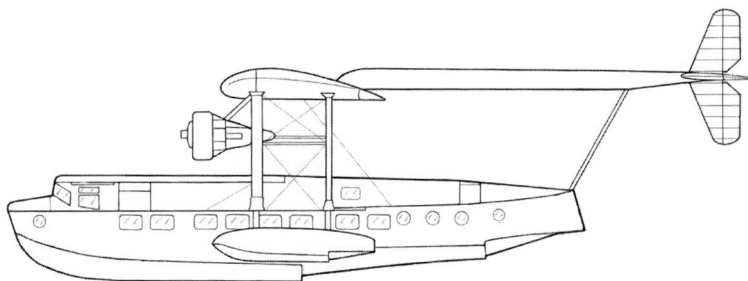

Sikorsky S–40.

In contrast, the elegance of Mitchell's sweeping lines was emphasised and com-plemented by the redesign of the Swan fins which were now swept back in a single curve, resulting in the new Southampton being regarded as 'probably the most beautiful biplane flying boat that had ever been built' and 'certainly the most beautiful hull ever built'. Control lines to the tail unit now ran within the hull rather than externally and untidily.

We can be sure, however, that the aeroplane's ability to maintain height on one engine as well as its maximum range of 500 miles weighed the stronger in Air Ministry minds than any aesthetic considerations. No doubt the Air Ministry advisers also appreciated the extreme practicality of the design: as with the Swan, Warren girders, apart from reducing drag, separated the centre section of the wings without the need for wire bracing and so enabled a change of engine or servicing to take place unimpeded and without interference to the airframe. This centre-section was plywood covered, again for ease of operation by mechanics; the leading edges of the outer panels were also plywood covered to ensure a smoother aerodynamic entry.

The lower wing roots were not incorporated into the boat hull, as was com-mon practice elsewhere at the time; instead the wing superstructure had attach-ment points on the top of the hull (*see* photo, p.323) and was braced by struts from the lower-wing centre-section spars to reinforced frames in the hull. In this way, Mitchell retained as much flexibility as possible in the Linton-Hope type hull and reduced the possibility of cracking around the wing fixing positions. In the Swan this arrangement had also created an unencumbered and roomy pas-senger space with adequate headroom and, in the Southampton, it also had the advantage of enabling good communication between crew members which was especially appreciated at a time before radio contact was available.

Ahead of the pilot was a bow cockpit for a forward gunner and, a little further back from where the Swan crew had been located, were two staggered cockpits for rear gunners, one on each side of the centre-line. Hammocks, basic cooking, and lavatory facilities were also provided – thus beginning the tradition of providing the RAF with maritime aircraft which could be reasonably self-sufficient for prolonged periods of time. As with the Scarab, the siting of the petrol tanks in the upper wing centre-section was a contributory factor in the improvement in crew facilities and communication and also gave a simple and reliable gravity feed to the engines.

Officials must have also been impressed by the efficiency with which the first Southampton was delivered. As Supermarine's publicity recorded: 'Something of a record in design and construction was achieved with the first machine of this class, for it was designed and built in seven months, was flown for the first time one day and delivered by air from Southampton to the RAF at Felixstowe the next day' (11 March, 1925). Its cause could not have been harmed when, after being damaged there in a collision with a breakwater, it was taxied all the way back to Woolston for repairs. Pilots subsequently reported that it 'never gave the slightest trouble … and was a joy to fly', 'a great step forward, a delight to fly and operate' – summed up by Penrose when he reported for the year 1925 that 'it was the beautiful new Supermarine Southampton flying boat which was receiving unstinting approbation from RAF pilots'.

Three Southamptons over Southampton docks (Mauretania in foreground) and about to overfly the River Itchen and the Supermarine factory at Woolston.

The combination of practicability, reliability, range and 'friendliness' resulted in the RAF undertaking a series of long-range proving flights as soon as deliveries to 480 Coastal Reconnaissance Flight began – in the summer of 1925. Four Southamptons flew a twenty-day cruise of 10,000 miles around the British Isles, including exercises with the Royal Navy in the Irish Sea, and a single Southampton – N9896, the first to be completed – made a three-day round trip from Felixstowe to Rosyth, followed by a fourteen-day exercise with the Scilly Isles as base, and then by a week's cruise around coastal waters.

It can be no exaggeration, therefore, to say that the advent of the South-ampton, which was first flown by Biard on 10 March, 1925, marked the real point at which Supermarine finally achieved economic stability and prosperity. In the Supermarine publicity (*see* overleaf), confidence is expressed in features that should appeal to potential customers: ruggedness, manœuvrability, and unre-stricted (by contemporary standards) fields of fire. Mention is also made of the metal hull version of the Southampton that the company was contracted to ex-periment with. The original order of six machines was eventually increased to a total of twenty-five – including the required experimental metal-hulled machine. And this total was later increased to 83 when the metal-hulled Marks II to IV were ordered and when sales were extended to Japan, Argentina and Turkey (*see* p. 127). As a result, the company took out a lease on the Air Ministry's large flying boat assembly units at Hythe on the opposite side of the Solent from Wooltson, for final erection and testing of their new Southamptons.

The penultimate wooden hulled Southampton.

Significantly, the Supermarine entry in *Jane's All the World's Aircraft* for 1925 records, for the first time, the identity of the company's Chief Designer: 'The firm has a very large Design Department continually employed on new designs, under the Chief Designer and Engineer, R. J. Mitchell, who has estab-lished himself as one of the leading flying-boat and amphibian designers in the country'. A more independent view of Mitchell's achievement in the field of sea-plane design came from the caption to a picture of a Southampton I flying boat at the beginning of *Jane's* for the same year: 'one of the most notable successes in post-war aircraft design'.

Mitchell had just passed his 30th birthday.

* * * * *

The efficiency of the Southampton and the elegance of its hull design could not have been clearer signs of the emergence of a designer in his own right but nothing could have prepared one for his next Schneider Trophy aircraft, the S.4. And whilst the Southampton represented a real advance on the current biplane flying boat formula, the boldness of the S.4 design showed Mitchell also moving ahead of other *racing* aircraft designers and beginning the establishment of his reputation in this high-speed field, as will be recounted below.

Supermarine publicity, which draws attention to the main technical features of the Southampton and its many up-to-date features likely to attract potential customers:

THE SUPERMARINE 'SOUTHAMPTON'.

This machine, which appeared for the first time in the early part of 1925, is now the standard twin-engine reconnaissance flying-boat of the Royal Air Force, and has been ordered in large numbers for duty with the Naval Cooperation Squadrons.

Recently, a most instructive and successful cruise has been carried out around the British Isles by five of these machines, the total distance covered during the cruise being approximately 10,000 miles. Throughout, the weather was distinctly bad, yet the boats carried out the programme previously drawn up, and demonstrated that they can function successfully quite separately and independently of their land bases. Refuelling at sea was carried out on all occasions without a hitch.

This machine is fitted with two 470 hp Napier 'Lion' engines and is a biplane flying boat of very clean aerodynamic design. The empennage consists of a monoplane tail with three cantilever fins above it. This design of tail is one of the many novel features of the machine, and, in conjunction with the positioning of the rear guns, enables an unrestricted field of fire to be obtained. The design of armament on the 'Southampton' machine has proved that a flying boat can defend itself most efficiently, and has revolutionised all previous beliefs on this matter. The petrol is carried in the top planes, which not only permits a gravity feed, but also reduces the risk of fire to a minimum.

The machine is extremely seaworthy. It is capable of riding out the roughest of seas, and can be taken off and landed with safety under these conditions. The hull is very roomy and efficiently fitted out for the crew. There is a through passage from bow to stern, and no petrol is carried in the hull. The standard arrangement of crew is as follows: In the bow is the gunner and bomb operator. Behind him are the two pilots, in tandem, with complete dual control. The after pilot also has a complete navigating compartment. Aft of this is the W/T compartment, which opens out into two rear gun-ring positions behind the main planes. A crew of five is normally carried. The accommodation for the crew is very comfortable and efficiently planned, and is unusually free from noise and draught. Hammocks can be easily fitted, so that the crew can sleep on board and remain afloat for long periods.

The machine has been flown continually on one engine, and can be manoeuvred and turned against the pull of the one engine without difficulty.

The well-known qualities of the Napier 'Lion' engine have been used to the fullest extent by an efficient installation, with the result that not the slightest troubles have been experienced from the power units throughout the many thousands of hours of flying these machines have carried out.

The hull can be supplied either in wood or metal. The first duralumin hulls for the 'Southampton' have already seen considerable service and have proved themselves to be extremely robust and capable of standing up to heavy usage. They are 450 lbs (204 kgs) lighter than the wooden hulls, and this weight can be used either to increase range or military load.

S.4 – Mitchell's revolutionary Schneider floatplane

It will be recalled that Mitchell's upgraded Sea Lion III was no match for the American Schneider Trophy machines; indeed, the other British aircraft which also competed, the Blackburn Pellet flying boat, had been noticeably cleaner in design than the Supermarine entry, but it was the American floatplane, the Curtiss CR-3, which had set the new standards of streamlining for the 1923 contest. By abandoning the flying-boat configuration of the winners of the three previous contests, the Americans were able to shape their fuselages more specifically to the cross-section of their engines and avoid any aerodynamic uncleanness of engine-mounting struts and wires. As the American D-12 engines only produced 465 hp, compared with the 525 hp of the Napier Lion engine in the outclassed Sea Lion, something more fundamentally different was needed if Supermarine were to compete successfully next time. (The Fairey Aviation Company also recognised what had been demonstrated by the 1923 Schneider Cup winner: as its chief test pilot, Capt. Norman Macmillan, said, 'Fairey saw that the American success was primarily due to a clean engine of small frontal area mounted in a well-streamlined seaplane'. As a result, the company used the American engine and its small frontal area in its Fairey Fox bomber which, when it came into squadron operation in 1926, could not be intercepted by the RAF front-line fighters of the day.)

Once more, it was the lottery of external circumstances that made Mitchell's next Schneider design, the S.4, a possibility. At that particular time, the Gloster Aircraft Company, in the hope of attracting a military contract, had produced the very promising Gloster II floatplane but it porpoised on landing after its first test flight and sank when one of the float struts collapsed. The Italians had purchased two of the Curtiss D-12 engines for experimental purposes but no plane materialised in time for a 1924 contest. And so the American National Aerobatic Association, responsible for hosting the next event, declared it void and the Royal Aero Club cabled its 'warmest appreciation of this sporting action'– a second win, by a flyover, would have put America in a strong position to win the Schneider Trophy outright, on home ground, in 1925. This unexpected turn of events gave time for the British government to be persuaded to offer the substantial assistance required to produce a serious challenge to the Americans and the necessary breathing-space for Mitchell to produce, for the first time, a dedicated and competitive racing machine that was not dependent on straitened company resources and recycled airframes.

The result of the new-found government backing was that Napier was given a contract for a 6:1 compression ratio development of the well-proven Lion engine which, it was hoped, would deliver 700 hp; and an order was placed with Gloster, as well as with Supermarine, for new machines, for 'technical purposes' – it being understood that, if the machines proved to be suitable for the competition, they would be loaned back to the manufacturers for the next contest. Flying was to be at the Air Ministry's risk, with insurance of personnel the responsibility of the firms involved.

Supermarine, unlike Napier, could not breathe new life into their previous offering and so Mitchell was now faced with the necessity of producing an aircraft which was to be a significant departure from all the Supermarine aircraft which had preceded it.

As we shall see with the Spitfire, his first response was not, however, dramatically original; indeed, his proposed Sea Urchin still looked towards the flying-boat approach (perhaps not surprisingly since almost all the Supermarine design effort had been directed into this type of seaplane) and might well be regarded, essentially, as an improved Savoia S.51 Schneider racer.

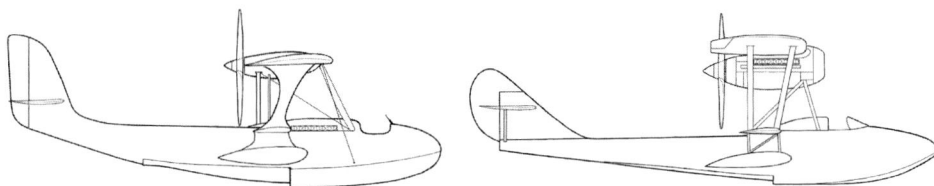

Sea Urchin.   *Savoia S.51.*

Whilst a similar sesquiplane arrangement was proposed, it can be seen that Mitchell's hull revealed somewhat similar styling to that of the slightly later Southampton, particularly in respect of the upswept rear fuselage. Indeed, it was to incorporate an integral fin and unbraced tailplane and, additionally, the drag penalty of the high-mounted engine was to be reduced by situating the engine in the hull and driving the propeller through bevel-geared shafting – some response, at least, to the successful in-line engined CR-3.

But nothing eventually came of the Sea Urchin for, as Alan Clifton said: 'The next design R.J. got out was never built because of doubts about the shaft drive'. Meanwhile, the extra time presented by the American postponement allowed Mitchell time to move comprehensively from the well-tried flying-boat approach to a float monoplane with 'breathtakingly clean lines, which caused a sensation when photos were released' (Clifton). As *Flight* put it when its shape became known outside of the Supermarine works:

> One may describe the Supermarine Napier S.4 as having been designed in an inspired moment. That the design is bold no one will deny, and the greatest credit is due to R. J. Mitchell for his courage in striking out on entirely new lines. It is little short of astonishing that he should have been able to break away from the types with which he had been connected, and not only abandon the flying boat type in favour of a twin float arrangement, but actually change from braced biplane to the pure cantilever wing of the S.4.

The dramatic leap from the type of aircraft he had modified for the previous two Trophy competitions to the new design can be readily appreciated from the following side-views:

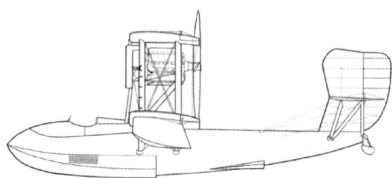

Sea Lion II (1922).   *S.4 (1925).*

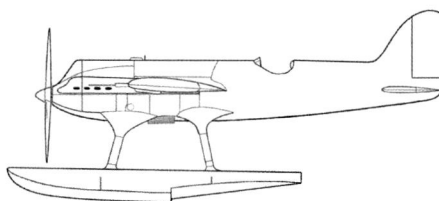

The time available to construct this machine, with all the attendant problems of building such a novel aircraft, was not great as Supermarine only received approval to begin building on the 18th of March, 1925. The allocated Air Ministry serial number of the new machine was N197, although this was never carried; Supermarine referred to the new machine only as the S.4: 'S' presumably referring to Schneider and '4' indicating that it was the successor of the Mark III Sea Lion.

Even his Spitfire was preceded by another monoplane fighter and by the high speed S.5/6 series of trophy racers; by comparison, the S.4 was far more dramatically original although it ought to be noted that nearly a year earlier the French speed record holder, the Bernard V.2 landplane, had displayed some features which might have prompted Mitchell's design: its Hispano-Suiza engine was a broad arrow design similar to the S.4's Lion engine and was incorporated almost identically into the fuselage and wings, with similar aluminium engine fairings; it also had cantilever flying surfaces, under-wing radiators and a similar pilot's cockpit position. Nevertheless, when Harald Penrose of Westlands later wrote of 'the startlingly novel and beautiful Supermarine S.4' he was at least reflecting the dramatic appearance of a entirely new floatplane design and was surely right in responding to its fine lines; in comparison, it might not be unduly partisan to consider that the Bernard landplane had a much more clumsy appearance.

Clifton also drew attention to the undoubtedly unique attachment of the floats: 'It was an exceptionally clean design, with a central skeleton of steel tubing which included daring cantilevered float struts.' This 'central skeleton'

The S.4 at Calshot.

was, characteristically, a deceptively simple arrangement of two very strong 'A' frames which related directly to the three sections of the fuselage: the engine mounting was bolted to the front frame and the rear monocoque fuselage section to the rear one; between the two, the wing was fixed and the floats were attached to the feet of the frames. This bold mid-fuselage structure was known in the works, less reverently, as 'the clothes horse'. The struts were carefully faired into the tops of the floats and into the fuselage which, in its turn, was tailored to the contours of the engine cowlings.

The move from flying boat to floatplane brought the new requirement for floats and this led Supermarine prudently to subcontract this item to Shorts who had recently installed their own testing tank that not only saved the delay of using the National Physical Laboratory ship-model facility at Teddington but was also designed more specifically to simulate seaplane conditions. As a result, the British contenders for the next competition had state of the art floats which were clean-running and low in drag (whilst their single-step design was a gratifying confirmation of the more intuitive design philosophy of Supermarine's flying-boat hulls).

The wing also represented a striking departure from earlier company structures. Its cantilever structure was given the rigidity that bracing wires normally provided by the addition of spanwise stringers rebated into the ribs, covered with load-bearing plywood sheeting top and bottom, gradually decreasing in thickness towards the tips.

Thus it was that Mitchell felt able to take the, then, radical step of dispensing with struts and wire bracings for the wings and tail surfaces; he also did away with bracing wires for the floats, although their struts were braced by two thin-section cross members. An appreciation of the conceptual leap represented by the S.4 can be gained by a comparison of its cantilevered structure with that of the previous Sea Lion which had thirty-three struts and forty-two bracing or external control wires; it also had struts between hull and lower-wing centre-section joints, and no less than five struts each side of the fin.

Further streamlining was now achieved by the use of the newer Lamblin radiators which were mounted on the underside of the wings. These, and the oil

Supermarine S.4 (1925)

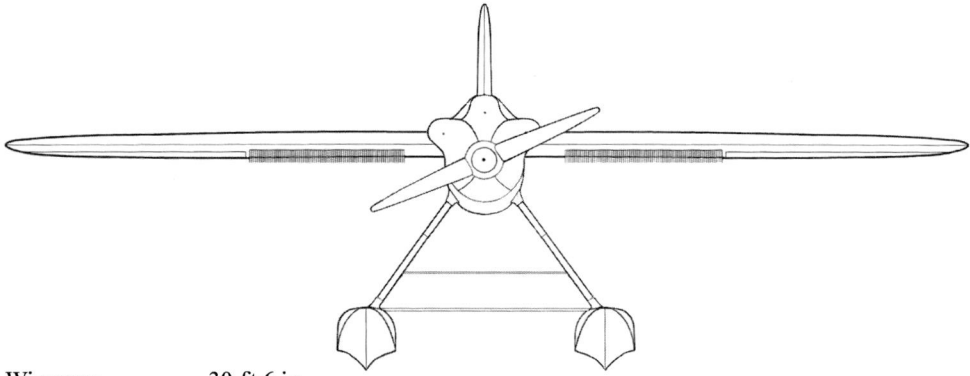

Wingspan	30 ft 6 in.
Wing area	139 sq ft
Loaded weight	3,191 lb
Maximum speed	239 mph

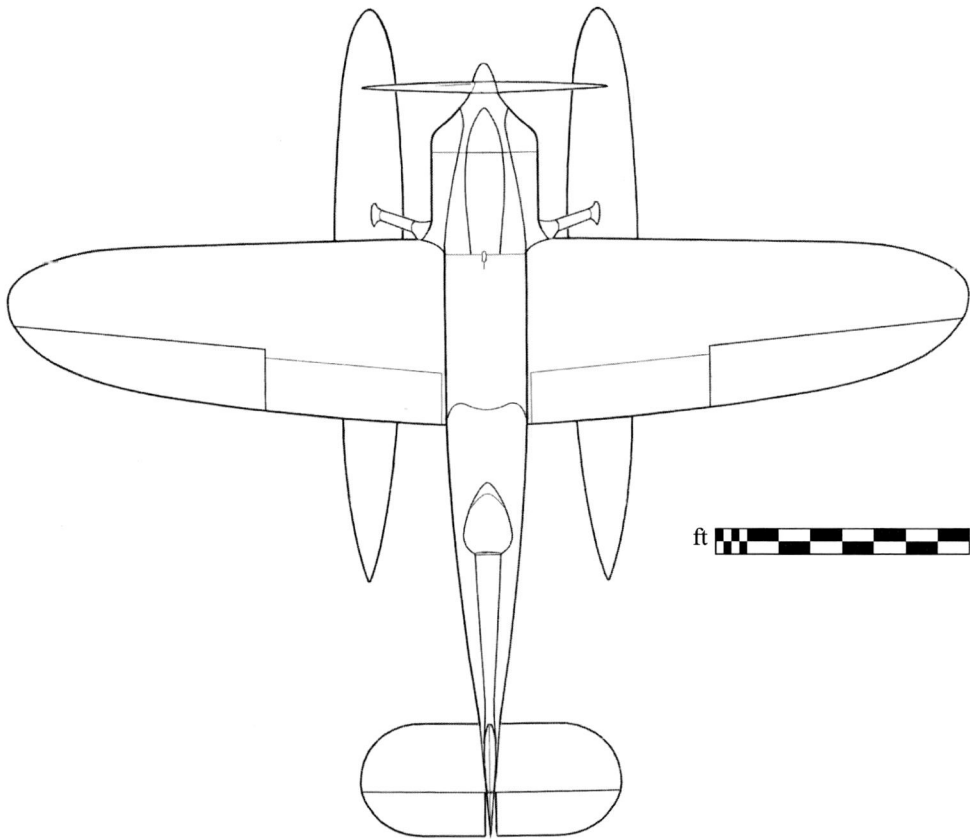

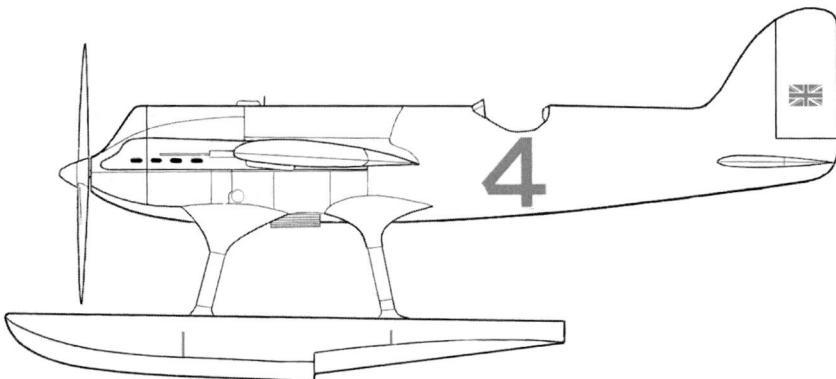

ft

The S.4 at Woolston prior to departure for America. Strut fairings incomplete.

Biard and Mitchell with S.4 under construction.

cooling fins on the underside of the fuselage, were the only significant protuber-
ances exposed to the slipstream – the coolant water was carried to and from the
engine via troughs buried in the underside of the wings and the interconnected
flaps and ailerons and the tail surfaces were activated also from within the structure via rods and torque tubes. The streamlining aluminium fairings from the
engine section extended almost to a point level with the wing trailing edge – also
giving protection from the searing heat from the Lion's stub exhausts.

Mitchell's design was built in five months and Biard first flew the machine on
the 24 August, 1925. According to Biard, Mitchell had put on a bathing costume,
assuring his test pilot that 'if anything happens, I'll dive into the water and pull

> The following reminiscence of Cozens reminds one that the S.4 was, despite its futuristic shape and polished metal cowlings, built with traditional woodworking techniques. Its sound, however, was something else – no doubt the result of its ungeared racing engine producing propeller tip speeds around the speed of sound (2,600 rpm turning an eight and a half foot airscrew – go figure):
>
> After being beaten in 1923 by trying to make the best of an outdated machine, Mitchell went to the other extreme and produced something that was far ahead of its time. It was, of course, of wooden construction, not surprising as neither the designers nor the workforce were capable of building a sophisticated metal machine, and it exploited the Linton-Hope technique to the limit, and the cantilever wings and float struts were faired into the fuselage so that the whole was clean and the flying characteristics were bound to be the matter of some speculation. When considering it today, fifty-six years later, one must concede that it was a daring gamble. It was built in great secrecy which gave rise to even more curiosity and expectancy than was usual for a Schneider Trophy, always a sensitive subject and it was guarded like a racehorse in a training stables.
>
> By raising the compression ratio the Napier's power was increased and the metal Fairey Reed propeller was quite new in design and construction, so that when it finally emerged and the engine was run up a new sound came, something that the local people had never heard before, and indeed, very few people ever heard in all their life, a sort of high pitched scream of immense power.
>
> Radiator drag was reduced to a minimum by building it flush into the wings, and because it was made of thin copper plates soldered together it had to be very carefully formed and fitted ... Plainly, this aeroplane needed very skilful handling, more so because the mid-wing shape made visibility poor, and Captain Biard was the only man who could fly it, and an eyewitness said that even he made an airborne hop of a mile before he finally got off between Lee-on-Solent and Calshot ...

you out'. In fact this first flight nearly began and ended in disaster owing to the pilot position. Centre of gravity considerations necessitated the cockpit being situated well back behind the trailing edge of the wing and the high position of this wing thus created a blind spot ahead when taking off and landing. Biard claimed to have nearly collided with the liner *Majestic* on take-off – having not seen it at all until the last minute – and, when he came in to land, he nearly hit a dredger. However, the S.4 thereafter proceeded to gain the World Speed Record for Seaplanes and the outright British Speed Record by registering 226.75 mph – nearly 40 mph more than the Curtiss CR-3 record established the previous October.

In contrast to the Supermarine design, the Gloster III, whilst featuring the uprated Lion engine and a metal propeller, stayed with the current orthodoxy of the wire-braced biplane formula. The new Italian entry, the M.33, also retained many of the previous features – it was a flying boat which necessitated an engine mounted on struts above the fuselage and therefore unable to be effectively streamlined – but, on the other hand, its cantilever monoplane approach showed that not just Mitchell was moving in this direction, although far less dramatically (*see* drawings p. 115). Nor were the Americans making bold advances, being restricted to developing the previous CR-3 by fairing the upper wing into the fuselage and installing a relatively untried Curtiss V.1400 engine developing 610 hp (compared with the British Lion engines, now developing 700 hp).

Owing to the short time between gaining Air Ministry backing for the building of the aircraft and the race itself, neither the S.4 nor the two Gloster IIIs had many suitable opportunities for test flights before planes and personnel had to be transported to America, along with a practice machine. On the other hand, they were the first to arrive at the proposed venue, Chesapeake Bay near Baltimore, on 5 October – further evidence of Britain's new-found determination to compete successfully.

Mitchell (centre) in front of completed S.4.

The actual base of operations was Bay Shore Park, a beach area fourteen miles south-east of Baltimore, where tented accommodation for hangars and workshops was to be provided. This provision was found not to be ready and it was only possible to begin erecting the aircraft on 12 October; even then, Mitchell and Folland, Gloster's designer, were appalled at the conditions which their ground crews had to work under. Several days elapsed before any flying was possible, the Gloster III going first, followed by the S.4 on the 16th of October. Then Biard went down with influenza and the weather worsened to gale conditions, causing the collapse of some of the tents. A heavy pole fell across the tail unit of the S.4, which necessitated hard work by four Supermarine rigger/engineers in order to get it ready in time for the navigation tests on the 23rd of the month.

By this time, Biard was up and about, although not his usual self, and the British team leader, Capt. Charles B. Wilson suggested that the reserve pilot, Bert Hinkler, should take over the S.4. This Biard resisted as only he had had experience of handling this advanced machine. Unfortunately, the S.4 story was not to have a *Boys' Own* ending as it crashed into the bay following a steep turn which appeared, perhaps, to have been caused by a high speed stall; flutter or wing distortion was also suggested by contemporary observers. Whatever the cause, it luckily occurred at low level and Biard survived. Mitchell, always concerned for the safety of his pilots, had set out to rescue him but his boat had engine failure and Biard was only picked up after some time in very cold water.

Flight had described how Biard took off and circled the tented area but 'coming back over the pierhead … at a height of about 800 ft he seemed to make a steeply banked turn, which at first led the spectators to believe he was stunting, but it was soon realised that the machine was in difficulty and not under proper control'. *Aviation* magazine reported that 'the machine appeared to stall and sideslip first one way and then the other from about 500 feet. After a half dozen of these right and left wing ups and downs, Captain Biard appeared to lose control completely and dropped about a hundred feet, pancaking into the bay.' The result was also described by one of the Schneider Committee: 'He decided to practise some sharp turns and completed one satisfactorily and then attempted another. It was noticed his aileron was hard down on the lower side. The machine seemed to

get out of control and did a falling leaf descent, making a huge hole in the sea.'
According to the Baltimore Sun, the S.4 'nosed into the water and … catapulted
over on its back'.

The S.4 taking off for the last time.

Eyewitnesses seemed certain there had been no structural failure of the wing,
which later examination seemed to verify, although the report of a depressed
aileron on the lower wing suggested an attempt to correct the effect of a wing
twisting out of correct incidence. On the other hand, the *Times* correspondent at
the contest said that the machine went out of control because the wings fluttered,
perhaps suggesting that they were not torsionally strong enough and the correspondent of the *Aeroplane* reported that 'Mr Biard, after getting off made several
sharp turns and put the machine on to vertical banks. After the last it seemed to
develop wing flutter and get out of control'.

Later, in England, Biard was found to have two broken ribs and damage to
some stomach muscles which later needed an operation. His own account of the
crash was that, on coming out of the turn at speed and diving down for a straight
run, the control column set up such violent side-to-side oscillations that he lost
control. Wing flutter was being experienced at about this time with military aircraft – Penrose quotes a Fl. Lt Linton Ragg of the Royal Aircraft Establishment
at Farnborough as experiencing similar stick behaviour: 'wing flutter had caused
trying [!] experiences, such as coming down with hand and knees badly bruised
by the control column as it played hide-and-seek round the cockpit'.

Thus, whilst torsional weakness or stall cannot be ruled out, 'flutter' might
very well have been the most likely cause of the crash – especially as the landplane version which preceded the Gloster Schneider entry had had an emergency
landing because of tail flutter and Biard had also reported a more minor tremor
of the S.4 wings before going to America (as, it was revealed later, had the Italian M.33). Certainly, as other aircraft caught up with the speed of the S.4, flutter
and aileron reversal were soon to emerge as something needing to be understood
and remedied: indeed, in 1926, Martlesham introduced 'terminal velocity' dive
tests as standard. Biard's description of side-to-side movement of the control
column, similar to that experienced by Ragg, points to aileron flutter and later

remarks at Supermarine confirm this conclusion: Mitchell, concerned about the need to avoid overbalancing of the Spitfire ailerons in a dive, wrote somewhat enigmatically, 'I believe this is the cause of several accidents involving ailerons' and Ernest Mansbridge, explaining the thickness of the later Type 224 wing being due to caution, was more direct: 'We were still very concerned about possible flutter, having encountered that with the S.4 seaplane'.

It was also suggested that Biard, probably still not sufficiently recovered from illness, had stalled through unfamiliarity with the effect of a very tight high-speed turn. In the course of his report to the Royal Aeronautical Society, Maj. J. S. Buchanan, the Air Ministry representative at Baltimore, merely stated that the S.4 stalled and crashed into the sea but, after his lecture, Biard made the rueful or joking rejoinder that 'I also note that Major Buchanan says, "High speed diving is not necessary [during turns] in the Schneider race" – I will take this to heart but wish he had mentioned it before we went to America'.

The Eighth Schneider Trophy Contest, 26 October, 1925, at Baltimore
After the crash of the S.4, the second Gloster III was hastily prepared for, un-accountably, the Royal Aero Club had only entered two machines, instead of the three permitted. The navigation tests this year consisted of flying two laps of the course, during which the pilots had to alight twice and taxi for half a nautical mile at not less than 12 kts in both directions. The Italians, Giovanni de Briganti and Riccardo Morselli, the three Americans, Lieutenants G. T. Cuddihy, J. H. Doolittle and R. A. Ofstie, and the Briton, Hubert Broad, completed these navigation tests by the end of 23 October and their machines were anchored out in the bay for the watertightness test. Bert Hinkler also finally got away in the second Gloster but soon landed with a broken flying wire. By the time that repairs had been finished, it was judged that the failing light had made conditions too dangerous for flying, despite Hinkler's protests. The other competing pilots sportingly requested that he be allowed to try again next morning and the judges agreed.

In the event, Hinkler was unable to do so because of another severe storm. Of seventeen US Navy Glenn Martin SCI floatplanes which had arrived to perform in a Naval Air pageant that was to precede the contest, seven were wrecked and the rest damaged. It was 26 October before conditions were judged suitable for Hinkler to resume his navigability tests. He completed the required two laps of the course but, on alighting to begin the two taxiing tests, he hit some rough water resulting from the storm; his front struts collapsed, and the propeller irreparably damaged the floats. In addition, an Italian M.33 had to be withdrawn before the contest proper was to begin, owing to engine trouble, and so the contest was left to five aircraft, three American, one Italian and one British.

The main flying competition was now able to start at 2.30 p.m. on the 26th. Doolittle in his R3C-2, went first, followed by Broad in the Gloster III; after him came Cuddihy and then Ofstie in the other R3C-2s. Finally, de Briganti got away in the remaining M.33. It soon became clear that Doolittle was putting up the best times, which was not unexpected as he was known to be a brilliant and perfectionist pilot who had been making studies of 'G' forces on pilots for the Army. This experience was now being put to good effect on the seven lap course and his tight turns around the three pylons at low level were in contrast to the slower, wide turns of the other two Americans which had been worked out for the previous contest. Broad, whose Gloster aircraft had 100 hp more than the R3C-2s, tried to emulate Doolittle but his machine's directional stability left something to be desired – the problem had been identified in England but, as was so often the case with Schneider Trophy entries, there had been insufficient time to sort it out, beyond an ad hoc increase in fin area. The pilot experienced side-slipping in wide arcs at the pylons which he later compared to 'the back wheels of a car on an icy road'. He was also averaging 30 mph less than the leading American. In contrast to all the others, de Briganti employed the classic Italian technique of climbing turns on full power, followed by a dive into the following straight but he was, nevertheless, slower than Broad as the Curtiss engine of the M.33 was by now showing the result of earlier extensive testing by Fiat.

The different cornering techniques were an impressive sight for the spectators who saw that the American pilots were doing well. Then the new Curtiss engines of the American team began to fail. On lap six, Ofstie was forced down with magneto failure and then, on the seventh and last lap, Cuddihy's engine ran out of oil; the aircraft caught fire but the pilot managed to hurriedly alight. The American spectators could well have done without this sort of excitement but at least they were able to see Doolittle win the trophy at an average speed of 232.562 mph, followed by Broad at an average of 199.167 mph, with de Briganti a disappointing third – well over 60 mph slower than Doolittle.

* * * * *

The competition showed clearly that the newer version of the Curtiss machine was much sleeker, and therefore faster, than the more powerful Gloster float-plane. The latter, apart from its controllability problems, had been handicapped by its protruding leading-edge radiators which had had to be fitted when their wing surface radiators could not be readied in time. As both these aircraft were floatplanes it was now clearly evident that the flying-boat formula, still favoured by the Italians with their M.33, was now outdated and in the paper given on January 21, 1926, to the Royal Aeronautical Society by Maj. Buchanan various conclusions were drawn. He stressed the need for reducing fuselage drag, the need for wing surface radiators and the need for ample time to test the efficiency of different propellers. And he also called for the use of pilots trained for high speed flight – as per the American Navy and Army teams of the last two contest wins. Despite civilian pilots' evident skill and willingness to take risks, it still remained a fact that their usual flying experience was of much slower machines and that their comparatively isolated experiences of Schneider speeds was compounded by the very limited amount of practice time that was usually available – as we have seen, the Schneider events were characterised by late go-ahead decisions and, therefore, late delivery of new machines, not uncommonly coupled with curtailed flying at the race sites, either because of mechanical problems or because of weather unsuitable for specialist racing machines.

The case of Henri Biard was not untypical: the S.4 was first flown by him on 25 August and, despite reporting slight wing tremors, he had to leave soon afterwards for the Schneider competition in America; influenza and damage to the floatplane resulted in his being only just ready for the Trophy navigability trials on the 23rd. It is to Biard's credit (or belief in his own immortality) that he was prepared to attempt to race in a somewhat suspect machine, with poor forward visibility for take-off and landing, and with little time to familiarise himself with what the other competitors had already surmised to be the fastest aircraft in the field. Not only had his experience on the revolutionary S.4 been very limited but, between his flying the Sea Lion III at a maximum speed of 175 mph in 1923, and achieving 239 mph in the new machine two years later, his day-to-day flying experience with Supermarine was with the Swan passenger amphibian, the Scarab reconnaissance amphibian, the Sparrow I light landplane, and the Southampton I flying boat whose top speeds averaged out at something less than 100 mph.

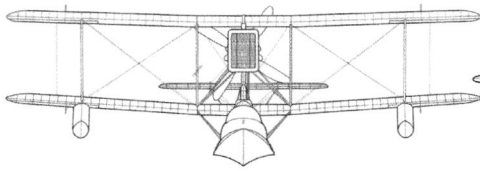

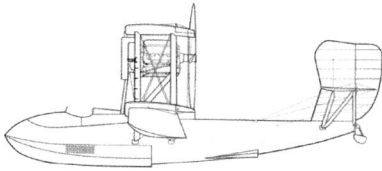

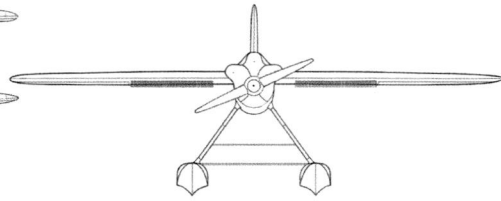

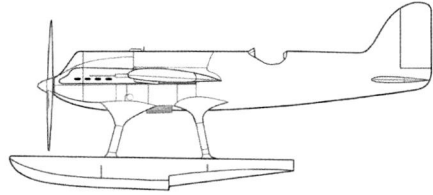

Supermarine Sea Lion II (1922).   *Supermarine S.4 (1925).*

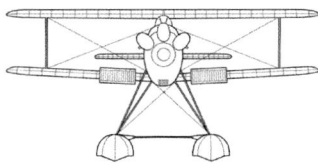

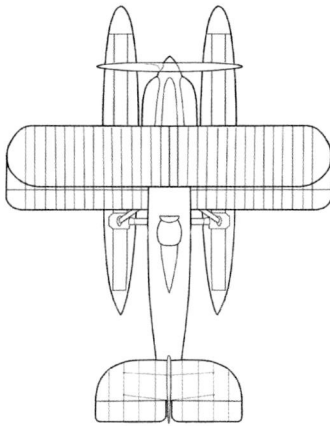

Gloster III (1925).   *Macchi M.33 (1925).*

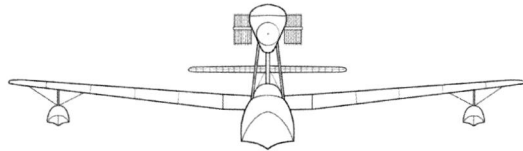

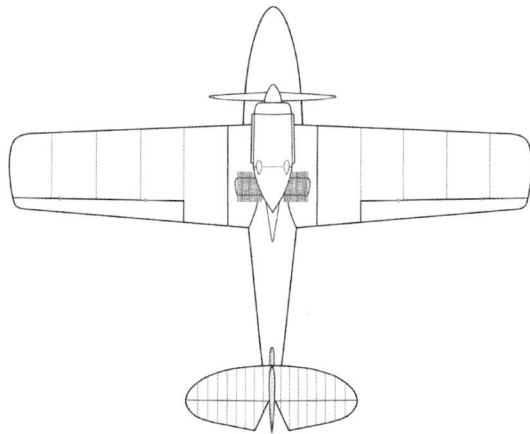

* * * * *

Before the year of the eighth contest was over, it was announced that the next Schneider Trophy was to be held in the following year in the week beginning 24 October. The need for a radical overhaul of the British effort, prompted the Royal Aero Club to ask for a one year postponement but, this time, the NAA refused. After all, they had now achieved two wins in a row and could very probably manage a third and final success without costly new designs, bearing in mind that Doolittle had now set up a new world record for seaplanes at 245.71 mph in the existing R3C-2.

Additionally, a delay might not be wise as American governmental priorities were hardening towards the development of air commerce and transport: the last Pulitzer Cup race had been run in 1925 and barnstorming was not being encouraged. It was now more important to promote a public appreciation of safe commercial flying; racing had been valuable but the development of military aircraft could now directly benefit from the currently existing experience of high speed flight without diverting funds to dedicated, and probably, esoteric new racers – especially floatplanes.

Despite the distinct possibility of a third, and outright, win by America, the Royal Aero Club announced that it would not be competing. Italy had also asked for a postponement and, had their response been the same as that of Britain, it would have resulted in an American flyover and the end of the Schneider Trophy. Fortunately for the future public standing of Mitchell and, arguably, for the later development of the Spitfire, events in Italy intervened. Mussolini had come to power in 1922 and by now felt it needful to demonstrate the success of his dictatorship – in particular, he decided that Italy must win the next Schneider Trophy at (literally) all costs. It was decreed that the state would provide all necessary financial and other assistance to create both a suitable airframe and a matching engine. Three aircraft were ordered, finished to contest standard, another two were for training and practice flying, with an additional airframe for structural test work; and an approach was made to the United States for six of the Curtiss engine which had brought victory to the R3C-2 at Baltimore.

Not surprisingly, Macchi and Fiat were the Dictator's chosen instruments. The 1923 and 1925 winners, as well as Mitchell's revolutionary but ill-fated design, had shown that the floatplane approach was the future. The American machines, however, had been biplanes whereas the Italian M.33 had represented a movement towards the S.4 monoplane approach. Macchi now went further, producing a floatplane, instead of the previous flying-boat design, and continuing the monoplane configuration. However, because the M.33 had experienced flutter and because of the strong suspicion of flutter as the cause of the S.4 crash, the new wing had wire bracing. And, as floats were now to be a new feature, it was not surprising that those of the new Macchi machine followed the general shape of other 1925 aircraft. Nevertheless, having been preceded by Mitchell in certain respects, Mario Castoldi must be credited with producing a quite distinctive and elegant Macchi design, the M.39, with a cruciform tail-unit and a low

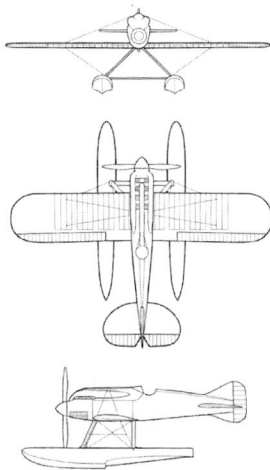

Macchi M.39.

wing with slight sweepback – this last dictated by weight distribution rather than by any futuristic aerodynamic consideration.

Meanwhile the Italian Air Ministry had been unable to obtain approval for the purchase of the American engines but Fiat, having profited from the earlier study of the Curtis D-12 engine, was able to offer a 12 cylinder 'V' that was capable of developing over 800 hp. The new AS.2 engine eventually produced 880 hp on the test bench and, by mid August, the first competition aircraft was ready for testing. The remainder were delivered in the next few weeks and it was confirmed to the American club that Italy would definitely compete in the 1926 Trophy competition. But engine heating and carburation problems soon began to emerge and one pilot was killed in training: so an application was made for a short delay, to which America sportingly agreed. Despite British protests at this, the race date was put back to the 11th of November and, on the 12th of October, the Italian team set off for prohibition America with four of their new M.39s and with a plentiful supply of Chianti smuggled in their floats.

The Ninth Schneider Trophy Contest, 13 November, 1926, at Hampton Roads

In view of the future successes of Supermarine, powered by very reliable British engines (*see* Chapter Six), it is instructive to consider the very mixed fortunes of the 1926 aircraft.

For the contest, the Americans fielded an R3C-2 with the 1925 600 hp engine, another Curtiss floatplane with a 700 hp geared Packard engine – designated R3C-3 – and also an R3C-4 which had a 700 hp Curtiss powerplant. And there was a Curtiss Hawk fighter equipped with floats for team practice and as a reserve. Against them the Italians now revealed their M.39s which looked to represent a formidable challenge and were known to have considerably more horsepower than the R3Cs.

Poor weather was again a factor and so both teams had limited testing time before the contest date; in addition, the R3C-3 engine was not giving full power and R3C-4 engine overheated during practice and was seriously damaged. Meanwhile, the Italians also had their engine problems: one M.39 had to be force-landed because of an engine fire and the engine of another had a connecting rod failure; nor had their carburation problems been overcome. The earlier call for a year's postponement by both Italy and Britain began to seem rather sensible.

The revised date of the competition was also set back because of the weather and so the navigation tests did not start until late on 11 November, with a possible extension into the next day. Following the completion of the navigation tests, the flying competition was able to start on the next day, the thirteenth of November, before a crowd estimated at 30,000. After a 2.30 start, the first lap times revealed that Tomlinson in the stop-gap Hawk had achieved a not unexpected average of about 137 mph, and that Bacula had posted a surprisingly modest 209.58 mph. The crowd was not to know that the Italians had planned to have Bacula beat the slow Hawk but, otherwise, nurse his engine in order to ensure that he might do well if the two R3Cs dropped out. Then came the main contenders: Cuddihy's average speed was timed at 232.427 mph but this was slower than the next two Italians. Ferrarin had delayed his start in order to shadow Cuddihy who was regarded at the main threat to the Italians, but it was found that the Italian was achieving the better time of 234.61 mph.

Then, on lap four, Ferrarin's repaired engine failed, as a result of a fractured oil pipe. The competition then settled down to a contest between Cuddihy and de Bernardi, with Schildt's R3C-2 lapping at about 230 mph, hampered by a float-wing wire that had parted and was causing considerable wing and aileron flutter on approaching top speed. Then, Cuddihy dropped out of the competition on the seventh lap as he had done in the previous contest; this time his petrol supply had failed. He landed safely but with a severely blistered hand from furiously operating a handpump for lifting fuel from the floats. And so De Bernardi came in first at an average speed of 246.496 mph, followed by Schilt with a figure of 231.363 mph; Bacula came third, having flown a circumspect 218.01 mph but beating, as planned, the slower Tomlinson, whose Hawk could only achieve an average of 136.95 mph. De Bernardi duly sent a cable to Mussolini, stating: 'Your order to win at all costs has been obeyed.'

* * * * *

Whilst the American phase of the Schneider Trophy competitions had brought no real luck to Supermarine, with the gift of hindsight, it can be seen as a most important milestone in Mitchell's career. Later, when Mitchell returned to the design of racing floatplanes, he turned from the wooden airframe of this 1925 aircraft to embrace the metal structures that were to become a feature of the future generations of fighter aircraft. We shall see the successful outcomes of the new technology in the next three Schneider Trophy competitions and it is these which established his reputation beyond the aircraft industry; but, when one considers the quantum shift from the Sea Lion II of 1922 to the S.4 of 1925, a special place should be reserved in British aviation history and in Mitchell's design career for the ill-fated but beautiful S.4 – as E. Bazzocchi of Aeronautica Macchi said, 'the real revolution of 1925 was the appearance of the Supermarine S.4: its very clean design set the pattern for all subsequent Schneider racers'.

The S.4 was a failure but it marked out the emergence of a notably innovative designer dramatically pushing forward the frontiers of high speed flight. Supermarine's publicity in 1926 (*see* below) draws particular attention to this and to the design features which were employed; it also points out that the previously quoted top speed of over 226 mph was later increased to an impressive 239 mph and makes the claim that it was 'the fastest British aircraft of any type':

Supermarine publicity from 1926:

The Supermarine-Napier S.4 is a twin-float cantilever monoplane of high performance. The machine was built as part of the Air Ministry's programme of high speed development, and was loaned to the Supermarine Company for entry in last year's Schneider Trophy Race. Instructions to proceed with the construction of this aircraft were issued on March 18, 1925, and the first flight was carried out on August 25, 1925. In view of the extremely novel type of design and the large number of experimental features incorporated in this machine, this may be fairly considered a remarkable achievement.

The wing, which is of a new high-speed section, is built of wood and constructed as one unit. No fabric is used for covering; three-ply is used throughout, and in such a manner as to take its share of the load. The trailing portion of the wing can be used as a flap to reduce landing speed, and the ailerons are also geared in with the flap mechanism.

The chassis consists only of four high-tensile tubes, with two light horizontal bracing tubes. The main support tubes are braced together within the fuselage, and thus form a complete structure on which the remainder of the machine is erected. The engine-bearers are built on forward, the wing is attached to the top, and the rear fuselage is bolted on to the aft end of this central section. It is well to note that this machine is not only based on excellent aerodynamic design, but the floats are admitted to represent a very great advance on anything previously achieved. A minimum of spray is caused when taking off. It will be recalled that on September 13, 1925, this machine set up a World's Speed Record for Seaplanes, covering the 3-kilometre course at a speed of 226.6 mph. Since this date the performance has been considerably improved by special tuning of the engine and fitting of a propeller of greater efficiency. At the same time, the S.4 achieved the distinction of being the fastest British aircraft of any type, achieving a maximum speed figure of 239 mph.

[*One might not be too surprised that the company saw no reason to mention the final fate of the machine.*]

— *Chapter Five* —

1926 to 1927
Consolidation and
International Successes

The previous chapter showed how Mitchell, in 1925, had set a new standard in reconnaissance flying-boat design and had produced a revolutionary floatplane racing machine and both had shown the emergence of a creator of elegant shapes and a designer who had fully transcended the design precedents of his company.

But, however far in advance of contemporary practice such products might be, they only formed part of a busy designer's overall responsibilities. It is instructive therefore to note that, as late as 1926, Supermarine was still advertising the 1919 Channel flying boat in its four-seat passenger-carrying and dual-control trainer versions – with a photograph of the last one sold, to Chile in 1922 (modified, interestingly, with the later Seal type hull). It is clear from the accompanying text that the company was still hoping for commercial contracts: 'A machine of this type was used to demonstrate to representatives of the Port of London Authority, the Trinity Brethren, and Scotland Yard that a flying boat could be handled in a busy waterway, and that it was possible to use the Thames as an air port.' (Note that the now familiar 'airport' was not yet in common use – *see* p. 72.)

In the years which followed, Mitchell's varied work pattern intensified as the company began to prosper: apart from fulfilling the first orders for twenty-four Southampton Is – by far the most significant requirement that Supermarine had so far received – there was the need to improve upon the performance of this aircraft, there were specifications for larger flying boats to meet, it was necessary to start meeting the exacting demands for improved Schneider Trophy floatplanes, and there was now the important move from mainly wood construction to metal.

SPARROW II

A good example of the diversity of the Supermarine output (and of the fundamental insecurity of the aero industry) was that the small Sparrow landplane still continued to occupy the time of the company. In September, 1926, a second light aircraft competition was organised, again at Lympne, over a series of courses totalling about 2,000 miles, and with an engine size restriction in common. This time, the Supermarine machine was fitted with the type of engine which had powered the two most successful aircraft of the 1924 Competition, a Bristol Cherub III of 32 hp.

Mitchell also replaced the biplane structure with a parasol high-wing monoplane superstructure and the resulting aircraft was now designated Sparrow II. The new layout must have vastly improved upon the pilot's vision, which had been criticised at the earlier competition. The one piece wing was smaller in total area than the biplane surfaces of the Sparrow I, and its thick aerofoil section gave what Biard described as 'an exceptionally low landing and starting speed, which

Sparrow II.

would have been most useful in a machine meant for small aerodromes'. He also recalled how, when Sir Sefton Brancker, the then Director of Civil Aviation, was a passenger in the Sparrow, he had become more than a little interested in the true airspeed when it dropped to about twenty miles an hour during a landing into a strong headwind. It was perhaps fortunate for the company's reputation that he was not aboard a few days later:

> ... we went up in fine style, circled round, dived and so on, and then I came down, vaguely aware that there had been some sort of commotion among the Directors who were watching below. Mr Mitchell came running up as I climbed out of the cockpit. 'Didn't you see the wings? Couldn't you see the wings?' he asked in a very agitated voice. It happened that I couldn't properly see them from the cockpit, because they were away up above my head. But he told me the whole time I was flying the wings had been trying to swing round, first one way and then the other.

Suitably stiffened up, the machine was duly entered in the 1926 *Daily Mail* Two-seater Light Aeroplane Competition which required six days of out and return circuits from the Lympne airfield where the competition was based. When Biard took off on 12 September, the first day of flying, the weather had worsened and, after less than thirty miles outward bound for Brighton, he decided that battling the strong headwind would not allow him sufficient petrol to complete the circuit. He returned to Lympne and, hopeful of better conditions, refuelled and set off again. This time, when he had reached Beachy Head, his passenger, one of the Supermarine mechanics, pointed out that one of the pins holding the wing struts in place had nearly worked itself out. To avoid the 'very annoying' prospect of the wings again coming off, Biard hastily landed on the Head where the aircraft was promptly blown on its side.

By the time that it had been righted, it was too dark to attempt the return flight to Lympne and so an uncomfortable night was spent beside the machine. The next morning the engine had to be run flat out whilst Biard and his mechanic guided the Sparrow several hundred yards up the slope of Beachy Head; then, leaving behind his passenger who insisted on safeguarding the lead ballast that the competition handicappers had required, Biard turned downhill, made a successful take-off in the lightened plane and finally returned to base.

Unfortunately, the rules of the contest required that each of the six circuits had to be completed in the day allotted and thus the Sparrow II was eliminated on the first day of the competition (eventually won by Sidney Camm's Cygnet). Supermarine entered the Sparrow five days later for the Steward's Prize for the eliminated aircraft and in the Grosvenor Cup Race on the same day. It was un-placed in both.

No doubt the promise of an Air Ministry contract to test various aerofoils had influenced the company to persevere with the type and to redesign it for this

Supermarine SPARROW II (1926)

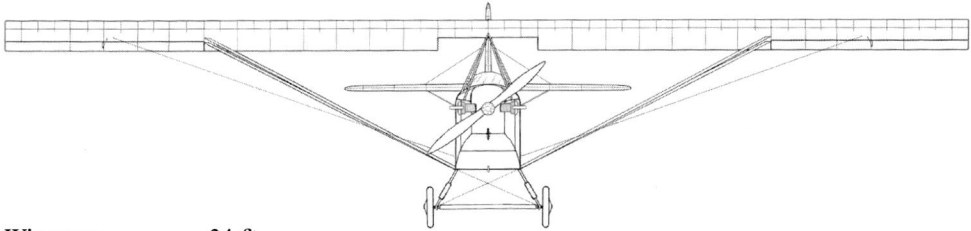

Wingspan	34 ft
Wing area	256 sq ft
Loaded weight	1,000 lb
Maximum speed	65 mph

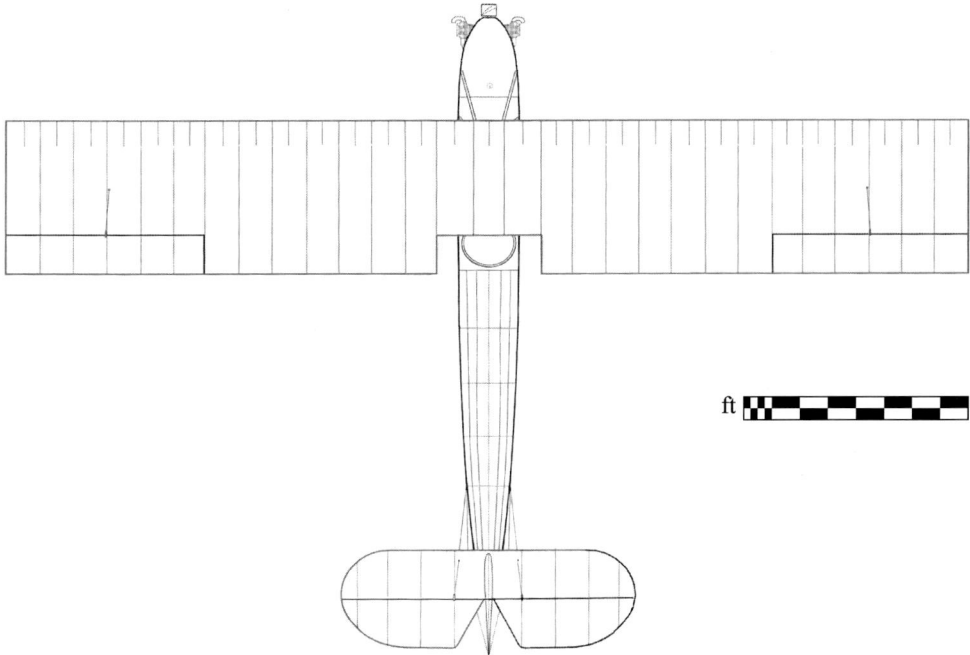

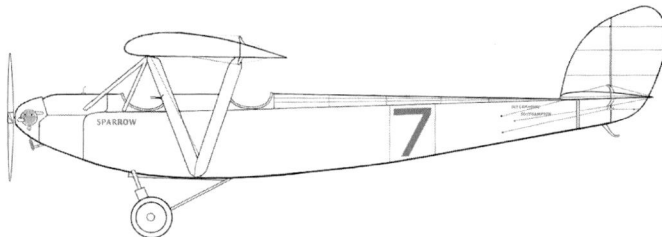

ft

purpose for, in view of Supermarine's specialisation upon seagoing aircraft, the two Sparrow episodes cannot be seen as particularly significant to the development of the company – although the need for companies to seek possible lucrative orders wherever possible (in this case the hoped-for increase in private club flying) was always an important consideration.

Whilst no orders for club aircraft materialised, the Sparrow II, now financed by the Air Ministry contract, was usefully employed for flight comparison trials of identical area wings with different aerofoils, with the parasol wing layout reducing interference effects of the fuselage to a minimum. An SA.12 aerofoil proved to be the best, giving the machine an excellent balance and making it easy to fly 'hands off'. It also gave the shortest climb time to the 5000 ft test height and, as a result, was used on the Nanok/Solent machine, to be described below. Thereafter, the Sparrow II was stored in a shed at Hythe until May 1929, when it was given to the Halton Aero Club and registered G-EBJP. It may have survived until as late as 1933 but there is little evidence of its being flown by members.

There were to be no further excursions by Supermarine into the light aeroplane field as the de Havilland D.H.60 Moth aircraft, sensibly powered by a more powerful engine, was accepted by the Air Ministry in 1925 as the basis for Britain's first five civil flying clubs (and thus it was in a D.H.60G Gipsy Moth that Mitchell gained his pilot's certificate in 1934). By 1939 there were sixty-six clubs in Great Britain and so Supermarine and all the other light aircraft competitors had not been wrong in putting in their bids.

SEAGULLS III and IV

Meanwhile the number of Seagulls sold was increased when the Australian Government decided that their Air Force should assist in the hydrographic survey of the Great Barrier Reef.

No. 101 (Fleet Cooperation) Flight was formed on 1 July 1925 and six Supermarine **Seagull III** amphibians were ordered (A9-1 to A9-6). These machines were essentially Mk IIs, but tropicalised with larger radiators, and the first of these was ready by February, 1926. By this time, six of the RAF aircraft had served a tour of duty with HMS *Eagle* and the type had then been pronounced as having 'no potential naval use', particularly because of their habit of porpoising on take-off. They were confined to coastal (non-carrier) reconnaissance duties

A Seagull III returning to the Supermarine works after a launching ceremony performed by Lady Cook, wife of the Australian High Commissioner.

and so the type did not come into contention as a future replacement for the long-serving Fairey III series, although it was given a place in the popular final set piece of the fifth RAF Pageant where it summoned Flycatchers, followed by Blackburn Darts, to destroy two large replica warships.

In sharp contrast, the Australian Seagulls were used more thoroughly, as their survey work extended into 1927 and continued on northwards to include some 10,000 square miles of Papua New Guinea and one staged flight of 13,000 miles. Referred to by the natives as 'the canoe that goes for up', it was also pronounced a 'delightful' aircraft to fly by one pilot, Commander F. J. Crowther – although he did note that, in these tropical regions, it took more than an hour to reach 8,000 ft. (But a Vickers Victoria transport, at about the same time and also in a hot climate, took nearly two hours to reach 10,000 ft.)

Traditional Supermarine ruggedness was also evident after the survey work was completed, as the Seagulls were assigned to the newly constructed seaplane tender HMAS *Albatross*, commissioned in 1929; and they continued in carrier use until 1933 when this vessel was placed in reserve. The Seagulls were then transferred to RAN cruisers. As they were not easy to deck land, they had been lowered and hoisted aboard, like the Spanish Scarabs before them. But, in spite of the Seagull's various limitations, three RAF Seagulls, engineless, were subsequently acquired at the scrap price of one hundred pounds each and were intended to be used for spares; however, they were found to be in such excellent condition that they were restored and put into service use.

Two views of HMAS Albatross *with a Seagull III.*

Seagull III at Hobart, 1930.

Because of its habit of porpoising on take-off, the type continued to occupy the minds of Mitchell's design team even until 1928: fitting hydro-vanes was considered and various permutations of the hull step position were tried out on N9565 and on N9606. And one aircraft, N9605, was fitted with Handley Page wing slots and a new tail unit with twin fins and rudders. This aircraft, designated **Mark IV**, was converted to take passengers in 1929, as the Supermarine company was looking forward to a small fleet of this later model resuming the previous Sea Eagle Southampton–Channel Islands routes. A pilot service was begun in July by the prototype five-passenger conversion (G-AAIZ) but most of August was void owing to serious damage to the hull caused by its hitting a submerged rock. Then, on September 2, the short-lived business ceased when the aircraft ran into engine trouble.

Two other Seagulls, N9653 and N9654, were converted for civilian use. Registered as G-EBXH and G-EBXI respectively, they began a coastal service at Shoreham but this also failed, owing to inadequate public response. However, two other modifications of the Seagull were of great significance to Mitchell's team. One was concerned with equipping a Seagull to initiate the testing of catapults for launching aircraft and the second was the exchanging of the usual water-cooled Napier Lion engine for an air-cooled radial engine in a pusher configuration. As we shall see later, when the Seagull V/Walrus appeared, it was as an aircraft engined in this particular way and stressed for catapult launching.

Another influence upon the eventual Seagull V design was the more successful hull planing arrangement of the Sheldrake. It has been mentioned earlier how this aircraft was first conceived in 1923 but was quite obsolete when it made its only appearance in public at the 1927 display at Hamble. The other design activities of the time, notably the Southampton and Schneider Trophy aircraft development programmes and the intermittent work on the Sparrow and Seagull types, must have contributed to its neglect – as with another machine of this time, the unimpressive Seamew.

SEAMEW
Perhaps its lack of promise accounted for Supermarine appearing to give the Seamew low priority, despite an awareness of the Air Ministry's pressing need for better types to serve the Royal Navy. Even while the orders for the Blackburn R.1 Blackburn and Avro 555 Bison three-seat gunnery spotters were being fulfilled, Specification 37/22 was issued for a replacement; the efforts of Hawker, Blackburn and Fairey all came to nothing when this last requirement was cancelled because of the poor performance of the contenders when ready by 1925. Thus Supermarine clearly had motivation, as well as previous experience, to produce a successful contender when Specification 29/24 was issued.

In response, the relatively small shipborne Seamew was conceived by Supermarine and it was to be powered by two engines in order to carry the requisite three crewmen, the additional weight of gun positions both fore and aft, retracting undercarriage and folding wing mechanisms. In view of the Ministry's concern with slow landing speeds for its deck-landing types, it is no surprise that Mitchell was also proposing thick, high-lift aerofoils for the wings as the design of arrester mechanisms at this time was none too advanced for coping with fast 'arrivals'.

Thinking was sufficiently advanced by the next year for the Ministry to issue Supermarine with a contract for two machines of this type. With the Southampton developments, especially in respect of new metal hulls and metal wing structures, the urgent requirement to complete the S.4 and then the S.5 Schneider Trophy racer programmes, the Australian Seagull orders, the Sheldrake and

Supermarine SEAMEW (1928)

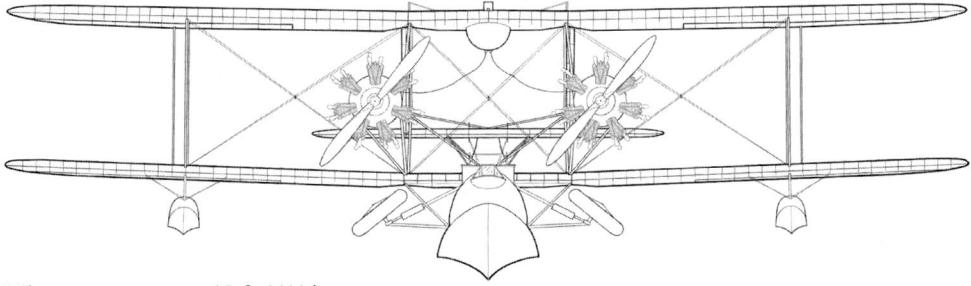

Wingspan	45 ft 11½ in
Wing area	600 sq ft
Loaded weight	5,700 lb
Maximum speed	94.4 mph

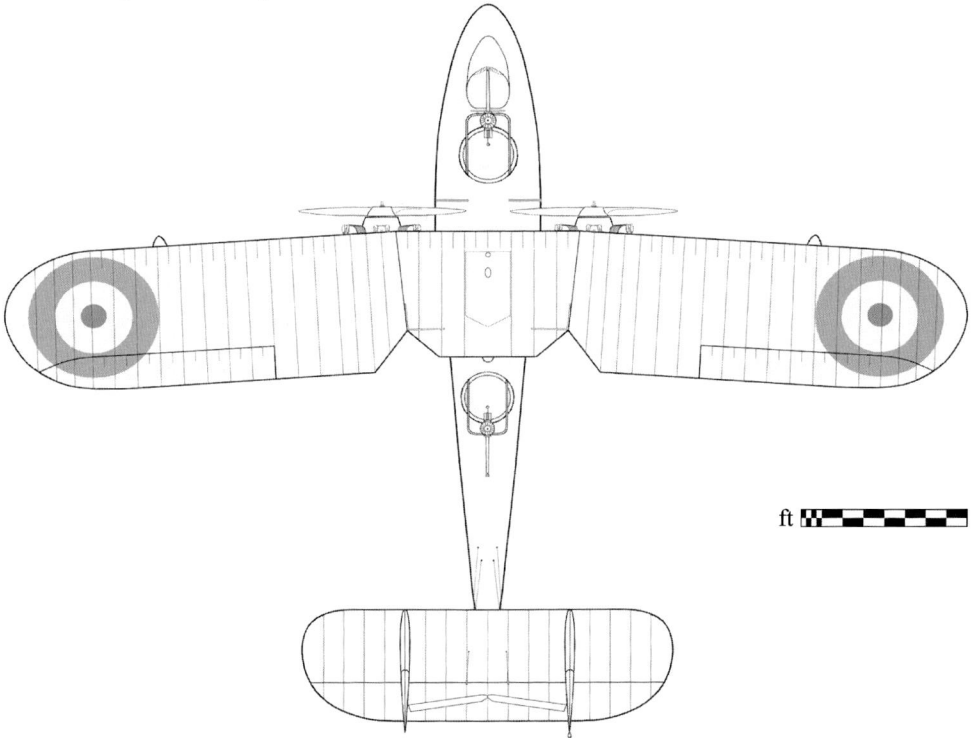

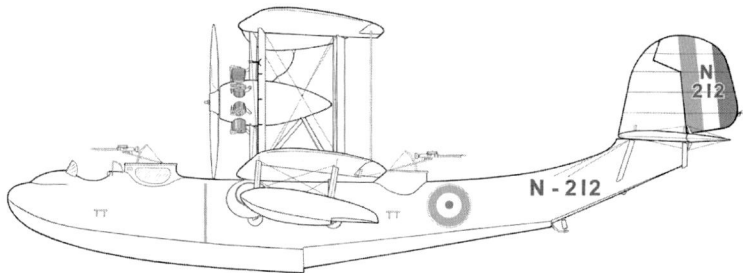

ft

Sparrow testing, and the Seagull IV activity, it is perhaps not surprising that the Seamew was slow to materialise.

When it did finally make its first flight, on 9 January, 1928, it might have been regarded not as a scaled-up Sheldrake or Seagull but rather as a scaled-down Southampton, particularly in view of the graceful shape of its hull. No doubt because of its early gestation, it was still, like the early Southamptons, of wooden construction although the wing structures made far more extensive use of stainless steel than the larger machine.

The Seamew and Mitchell.

Unfortunately, flight testing revealed the Seamew to be one of the few aircraft designed by Mitchell which did not live up to its design projections. The first to fly, N212, was found to be very nose heavy and the forward facing propellers had only a very short life, due to water impact during the take-off run. The second aircraft ordered, N213, was fitted with smaller diameter propellers, no doubt to try to overcome the problem of water ingestion, but this expedient then affected the amphibian's rate of climb. By 1930, balanced rudders had been fitted and the tailplane had been given more negative incidence to counteract the nose-heaviness problems.

The Seamew, fitted with balanced rudders.

Additionally, it was now found that the stainless steel fittings of one of the mainplanes were in need of replacing, owing to the inferior quality of the materials. Similar problems had appeared on operational Southamptons but the more extensive use of the material in the Seamew suggested the need for a more radical rebuilding of the airframe than simply replacement of parts. However, such a course of action was not justified by the overall performance of the aircraft and so the type was not proceeded with. The problem of spray affecting the forward-facing propellers of the smaller type of flying boat must have convinced Mitchell of the advantage of a single, more powerful, pusher engine configuration, further protected by the forward chines of a relatively wide hull – another factor in the eventual design of the Seagull V/Walrus.

There were various Air Ministry specifications from 1923 onwards for naval torpedo, fleet-spotter, or intercepter aircraft and for aircraft having interchangeable wheel and float undercarriages (with the Fairey IIIF emerging as a very successful contender in most of these roles). The Southampton firm, by now, had had experience of the float-equipped S.4, as well as previous interests in torpedo or deck-landing machines; also, as we know, they had for a long time cherished notions of a small manœuvrable fighter flying boat. Nevertheless, the Seamew represented the last small or medium-sized Supermarine aircraft built specifically for such Air Ministry requirements, giving way to machines from Hawker, Blackburn and, of course, Fairey. (As we shall see in Chapter Seven, the original requirement for the later Walrus did not come from Britain.)

Southampton Development
Whilst the Seamew was proving a disappointment, the Southampton I had fulfilled the expectations of the RAF in their cruises around the British Isles and these were followed, in 1926, by a cruise of two Southamptons from Plymouth to Egypt and back, via Bordeaux, Naples, Malta, Benghazi and Sollum to Aboukir, calling at Athens and Corfu on the return flight. It is not surprising, therefore, that the order for the first 1925 batch was not the last. A Mark II version, with a metal hull, was ordered and so a total of 79 production machines were eventually completed between 1925 and 1934.

A metal-hulled Southampton II moored on River Itchen, Southampton, in front of a wooden-hulled Mark I.

The Southamptons first equipped No. 480 (Coastal Reconnaissance) Flight at Calshot – which later became No. 201 Squadron – and afterwards four other squadrons were also supplied with Southamptons: No. 204 Squadron at Plymouth, No. 210 Squadron at Felixstowe and Pembroke Dock, No. 203 Squadron in Iraq and No. 205 Squadron at Singapore. Four experimental prototypes were also ordered and separate metal hulls were manufactured for replacement of damaged hulls as well as for the retrofitting of all **Mark I** Southamptons. The fact that Mitchell had designed the Southampton so that the entire wing structure could be removed as a single unit was an important factor in the decision to upgrade the Mark Is with the new hulls. And during this time there was also continuous detailed development of this flying boat. In view of the effect of this work on the Southampton upon other Supermarine projects, it is worthwhile giving a resumé of the various Southampton orders and developments undertaken from 1926; the development details also remind us of Mitchell's day-to-day concerns between the highlights of new design first flights or Schneider contests (*see* opposite).

The sixth of the 1929 Argentinian batch.

Harry Griffiths, (*see* Introduction) recorded an incident involving one of the Southampton developments, which gives an insight into the sort of unanticipated problems that would arise after a design had left Mitchell's drawing-office:

> The firm received an order from the Turkish government for a number of Southampton flying boats, but it was specified that they were to be fitted with Hispano Suiza engines instead of the Napier Lions which were standard.
>
> In fact the installation looked very much neater but on test flights the pilots reported heavy vibration on the control column during take-off and climb to cruising height.
>
> Arthur [Black, Chief Metallurgist] obtained a Vibrograph which produced traces on a celluloid strip – a cumbersome device. Being rather small I could get into the rear of the hull so I was deputed to use it during ground engine runs and subsequently on a test flight. I had to kneel on the cockpit floor and hold it against the control column during take-off, and then crawl down to the rear and get a number of readings on the way back.
>
> After that the theory was that I could go back to the rear gunner's position and enjoy the rest of the flight.
>
> I got the readings and then was horribly airsick all down the outside of the hull – fortunately the spray on landing cleaned everything up, but after we had come ashore I went to first aid who gave me a dose of Sal Volatile which made me feel worse. What's more, not having a flying helmet, I was deafened for several hours afterwards from the engine and propeller noise.
>
> We projected the traces onto a screen and measured frequencies and amplitudes but they really told us nothing except that the aircraft was shaking like mad under certain conditions.
>
> So there had to be another approach, and Oscar [Sommer, in charge of structure testing] came up with the idea that if he sat in the cockpit and held the control column while I thumped on various parts of the tailplane we might find out where the trouble came from.
>
> I was on top of a trestle thumping away when I heard a voice below. It was Trevor Westbrook, the works superintendent, who called up: 'Griffiths, if you are trying to smash that bloody aeroplane go to the stores and get a sledgehammer, don't use your fist, you might hurt yourself!'

Details of Southampton design activity:

- increased incidence to wingtip floats (they tended to dig in) and later redesign of them;
- N9896 experimentally fitted with alternative fuel tanks to replace the normal external underwing tanks; later fitted with a fore and an aft gun turret on the centre-line of the top wing – the Mark III;
- N9900, from the original order, modified to carry torpedoes; formed the basis for the Danish Nanok order (*see* below);
- N218 with an experimental metal hull, also from the original order; later used as a test-bed for the Bristol Jupiter IX engine; also fitted with Handley Page leading-edge slots;
- twelve aircraft (S1036–1045 and S1058–1059) built in 1926; S1059 fitted with a canopy over the two pilots' cockpits. All the Southamptons supplied for No. 203 Squadron in Iraq were so modified – known as the 'Persian Gulf' type;
- eight aircraft (S1121–1128) ordered in 1926 with instructions that the last two should be fitted with metal hulls. These last two became the first Southampton Mark IIs; S1122 fitted with Kestrel engines;
- four (S1149–1152) built for the Far East Flight (*see* below) with modified fuel tanks of increased capacity and with increased radiator surface areas;
- five (S1158–1162) built during 1927; S1159 went to Australia;
- nine more (S1228–1236) built in 1927;
- eight (HB1 to HB8) – for the Argentine Naval Air Force, fitted with Lorraine 12E engines, 1929;
- three experimental airframes ordered in 1928: N251 to be fitted with a special hull built by S. O. Sanders Ltd.; N252 to be converted to take three Jupiter XFBM engines – the Southampton X (*see* later); N253 – the fitting of Kestrel IV engines to S1149 and the fitting of an all metal airframe with Frise balanced ailerons;
- three batches of five each ordered: (S1298–1302), (S1419–1423), and (S1643–1647); additionally S1464 ordered;
- S1648 ordered in 1931 as an 'Improved Southampton Mark IV' – renamed Scapa (*see* next);
- two aircraft (K2964–2965) ordered;
- six (N3–8) fitted with Hispano-Suiza 12Nbr engines – which produced vibration necessitating the strengthening of the rear part of the hull and tail surfaces. These were delivered to Turkey in 1934.

Cozens again provides information and anecdote about the Southampton:

Metal frames were becoming common on aircraft but it took a long time before there was an attempt to use metal for the skin, especially if there was the risk of contact with salt water and the danger of corrosion … Some wooden Southamptons were fitted with stainless steel bottoms and some were built with stainless steel bottoms and duralumin hulls, and of course the expected corrosion did appear, and like other companies building flying boats, Supermarine set up a metallurgical section to try to overcome it. [*See* appointment of Arthur Black, p. 15.] … the men on the workshop floor had to change to metalworking, but they were practical men and times were difficult, the General Strike was hardly over, and most of them stayed, in any case woodworkers were still needed for building wings and control surfaces …

Mr S. F. Tilman, who was concerned with Southamptons while in the RAF, wrote in the *Southern Evening Echo* 'Letterbox'

'I think the City of Southampton should be proud of the Supermarine Southampton flying boats. The "Swampton" as we affectionately called it, was the mainstay of Coastal Command for at least twelve years, and saw service in Hong Kong, Singapore, Iraq, Gibraltar, Egypt, and Pembroke Dock, Mount Batten, Calshot and Felixstowe …

'Southamptons made many other memorable flights including escorting the Prince of Wales on a tour of the Baltic and Scandinavia. This particular flight was made by 201 Squadron from Calshot, and the only thing that went wrong was when the C.O. of the Squadron, in spite of his own explicit orders that it should not be done, signalled with his arms from the centre cockpit and had his thumb chopped off by one of the propellers.'

Well you didn't shout back to the works superintendent from the top of a trestle so I climbed down to explain when, to steady myself, I caught hold of a strut which ran from the hull to the mid span of the tail.

Oscar stood up in the cockpit all excited and shouted, 'That's it! What did you do?' So I stayed halfway down the trestle and gave the strut a series of hard thumps, just by way of confirmation, before going down to the ground and explaining.

The drawing office stiffened up the strut and everybody was happy, including the works superintendent.

One development of particular significance was the completion of the experimental metal-hulled machine, N218. This requirement marked the end of the traditional Supermarine wooden hull of which the company had been justifiably proud and which had differed markedly from the girder-type, cross-braced, fuselage structures of the typical landplane – leaving a hull completely unobstructed within and thus enabling the crew the rare luxury of moving about with relative ease.

N218, the prototype Southampton Mark II.

Replacing the double-bottomed wooden hull of the Mark I machine with a single-skinned duralumin hull provided even more internal space and a saving of 300 lb in weight; a further 400 lb, caused by gradual water soakage into the wooden hull, was also saved. These weight savings, together with a change from the Napier Lion V to the more powerful Lion Va, increased the Southampton's range by 200 miles. Increased load testing was carried out at the Marine Aircraft Experimental Establishment at Felixstowe in 1927 and it was found that, at any weight up to 18,000 lbs, control, manoeuvrability and take-off were unaffected. Given the interest in developing an Imperial air route, one further advantage of the new metal hull was significant: the greater ability to withstand the rapid encrustation by barnacles and other marine growths encountered in tropical waters.

At this time, the only earlier long-distance flight by standard RAF machines which had exceeded the two Southamptons' 7,000 mile Mediterranean cruise was across the land mass of Africa (Cairo to Cape Town) using relatively small D.H.9s. The lack of prepared landing strips for larger aeroplanes in most countries and the short range of aircraft at this time had not encouraged long-distance military or commercial flying but the advent of the efficient and reliable Southampton, able to use the widespread landing areas provided by lakes, large rivers and the sea, gave increased confidence to the political consideration of the possibility of opening up air routes to the far-flung outposts of the British

Empire. (It may be recalled that one of the topics of conversation between the Prince of Wales and Henri Biard during the 1924 royal visit to Supermarine concerned the development of Imperial air routes. In the same year, the Government and the Air Ministry, among others, had sponsored surveying flights by Alan Cobham to India and Burma and, in 1926, his seaplane also made a commercial route survey to Melbourne.)

The Royal Air Force Far East Flight

The new Mark II machine duly encouraged the Air Ministry to order, also in 1926, four new Southamptons specifically to initiate a proving cruise to the far reaches of the Empire – as the Secretary of State for Air put it: 'it was our settled policy to show the Air Force, as the Navy showed the Fleet, in the distant parts of the Empire'.

The confidence in the Supermarine machine is evident from the fact that the cruise, by basically standard RAF machines, was to incorporate overflights of the countries only previously visited by the pioneering Cobham in his two separate flights and to go as far as Australia – which had only been visited, singly, by four previous aeroplanes. Additionally, it was to circumnavigate that continent – a feat which had been achieved only once to that date – by a Fairey IIID between the sixth and the eighteenth of May, 1924. Readers of the 1927 issue of *Jane's All the World's Aircraft* would therefore have been well aware of the ambition, and confidence, of Supermarine when the company announced:

> A number of the metal-hulled 'Southamptons' are now being completed to equip the RAF Far East Flight. These Southamptons will be flown out to India, via the Mediterranean, and then on to Singapore and along the Dutch East Indies to Australia, where an extended flight round the Australian seaboard, in conjunction with the Royal Australian Air Force, is contemplated.

The leader of the Southampton's earlier Mediterranean cruise, Squadron Leader Livock, was again chosen, as well as Fl. Lt H. G. Sawyer who, as a junior officer, had taken part in some of the early British Isles proving flights of the Southampton I. But, on the occasion of this much more extensive and important Far East Flight, a Group Captain was put in command – H. M. Cave-Brown-Cave (who, on arrival in Australia, became known as 'Home-Sweet-Home'). His orders were: 'to open the air route to Australia and the East, to select landing sites, to see how far flying boats and their crews were capable of operating away from fixed bases and under widely varying climatic conditions, and to show the flag'.

Thus, whilst there was a clear imperialist motive behind the proposed Flight, the other main concern was to prove the feasibility of reliable transport – with scheduled stops for servicing and for inspections to see how the aircraft were standing up to the very testing itinerary. The main cruise began from Plymouth on the 17th of October, 1927, and finished at Seletar, Singapore, on the 28th of February, 1928. The engines were replaced on arrival at Singapore and one of the aircraft, as prearranged, was dismantled and sent back to England for detailed inspection. The remaining three planes then proceeded to circumnavigate Australia and fly around the China Sea to Hong Kong and back. In all, the four flew 27,000 miles, in formation, at an average speed of 80 mph and in 62 timetabled stages of about 400 miles at a time. As might be expected, many minor running repairs had to be carried out, but during the whole cruise, the Southamptons only fell behind schedule three times – twice because of bad weather and once with engine trouble; one machine, additionally, was delayed by a cracked airscrew boss.

Efficiency

108,000
Machine Miles
"No Troubles"

Far East Flight publicity.

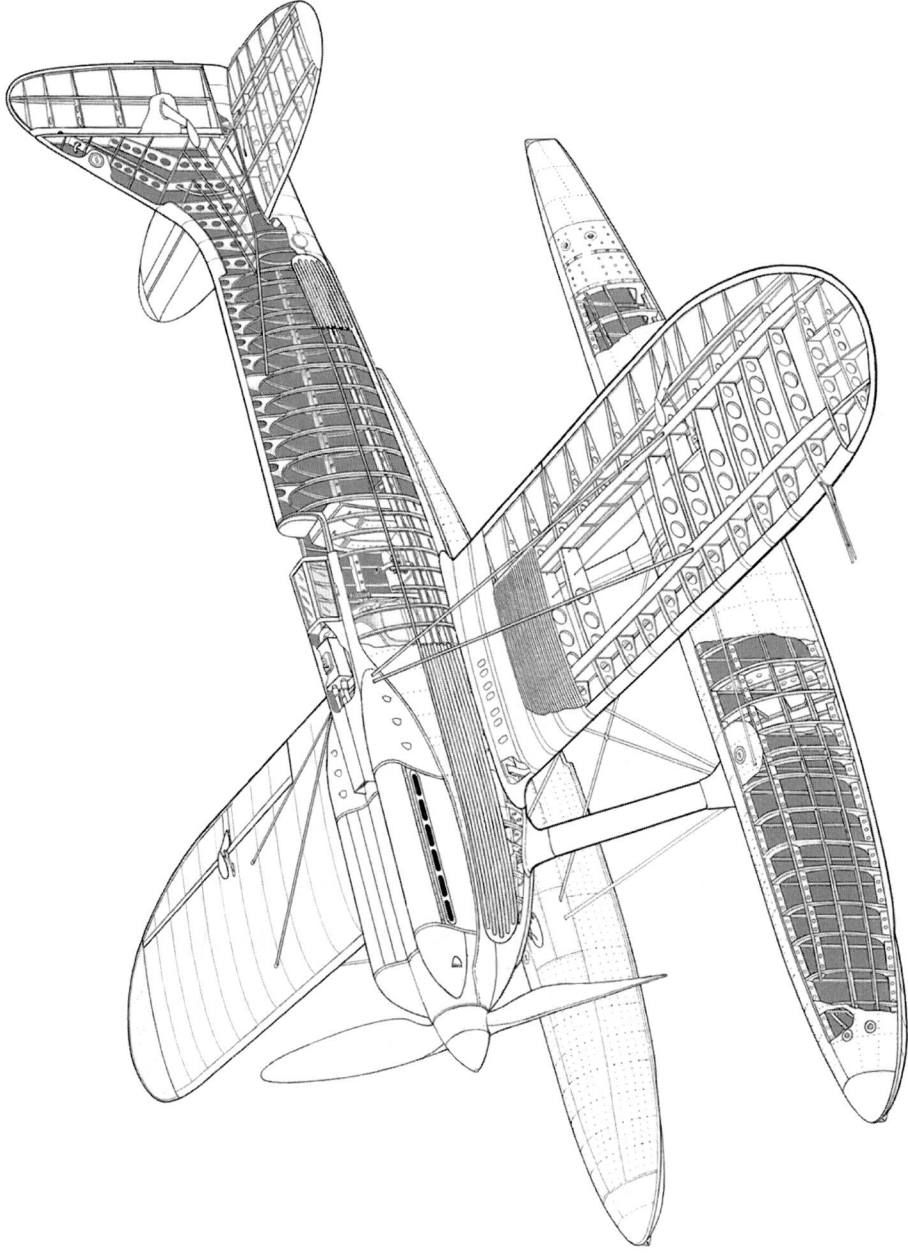

S.6B cutaway drawing, showing the all-metal development of the Linton Hope hull structure and foreshadowing the Spitfire fuselage.

The first flight to Australia, in a Vickers Vimy, had taken place nine years earlier, two out of a flight of four Douglas DCWs flew round the world during 1924, and other nations made more publicised formation flights in following years; yet the Southamptons' Far East Cruise, which was completed in scheduled stages by the *entire* formation, must be regarded as directly instrumental in the establishment of the Imperial Airways Empire routes of the 1930s and as one of the milestones in aviation history. The *Daily Mail* was in no doubt: 'As a demonstration of reliability, the flight will rank as one of the greatest feats in the history of aviation'. (n.b. Also in 1927–28, Cobham flew around the continent of Africa although this was a solo effort, in a rival Short prototype flying boat.)

The other extended formation flight carried out by Mitchell's aircraft, and mentioned above by Cozens, the Baltic flight, included the cities of Esbjerg, Copenhagen, Stockholm, Helsinki, Tallin, Riga, Memel, Gothenburg and Oslo. A total distance of over 3,000 miles was covered, again, without mishap. Squadron Leader Livock, who was second in command of the Far East Flight and leader of the formation, gives a full account of these flights in his autobiography, *To the Ends of the Air*, well worth reading for its accounts of the difficulties and frustrations encountered when pioneering air routes in areas where, understandably, there was little comprehension of aviators' special needs.

Southamptons of 201 Squadron during 1930 Baltic flight.

These cruises had taken place at the time when Supermarine was winning successive Schneider Trophy contests (*see* next chapter) and so Mitchell and his designs were becoming more widely known outside the British aviation community. The prestige of the Schneider wins reinforced the reputation of the Southampton which outperformed other European flying boats of the time and led to sales in Argentina, Japan, Turkey, Australia and Denmark. The US Government, usually a staunch supporter of its own native industry, also made enquiries which, perhaps because of the worsening economic situation there, did not materialise into orders.

The Supermarine publicity which follows gives a relatively comprehensive account of the physical features of the Southampton as a practical proposition and as a fighting machine, including the torpedo version, and offers wooden or metal hull and wing (some other nations might prefer the older technology):

TYPE – Twin-engined, five-seat reconnaissance flying boat.

WINGS – Equal-winged, unstaggered biplane. Top and bottom centre-sections of equal span, interconnected by vertical struts at their extremities, and four sets of struts in the form of a 'W' when viewed from the front, in between which are mounted the engines. One set of vertical interplane struts to each outer wing section. Normal structure of wood covered with fabric. A set of metal wings has been produced. The metal wings are approximately 200 lbs (90 kg) lighter than the wooden ones, with which they are interchangeable. Ailerons fitted to all four planes.

HULL – Can be supplied in either wood or metal. The metal hull is 300 lbs (136 kg) lighter than the wooden hull. Wooden hull of normal Supermarine circular-section, flexible construction, with two built-on steps. Metal hull of same form, built entirely of duralumin, with stainless steel fittings.

TAIL UNIT – Monoplane type, with three fins and balanced rudders mounted above. Tailplane of cantilever type and is adjustable. One-piece unbalanced elevator.

POWER PLANT – Two 470 hp Napier 'Lion' engines, on separate removable mountings, carried above the bottom centre-section. Each engine unit is self-contained and includes radiator, oil tanks and cooler, and all instruments, and may be removed without disturbing the main wing structure. Main fuel tanks (2) under top centre-section, giving gravity feed to engines. The 'Southampton' has [also] been fitted with two Bristol 'Jupiter VIII' geared radial air-cooled engines. With these engines the useful load was increased by 500 lbs (227 kg).

The 'Southampton' flying boats supplied to the Argentine Navy are fitted with 450 hp Lorraine-Dietrich water-cooled engines. The Rolls-Royce F type engines can also be fitted.

ACCOMMODATION – In nose is cockpit for gunner and bomber. Provided with Lewis gun, on Scarff mounting. Behind are two pilot's cockpits, in tandem, with dual control. The after-pilot also has complete navigating equipment. Below wings is the wireless compartment, which opens out into two staggered cockpits, aft of the wings, each equipped with Scarff gun-mountings. A crew of five is normally carried. Inside of hull, which is free from obstructions, may be equipped with hammocks and cooking apparatus, so that crew may sleep on board and remain afloat for long periods. Can be arranged to carry two 18 in. torpedoes, one on each side of hull. Winches for lifting torpedoes into position are carried under bottom centre-section.

NANOK/SOLENT; Mitchell's First Air Yacht

As a result of the widespread recognition of the exceptional qualities of the Southampton, there came a request from Denmark for a version of the type to carry torpedoes in a similar manner to that devised for N9900, one of the Southampton developments outlined above and alluded to in their publicity. Called 'Nanok', Inuit for polar bear, a Southampton was uprated to carry the heavier armament by being provided with a slightly larger thicker wing area and a third engine; and weight was saved by the provision for only a single rear gunner.

The engines selected were Armstrong Siddeley Jaguar IVAs and first carried the Nanok into the air on 21 June, 1927, whereupon it was found that the additional power at the high thrust-line caused the machine to become distinctly nose-heavy, especially at low speeds. Mitchells' expedient was to have an auxiliary elevator fitted, higher up between the three fins. However, the extra drag of the second unit lowered the flying speed some mph below that contracted for and the rate of climb was disappointing. Biard also records that the engines had a tendency to cut out and that the vibration experienced with the differently-engined Turkish batch of Southamptons (mentioned above) also occurred with the Nanok.

Nanok with torpedo seen under port wing.

Nanok with torpedo seen under starboard wing. Note auxiliary elevator.

In view of later developments (*see* below), Supermarine publicity in *Jane's,* 1927 is interesting in this respect for, under the heading of the 'Solent', it actually describes the torpedo-carrying 'Nanok':

> The first machine of this type was completed in June, 1927 and was built specially to the order of the Royal Danish Naval Air Service, who have renamed it 'Nanok'. The hull of the machine is very similar to that of the Supermarine 'Southampton' but the superstructure is entirely different. The three Armstrong-Siddeley 'Jaguar' engines are mounted as tractors about midway in between the top and bottom planes. A new thick wing section, which has been designed by the Supermarine Aviation Works, Ltd, has been used and has shown itself to have excellent all-round properties. Two 1,500 lb torpedoes can be carried, one on each side of the hull, suspended from the bottom centre section, which is fitted with all necessary accessories, including winches for lifting the torpedoes on to the carriers.

In fact, the main change to the engine arrangement was not the position of the thrust lines but rather the change from Warren interplane girders to more conventional struts in order to accommodate the third engine; the aerofoil section mentioned was the SA12, which had been tested out on the Sparrow II. But the positioning of the torpedoes, one on each side of the aircraft added to the other various difficulties associated with this machine: Biard reported control

problems which he experienced by the sudden change of trim when only dropping one torpedo – it being suspended fairly well out from the centre-line of the
flying boat. Such lurches, necessarily at low level, could hardly have induced
confidence in pilots assessing the Nanok.

In the end, the Royal Danish Navy took delivery of a standard Southampton
fitted with Jaguar VI engines and Supermarine were left with a modified but
unwanted flying boat. Thus the 'Solent' was announced as a civil version of the
Southampton. Although a Southampton was loaned to Imperial Airways for a
time, between November 1929 and February 1930, and a Japanese owned Southampton was converted for passenger carrying in 1930, the only Southampton-
type flying boat actually used for civil purposes at about this time was its
prototype, the Swan, and the Supermarine description of the envisaged 'civil
Southampton' is uncannily similar to that of the early aircraft:

> The passengers are accommodated in a roomy and comfortable saloon.
> It is fitted with spacious wicker chairs with deep and well-sprung
> cushions, carpets, hinged triplex glass portholes which can be opened
> by the passengers, cupboards and lockers, racks for lifebelts, etc. The
> whole saloon is beautifully finished and upholstered. There is no petrol
> inside the hull and the passengers may smoke without least danger. Being
> situated well clear of the engines, the saloon is free of engine exhaust
> gases, oil, vibration, etc., and is very quiet, thus enabling the passengers
> to converse quite normally. There is a through passageway from stem
> to stern of the hull. Adjacent to the saloon is the luggage compartment,
> which is fitted with racks for the stowage of passengers' luggage and
> freight, and is separated from the saloon by a partition. Access to the
> luggage compartment is by means of a door in the partition and a hatch in
> the deck can also be provided. This hatch enables the freight to be easily
> and quickly removed, and could be used as an emergency exit if required.
> Lavatory accommodation is fitted as desired.

The passenger comforts to which attention is drawn gives, by implication, an
interesting insight into the more spartan conditions normally experienced by air
travellers of the late 1920s and reveals how the relative bulk of a flying boat hull
was particularly advantageous.

The Solent Air Yacht.

Whilst hopes for the civil version of the Southampton never materialised, the
Hon. A. E. Guinness was subsequently persuaded of the potential of a flying boat
for comfortable travel. Despite his owning *Fantôme*, the largest and most spectacular barque-rigged yacht in the British registry, the unwanted Nanok variant
was sold to him, after conversion into an 'air yacht' with comfortable cabins
to carry up to twelve passengers in the luxury envisioned above. Now finally

named 'Solent' and registered G-AAAB, it soon became a familiar sight, flying for the next two years from the Hythe seaplane base on Southampton Water to Dún Laoghaire harbour, County Dublin, and thence to Lough Corrib, County Galway, close to Ashford Castle, its owner's home. The advantages of a seaplane for commuting were no doubt obvious to the new owner, given the terrain of the remote parts of Ireland where, as Biard recorded, 'even a train was a novelty there, and many of the peasants had never even heard of anything mechanical that could fly, for many of the older people could not read.'

Another Air Yacht was to follow in 1930, again a cancelled military prototype, but with a much briefer and more chequered career – *see* Chapter Seven.

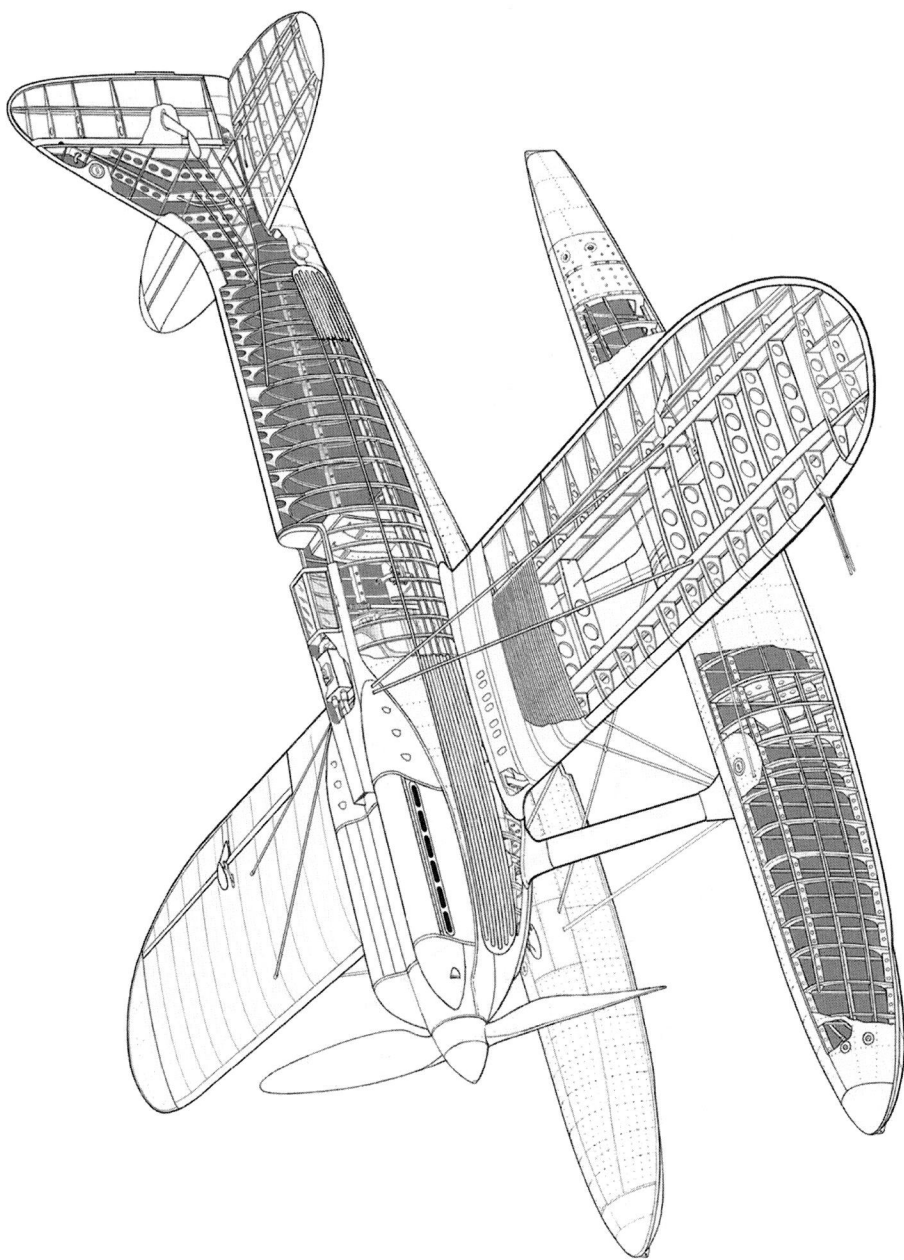

S.6B cutaway drawing, showing the all-metal development of the Linton Hope hull structure and foreshadowing the Spitfire fuselage.

— *Chapter Six* —
1927 to 1931
Schneider
Trophy
Domination

Whilst the Southampton and Seagull development, and the Sparrow II and Seamew testing were taking place, Mitchell again became involved with the Schneider Trophy contests that were soon to result in the Supermarine Chief Designer's becoming known outside the aeronautical community and Southampton for his outstanding contribution to high speed flight: the shy lad who joined his firm as the Personal Assistant to the General Manager in 1916 was soon to be honoured at Buckingham Palace, to give a talk on the BBC and to be elected a Fellow of the Royal Aeronautical Society. This recognition, while he was still in his thirties, was essentially due to his Schneider trophy aircraft of 1927, 1929 and 1931. His solid, everyday work on the various amphibians and flying boats that were sold to Imperial Airways, to the Royal Air Force, to the Royal Australian Air Force and elsewhere had resulted in his becoming one of the Directors of Vickers (Aviation) in 1928, but was unlikely to have placed his name in front of the general public.

Full Government Support
By now, Mitchell had gathered a group of men around him (*see* Introduction) who must share the credit for the later designs and he was also fortunate to be well served by British aero-engine manufacturers. Also, as we have seen, events themselves had conspired to present him with the platform for his more public successes: the damage to an Italian propeller in 1922, the sporting American cancellation of the 1924 contest, the demise of international landplane contests, the intervention of Mussolini in 1926, had all contributed to the Schneider Trophy remaining still to be competed for in 1927 as the most significant international aviation contest. Government backing for a British entry had already been in evidence in 1925 and was now so wholehearted that it had prompted *Flight* to comment that 'Never in the history of British aviation have we tackled an International speed race in so thorough a manner'.

The Trophy was always intended as an international competition by Jacques Schneider but it began in 1913 as a contest between the aero clubs of various nations, most of whose members were enthusiastic amateurs, as were their pilots. Whilst the Schneider Trophy continued to be organised by the clubs, by 1926 the character and costs of the meetings had produced the first confrontation of government subsidised teams with well organised military pilots and support staff. Equally, the demands of producing the sophisticated technology required of the modern winning entry was evidenced by the non-showing of Britain in that year and by the complete or partial failure of all the leading aircraft which did compete, owing to lack of adequate development time.

Clearly, if the Schneider competition were held on a bi-annual basis, a government would be able to spread its costs and America, in a reversal of the previous year's vote, now opted for the two yearly event. But Italy, currently holding the world seaplane record at 258.87 mph and contemplating further development of the winning Fiat engine and M.39 airframe, voted for a competetion in 1927. So did Britain, mainly thanks to Air Vice-Marshall W. G. H. Salmond, the Air Member for Supply and Research at the Air Ministry, and the Secretary for State for Air, Sir Samuel Hoare. The view had been formed that there was now a good chance of success in 1927 (assuming that America did not win the trophy outright with a third win in 1926). Irrespective of any Treasury decision about the costs thereof, funds from the existing Air Ministry development budget were utilised and specifications for new racing engines and aircraft had already been drawn up as early as March 1926 (!) and work was soon under way, involving careful appraisal by wind tunnel and tank testing of ¼ scale models.

Particular attention was to be paid to floats, flush wing radiators and airscrews. Gloster, Supermarine and Shorts were asked to design machines capable of speeds not less than 265 mph at 1,000 feet and Napier was responding by increasing the Lion's compression ratio to 10:1, with a view to approaching the 900 hp mark. (In passing, it ought to be noticed that this Air Ministry order for seven machines was the largest ever given for British Schneider Trophy entrance and was only matched by Italy in the following year.)

And, on the 1st of October, (before the January meeting to decide the date of the next competition) a High Speed Flight was formed, consisting of exceptional air force personnel, to test the new racing machines. As all this preliminary effort was dedicated to achieving a win in 1927, America's proposed postponement to an unpredictable contest in 1928 was not welcomed. Italy, as the previous winner of the Trophy, was to host the competition and chose Venice once again. No government-backed entries from the American Aeronautic Association were forthcoming.

The British effort was still to be spearheaded by the existing Napier Lion engine and faith in this reliable engine was justified as it was now made to deliver 900 hp in the ungeared model and 875 hp in a geared version. Just as metal propellers were found to be superior to wooden ones as tip speeds increased, so it was considered that any extra weight and loss of rpm because of reduction gearing would be compensated for by greater propeller efficiency.

In this respect, the Air Ministry were not following the previous American and current Italian concentration of effort on one aircraft type only but were clearly hedging their bets by supporting three Supermarine entries with geared and ungeared engines, a geared Gloster machine and two ungeared ones, as well as a Short seaplane powered by a more standard air-cooled radial engine – a considerable departure from the water-cooled, in-line type of engines which had resulted in the sleek, streamlined winners of the last three contests. As previous and current post-war RAF fighters were all powered by radial engines, there was a good reason for seeing what sort of performance the Trophy competition might produce with the Crusader. Its Bristol Mercury I radial air-cooled engine had individual 'helmets' fitted to the nine protruding cylinders – a compromise between streamlining and cooling.

As we shall see, Mitchell's forthcoming S.5 was to stay with his S.4 monoplane formula of 1925 and was followed in this respect by the Short Crusader. In contrast, the Gloster IV proposal was to favour the more traditional biplane approach although, for the new contest, the Gloster wing layout approached that of a sesquiplane. The resultant decrease in wing area was possible because of the extra power that the Napier Lion engine was now actually producing and its accompanying reduction in frontal area permitted an even sleeker model for the forthcoming contest. Streamlining was furthered by the removal of the top-wing pylon mounting in favour of careful fairing into the top of the fuselage, as had been the case with the Curtiss machines from 1925 onwards. In this way, an increase of 70 mph was achieved whilst adhering to the biplane formula whose shorter wingspan coupled with wire bracing was, at least, expected to produce a robust airframe that would be less susceptible to the problem of flutter which had been emerging since the 1925 contest.

Meanwhile, the Macchi opposition was to be equipped with an even more attractive development of the M.39. This, the M.52, also had reduced area flying surfaces because of the promise of more power being available – uprating of the previous year's Fiat AS.2 engine was expected to produce 1,000 hp, compared with the previous 800 hp. The engine was also to be lighter and so the floats were reduced in length and volume, and the wings more swept back to accommodate the backward movement of the centre of gravity. The Fiat engine also allowed a smaller frontal area to be designed in but its reliability was still to prove a problem, mainly owing to the new use of alloys, coupled with a higher compression ratio and higher rpm: it was reported that six out of twelve engines ordered for the development programme had been damaged beyond repair during tests.

S.5

For the new Supermarine machine, Mitchell's continuation of the newer monoplane approach was hardly unexpected and he was also content to use the reliable Napier Lion engine: indeed, in the discussion which followed the 1925 Buchanan lecture, he had remarked that 'At one time I thought that the "Lion" engine was at a disadvantage with the American engines, but I have changed my views rather, and certainly consider the "Lion" is capable of winning the Schneider Cup.' Indeed, Mitchell (and Folland for Glosters) had consulted with Napiers and the new Lion was designed whereby the engine's frontal area was reduced by repositioning the magnetos; and the cam covers of the three cylinder engine banks were even contoured to mate with the engine's streamlining fairings fore and aft.

Mitchell then continued these lines without reference to an ideal aerodynamic shape, as with the S.4, which widened to maximum thickness at about 30% of the fuselage length. The result was the slimmest fuselage of all the current and subsequent contenders – so drastic in fact that the pilot's cockpit was an extremely tight fit: there was no room for a conventional seat and the pilots sat on a cushion on the floor of the machine, their legs almost horizontal and their shoulders coming up to and pressing against the underside of the cockpit coaming. Fl. Lt H. M. Schofield, one of the pilots of the newly formed RAF High Speed Flight, described their visit to Supermarine 'for a fitting':

> The method of reaching the seat was to squeeze in sideways and down as far as possible so that the shoulders were below the top fairing, then turn to face the front, and in my case it needed no ordinary effort to get my shoulders home. There were many sighs of relief from the watching design staff when the last man had been 'tried-in', for it had been a near thing and it did look as though it was not going to be enough at times.

As there was now insufficient room for the fuel tank in the fuselage, the starboard float was used. This expedient had the advantage of giving the aircraft more stability in the air by lowering its centre of gravity. More importantly, it would also help towards counteracting the torque of the engine which, during take off, was expected to cause the opposing float to dig in and swing the aircraft offline before it gained sufficient airspeed to be effectively governed by the control surfaces. Mitchell also offset the fuel-loaded starboard float an extra eight inches from the centre-line as an additional response to this expected problem.

Supermarine S.5 awaiting cowlings for Napier Lion engine. Note forward hinged coaming for cockpit access.

Supermarine S.5 on engine test; R. J. Mitchell on left.

However, the most telling improvement, apart from a more powerful engine, was the estimated increase of about 24 mph by the proposed change from the Lamblin type under-wing radiators of the S.4 to a system akin to that adopted by the Curtiss racers and, subsequently, in the Macchi M.39. The new radiators were to be made out of copper sheets, eight and a half inches wide, with their outer surfaces formed to the contours of the upper and lower wing surfaces; thus the outer sheeting, exposed to the cooling airflow, offered no additional drag. Corrugations on the inner surface of the radiators formed channels for the coolant, which was taken along troughs behind the rear wing spar, through

Supermarine S.5 (1927)

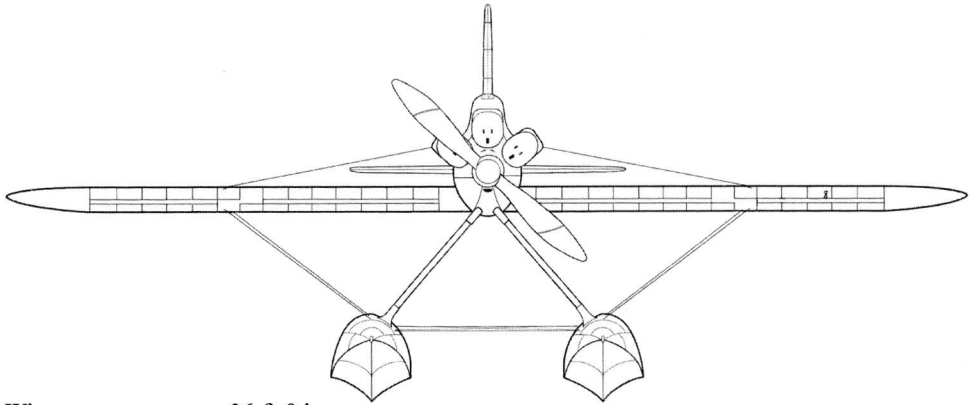

Wingspan	26 ft 9 in
Wing area	115 sq ft
Loaded weight	3,242 lb
Maximum speed	319.57 mph

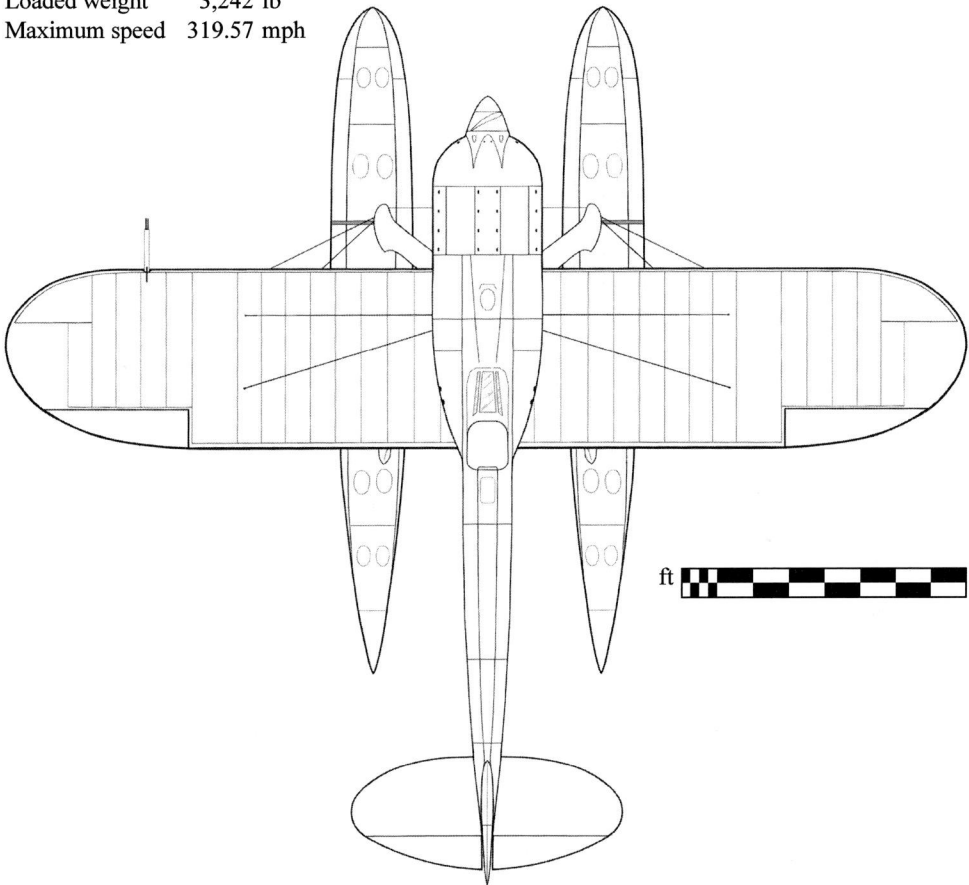

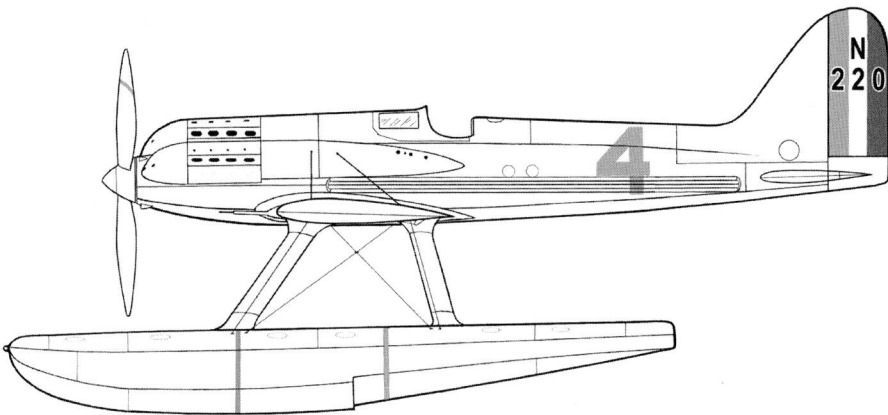

ft

N 2 2 0

the radiators and along the leading edge of the wing and then pumped to a header-tank behind the engine block. Attention was even paid to the effect of different paints on the effectiveness of the radiators. The lubricating oil was also cooled in surface radiators, consisting of corrugations which ran along the outsides of the fuselage and up to a header tank behind the cockpit.

In terms of structure, the Supermarine contender continued the move away from its S.4 predecessor. In line with the company's other developments at the time, the new machine was of mixed metal and wood construction, with the all-metal fuselage being a stressed-skin structure (which looked forward to that of the Spitfire) whilst the flying surfaces were, like those of the S.4, of wooden construction and ply-covered. It was designated S.5 as it represented a complete redesign of the previous monoplane and also incorporated the new information gained from meticulous work at the National Physical Laboratory test facilities, sponsored by the Air Ministry.

Mitchell had sent down three models for wind tunnel testing: one was a shoulder-wing design with wing roots cranked downwards and supported by streamlined struts from the floats; a second model had a low wing, similarly braced by struts; and the third configuration was an all-wire-braced proposal with a low wing position to give favourable bracing-wire angles. Biard's problems with forward vision during landing and take-off in the S.4 were no doubt an influence on Mitchell's considerations and eventually the low wing position was chosen, particularly as it had been found that wires offered less resistance than struts. The new wire bracing between the floats and, from them, to the bottom of the wing also allowed a wire 'cage' to be completed as the wires from the upper fuselage to the top of the wings were fixed immediately above the float bracing attachment points. Mitchell was clearly guarding against any wing flexing which might have contributed to the S.4 crash.

Whatever Mitchell's private thoughts were about the need to step back from the revolutionary concept of the cantilevered S.4, the pragmatic reversion to wire bracing also brought a further reduction of weight and drag which had been the result of the very sturdy float struts of the S.4. The balancing out of advantages and disadvantages attendant upon the wish to reduce frontal area and weight against the need to ensure adequate strength and pilot view was set out after the race in his speech to the Royal Aeronautical Society in 1927 whereby he had predicted that his design changes alone would achieve an increase of over 40 mph, irrespective of the increases that a more powerful engine would bring. (*See* details opposite.) It might also be noted that Mitchell took advantage of governmental support to devote considerable attention to float design at the National Physics Laboratory hydrodynamic tank. And Schofield mentioned that, with the second S.5 (which was to come first in the forthcoming contest) 'the hundreds of tiny rivets all over the skin were now flush with the surface instead of projecting like a mass of wee knobs as they had done' in the first S.5. (*See* a similar concern with the Spitfire, mentioned on p. 298.)

The result of all these design considerations culminated in a design which, when it went to Venice to compete in the 1927 Schneider Trophy competition, was seen by the Italians as a direct copy of the Macchi M.39 which had won the previous year. Penrose quotes an opposite view by Oswald Short:

> In 1926 I was invited to visit two Italian firms interested in my particular form of metal aircraft, and I was shown the M.39 undergoing its first engine test. I remarked that I could have sworn the floats had been made in our own shops. [Shorts had supplied floats to all the 1925 British Schneider machines.] One of the Macchi staff replied, 'Yes, when Mr Macchi Junior was in America for the 1925 Schneider, no doubt he kept his eyes wide open.' Undoubtedly Mr Macchi observed

the clean lines of the Supermarine and noticed that the attempt to produce a comparatively thin wing in pure cantilever led to trouble; he also saw the American float strut system was the cleanest. These facts no doubt led to the M.39 design and to avoid trouble the wing was braced to the floats and the top of the fuselage and this requirement alone, apart from the good view, would suggest dropping the wing to obtain reasonable angles on the bracing wires [as, of course, did Mitchell when he reverted to wire bracing]. The M.39 of 1926 was therefore a combination of the good points of the 1925 British and American machines and no one begrudges Messrs Macchi the credit of their achievement.

Extract from Mitchell's speech to the Royal Aeronautical Society in 1927:

(a) The primary object in lowering the wing on the fuselage was to improve the view of the pilot, which was never very good on the S.4 The higher position of the wing no doubt gave a lower resistance due to fairing in the outside engine blocks and thus saving a certain mount of frontal area. A loss in speed of about 3 miles per hour is estimated from this alteration. This loss is more than balanced, however, by the importance of the improved view.

(b) The system of wire bracing of the wings to the fuselage and floats was adopted for a number of reasons. The unbraced wings and chassis of the S.4 were very high in structure weight, and it was found very difficult to construct an unbraced wing sufficiently strong and rigid without making it very thick at the root, and thus increasing its resistance. The adoption of bracing was largely responsible for a reduction in structure weight of 45 per cent for the S.4 to 36 per cent for the S.5, with its corresponding reduction in resistance; also for the elimination of the two struts between the floats, and for the reduction in frontal area of the four main chassis struts. Against these must be set the addition of fourteen wires. It is not easy to estimate the final effect of a number of alterations of this nature, but from the analysis of the resistance of the two machines it is given on fairly good grounds that the overall effect was an appreciable saving in resistance, amounting to an increase in speed of approximately five miles an hour.

(c) The cross-sectional area of the fuselage has been reduced by about 35 per cent. This very large reduction was obtained through the redesign of the engine and the very closely fitting fuselage. This almost amounted to a duralumin skin in order to ensure that the very smallest amount of cross-sectional area was added. On several occasions during the construction of the fuselage the pilots were fitted, and much trouble was experienced through their being of varying dimensions … The reduction in body resistance was responsible for an increase in speed of approximately 11 mph

The floats were also reduced in frontal area by about 14 per cent. This was accomplished by using a much lower reserve buoyancy. The reserve buoyancy was 55 per cent for the 'S.4' floats and 40 per cent. for the starboard float of the 'S.5' [now being used for fuel tankage]. This figure is extremely low and called for very efficient lines.

The estimated increase in speed due to reduction in float resistance is 4 mph. These reductions in resistance of fuselage and floats are due to lower cross-sectional areas and not to improvements in form.

(d) Wing surface radiators were first fitted to the American machines in the 1925 race, and gave these machines a very big advantage in speed. The radiators added a certain amount of resistance to the machine due to their external corrugations increasing the area of exposed surface. As about 70 per cent. of the resistance of a high-speed wing is skin friction, and the corrugations almost double the area of surface, it is reasonable to suppose that an increase of at least 30 per cent. of resistance is added to the wing. It is evident that a saving in resistance would result if radiators could be made with a flat outer surface, and that they would give no direct resistance to the machine. After much experimental work, radiators with a flat outer surface were produced. The chief difficulty experienced was in sufficiently strengthening and supporting the outer skin to enable it to stand the heavy air loads without making the radiators unduly heavy. The estimated increase of speed due to their use in place of Lamblin radiators used on the 'S.4' is 24 mph.

Whilst Mitchell, like other engineers, was perfectly willing to profit from the successful design solutions of others (*see* particularly his Dornier-inspired Air Yacht – Chapter Seven), the Italian criticism had not taken into account Mitchell's trendsetting S.4 of 1925, how long Mitchell had been contemplating his latest design or how the wind tunnel tests had influenced his more pragmatic choice of this layout.

> **Supermarine's description of the new racer includes other details of the genesis of the S.5 machine and its increasing use of metal structures:**
>
> The S.5 is naturally a development of the S.4 and it may be interesting, therefore, to indicate the manner in which progress has been made. The shapes of all parasite parts, such as body and floats, have been arrived at as a result of lengthy wind-tunnel tests, and the cross-sectional areas of these components have been reduced to a minimum. In place of the cantilever float struts on the S.4, a peripheral system of wire-bracing has been adopted. The new Supermarine high-speed wing radiators completely eliminate radiator drag, as their surface is entirely coincident with the normal surface of the wings.
>
> WINGS – Low-wing, braced monoplane. Biconvex wing section, of medium thickness. Wing structure of wood, consisting of two spars and normal ribs, except for wider flanges necessary to secure the fixings for the wing radiators. Wing covered with ¼ in. plywood, over which are placed the wing radiators. Wings braced with streamline wires to top of fuselage and to floats.
>
> FUSELAGE – Oval section, of metal monocoque construction. Built up of a number of closely spaced transverse formers, covered with sheet duralumin, reinforced with longitudinal stringers. Front portion of fuselage acts as an engine-bearer, the two main bearers, of box-section, being secured direct to sides of fuselage and supported by reinforced cradles. The fuselage frames, to which wings and floats are attached, are strengthened and the skin in this region, as well as below engine, is laminated.
>
> TAIL UNIT – Monoplane type. Fin built integral with fuselage. All controls internal.
>
> FLOATS – Twin, long, single-step, streamline floats, of duralumin construction. Built up of one central longitudinal bulkhead, to which are attached transverse frames, which are interconnected by light longitudinal members, the whole being covered with duralumin sheet. The centre-section of the starboard float is built in the form of a petrol tank of steel, and to balance the machine laterally the whole chassis is slightly offset, relative to centre-line of body. Floats attached to fuselage by four struts, each pair meeting at a point under the centre-line of the fuselage.
>
> POWER PLANT – One special Napier 'Lion' racing engine, completely cowled-in. Either geared or direct drive engines may be fitted without alteration. Wing radiators header tank in centre cylinder-block fairing. Main petrol tank in starboard float, with auxiliary gravity tank in fairing of starboard cylinder-block. Total fuel capacity 55 galls. (250 litres). Oil-coolers set along sides of fuselage.
>
> ACCOMMODATION – Pilot's cockpit situated over trailing-edge of wing.

There was also to have been an American private entry to this year's Schneider competition, which deserves a mention, as it shows the determination of the British at this time to try for a victory: because of problems with the American machine, a request for a thirty day delay was made at the end of July – after all, the Americans had agreed to an 18 day postponement in 1926 – but Britain would not agree to the delay. The cost of re-arranging travel and accommodation arrangements or the much greater cost of keeping their Schneider team abroad longer than bargained for would have been an obvious reason for Britain's refusal but her comparative preparedness for the event on the due date was surely the main factor.

The word 'comparative' is used advisedly because the Crusader and the first of the two Gloster IVs were only delivered in May, with the first S.5 being ready in early June. However, the 14th saw the first flight of the S.5 and 284 mph was achieved; and, on 3 August, the Gloster machine put up 277 mph. Because of the straight-line development from the 1921 Mark I (Mars/Bamel), the new Gloster was unproblematic and early nose-heaviness in the S.5 was quickly corrected by an adjustment of the tailplane setting; the log of Webster (*see* later) recorded 'very very, nice. No snags'. Only the Crusader appeared not to be justifying faith in the British entries as it proved to be markedly slower and was afflicted by sudden engine cut-outs – 'with a whip that nearly took it out of the machine'. This was sometimes followed by the engine cutting back in just as unexpectedly which was not only alarming but also very uncomfortable for the pilot – as Schofield said, 'the backrest gave me a kick in the back that I shall remember for a long time. I literally felt quite silly'. The High Speed pilots were perhaps being even a little kind when they named the machine 'Curious Ada'.

S.5 at Calshot after release of the first photographs.

The tenth Schneider Trophy Contest, 26 September, 1927, at Venice

Notwithstanding the disappointments with the Crusader, it was decided to take this machine to Venice, along with the Supermarine and Gloster aircraft. Despite all the well-laid plans, there had been, as usual, little time for practice and certain necessary modifications before shipment but, at least, the travel arrangements had been designed to allow adequate practice time at the contest venue. One S.5, a Gloster and the Crusader left for Italy on 17 August on the SS *Eworth*; the remainder, with Mitchell and his wife, departed ten days later on the SS *Egyptian Prince*, which took them to Malta where they transferred to the aircraft carrier HMS *Eagle*, accompanied by four destroyers – more evidence that the British meant business. (Italy also fielded four battleships.)

Unfortunately, things did not start well: bad weather prevented test flying until 10 September, only thirteen days before the navigation tests were due; then, on the following day, the Crusader crashed. On take-off, a roll to starboard was seen to go past the vertical and the wing tip then dug in but, luckily, the machine had not obviously reached any great height, although the speed had reached 150 mph. Fortunately, the pilot, Schofield did not pay the ultimate penalty for not giving adequate attention to pre-flight checks – it was found that the crash had been due to crossed aileron controls on re-rigging in Venice.

The second batch of British planes arrived on the 11th but bad weather again prevented test flying until the 21st, when it was found that the problem of fumes in the cockpit still needed attention. Both Supermarine and Gloster had discovered this matter at Calshot but now, as a result, Fl. Lt S. M. Kinkead was confined to his room all the next day. His machine also required attention as part of his spinner had come adrift, causing severe vibration.

An Italian pilot was not as lucky as Schofield and was killed during practice. The new Italian engines were also proving unreliable and, when their pilots arrived on the 19th, they appeared not to want to fly flat out. This reticence had the effect of supplying no information to the British, except that the Italians still favoured their climbing turn technique at the pylons; as a result, the British team could devise no special tactics. The three competing entries were finalised as Fl. Lt S. N. Webster in the Supermarine N220, Fl. Lt O. E. Worsley in the Super-marine N219 and Kinkead in the better of the two Glosters. It was decided that N220, and the Gloster, N223, with the unproven geared engines were to fly flat-out, with the expectation that Worsley in the ungeared S.5 was likely to finish if, for any reason, the engines in the other two, more advanced, machines failed. They had also decided to continue with their technique of level flight with the turns as tight as possible.

The navigation and water-tightness tests were held on Friday, 23 September, as arranged and all duly completed them without trouble, except for Webster who had to make a second attempt as he was judged to have crossed the start-line incorrectly. This he made good on the next morning, in time for the contest proper on Sunday the 25th. Italy also had the time to replace a suspect geared engine with a direct drive AS.2 engine from the previous year.

Large crowds began to gather, and not just locals who were strongly supporting the 'local boy', Capt. Arturo Ferrarin: Maj. Mario de Bernardi was also a national favourite, having won the previous year, and he was therefore well supported by those brought in by the Italian State Railway on special half-fare excursions. Unfortunately, a strong wind and a heavy swell made conditions too problematic for the floatplanes. Slower, more seaworthy, flying boats would have coped with rougher seas and would have vindicated Jacques Schneider's aim to develop practical, as well as fast, seaplanes but, by now, the Trophy Competition had developed a breed of specialist floatplanes of a very different sort, with almost marginal buoyancy from the floats. Jacques Schneider, whatever his views on how his competition had developed, was not able to attend the event as he was recovering from an operation.

The crowds had to return on 26 September, when conditions had improved, and it was possible to commence the contest at 2.30 p.m. when Kinkead took off in the Gloster, followed by Webster and then de Bernardi. The new member of the Italian team, Capt. Frederico Guazzetti, was next, then Worsley and, finally, Ferrarin.

A new contest rule was that the start line should now be crossed airborne and this allowed an immediate assessment of likely performance over all the seven laps. British timekeepers made de Bernardi 15 seconds better than Kinkead in the Gloster, with Webster equal to the Italian. These calculations were rendered academic when Ferrarin disappointed his local supporters by turning off the course on the first lap with two pistons burnt through, followed by de Bernardi on lap two, suffering from a connecting rod failure. The third Macchi, with the older replace-ment engine, proved no match for the British with their new uprated Lions but then Kinkead retired at the beginning of the sixth lap when violent vibrations

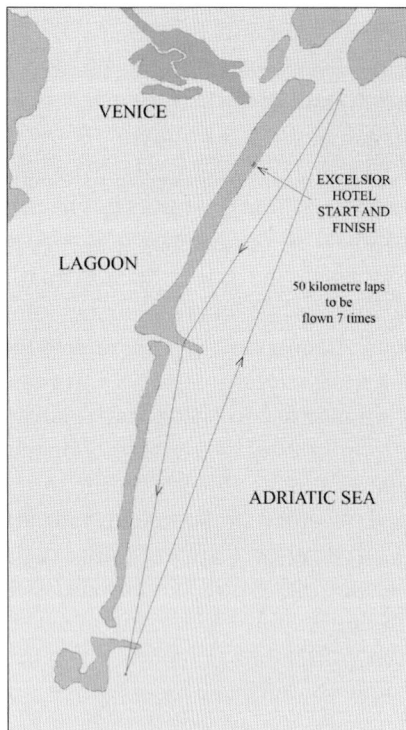

Schneider Trophy Course, 1927.

made it seem prudent to do so. This turned out to have been a wise decision as the previous vibrations were found to have caused a shear-line about three-quarters around the circumference of the propeller shaft.

Then, on the penultimate lap and in sight of being placed, the last Italian, Guazzetti, pulled out in spectacular fashion: Schofield (now able to walk with the aid of sticks) witnessed his 'hair-raising dashes in all directions when he was blinded by the bursting of a petrol pipe'. Luckily, he managed to get down safely although not before giving some of the spectators a scare, as Rodwell (later Air Commodore) Banks reported:

> I went to Venice, to see Webster of the RAF High Speed Flight team win; and, with others, including the Crown Prince of Italy, was nearly knocked off the top of the Excelsior Hotel on the Lido when … the pilot of one of the Macchis had engine trouble. He managed to turn his machine off the course but appeared to fly directly at us. However, he 'leapfrogged' the Hotel to land in the lagoon at the other side.

An S.5 over the Lido. HMS Eagle *in the background.*

Webster meanwhile had safely completed the required seven laps but carried on flying: his lap counter was a board with holes covered with paper and, after punching out the seventh, he found that he had been airborne for only 46 minutes instead of the expected 50 plus. Thinking that he might have miscounted some-how, he managed to complete another lap without having to force-land for lack of fuel. Watched by Mussolini, he thus led a British whitewash with an average speed for the actual distance of the contest of 281.65 mph, a new record for seaplanes and bettering by 3 mph the world speed record for landplanes. Worsley came second at 273.01 mph in the second, ungeared, S.5 – as Webster said after-wards, 'I was pretty sure it was in the bag. I knew I had the legs on Worsley'.

The skill of these specially selected pilots is demonstrated by the fact that the RAF front-line fighter at this time, the Armstrong Whitworth Siskin IIIA, had a top speed of only 186 mph and that, despite light rain and haze towards the end of the competition, both men, whilst having to negotiate two sharp turns per lap, had averaged about 87% of the maximum speed available to them. As Schofield said: 'It must be remembered that we were the first Service team, that our work … carried with it *an increase of speed of a proportion unheard of before*' [my italics].

It might be that Schofield's account of the Venice event was slightly col-oured by an over-compensation for his unfortunate part in it, but it is undeniable that the unprecedented leap in speeds took place in racing aircraft with extremely limited, almost non-existent forward views. Whilst the S.5 was slightly better, the Gloster contender had been introduced to Schofield as 'the blind wonder' and this did not just refer to the fact that 'it was impossible to see a thing anywhere over the top plane or indeed anything useful ahead at all, when sitting on the water' (or with the nose raised for landings); only when Schofield saw some reasonable height on the altimeter could he put the nose down a little and see the horizon ahead; 'a spot more height, and it was possible to get a generally helpful impression on either side, but it was still disconcertingly blind in most important points' – as he said of Kinkead in the same machine in the actual competition, 'he flew his deadly accurate course almost blind':

The S.5 flown by Fl. Lt Worsley and the Gloster IVB, flown at lower altitude by Fl. Lt Kinkead (in order to improve his peripheral view of the course).

* * * * *

Back in England, Mitchell was among those fêted by the Corporation of Southampton and his winning machine was put on display in London. A measure of the designer's increased status can also be gained by his being invited to address the Royal Aeronautical Society shortly afterwards.

But six weeks after the tenth competition, De Bernardi set up a new Absolute Air Speed record of 297.8 mph in the Macchi M.52, thus showing what this aircraft was capable of when its engine was running properly. And so, the following March, the Air Ministry, rather belatedly, had the reserve S.5, N.221 (not used in Venice), prepared for an attempt to recapture the record – which, it was hoped, would show the superiority of Mitchell's design, especially as its engine power was 125 hp less than that of the Macchi.

The attempt was to be flown by Fl. Lt Kinkead, a World War One pilot, particularly active afterwards in 'control without occupation', with a DSC and Bar, a DFC and Bar and a DSO, who had had to retire from the Schneider Trophy race in the Gloster machine. Unfortunately, the attempt was to end in tragedy. Poor weather had delayed the speed record attempt for ten days but, on the morning of 11 March, 1928, a test run in windy conditions suggested that a significant advance on the existing record might be achieved when the weather was calmer. On the next day, an oil leak had to be put right and then a snowstorm made low level, high speed, flying impossible. But, shortly after 4 p.m., conditions changed dramatically to calm and sunny although accompanied by a sea-mist.

Perhaps owing to the frustration of the previous days, Kinkead decided to go ahead. At 1700 hrs, he took off and, in accordance with the regulations, alighted as required – to prove the seaworthiness of the aircraft. (In fact, two landings were required, the previous day's test run was counted as providing the first.) He then took off again and climbed out to about 1,300 feet. In order that speed should not be built up in a dive, it was then necessary to descend to 150 feet, which had to be achieved 500 m before the start of the timed 3 km. course.

Precisely what happened then will never be fully established as the S.5 was barely visible to the witnesses on the land who, according to a *Flight* report, included several foreign Air Attachés, Biard and Mitchell. Some believed that Kinkead had decided to abandon the attempt because of the conditions and stalled on the landing approach; others maintained that, with the poor visibility, he never levelled out and flew straight into the sea whilst still intent on the speed record. Sun glare, coupled with an obscured horizon and a waveless sea, would have given precious little information to confirm altitude or flying attitude and it is known that he was suffering from a recurrence of malaria.

In these weather conditions, he had displayed his skill as a pilot by having just previously made a successful landing – and he had also previously overcome similar conditions when practising on the S.5 before setting off for Venice. Schofield, who believed that Kinkead died attempting a high speed run, remembered his previous flying when he wrote:

> One of his very first flights was on such a day as that on which he crashed.
>   Sky and sea blended together so ethereally that is was impossible to tell where one finished and the other started ...
>   Yet Kink insisted that he knew exactly what he was doing, and put up his usual faultless show. For the greater part of the time he cannot have known in what direction he was flying, and his 'landing' must have been accompanied by no small quantity of Faith and Hope.
>   It was on a similar day a few short weeks afterwards, whilst completing what looked like being a record speed run over the same waters, that he suddenly turned down and vanished for ever.

Nor can structural failure be ruled out – Kinkead was flying the aircraft which had not been used in Venice and, with a light fuel load and sprint tuned engine, would have been travelling faster than the other S.5s the year before. Certainly, the inquest verdict of a stall whilst landing, is the least convincing of the explanations (*see* Appendix Five). Biard, who was standing next to the Chief Designer at the time of the accident, is quoted as being quite certain that the crash was a result of structural failure and his reaction at the time might well have contributed to the designer's distress.

Mitchell was always known to be tense and difficult to live with before and during high speed and test flights. He had established a good rapport with the pilots and so he would usually stand apart from the other company watchers at these times, unsmiling and anxious. As Arthur Black had observed, 'When early test flights of a new aircraft were in progress his concern was so great that it paid not to attempt polite conversation'. Whilst the escape of Biard from the S.4 crash had not confirmed his worst fears, the sad death of Kinkead left him brooding over the tragedy for several days, finding it hard to accept the assurances of his colleagues that there appeared to be no reason why he personally should feel in any way responsible. But in a speech at a Rotary meeting in Southampton just after the winning of the Schneider contest, Mitchell had given a glimpse of the feelings that he felt in connection with this sort of design work:

> The designing of such a machine involved considerable anxiety because everything had been sacrificed to speed. The floats were only just large enough to support the machine, and the wings had been cut down to a size considered just sufficient to ensure a safe landing. The engine had

only five hours' duration; after that time it had to be removed and changed. In fact everything had been so cut down it was dangerous to fly. Racing machines of this sort are not safe to fly, and many times I have been thankful that it was only a single seater. The machine itself has been a source of anxiety to me right from the start, and I am pleased to know that at this moment it is safely shut up in a box.

Later, in June 1928, Fl. Lt D'Arcy Greig, took over the High Speed Flight, re-formed for the next Schneider Trophy contest, and N220 was 'taken out of its box', for an attempt to beat the Italian record speed, which had now been raised to 318.57 mph; however, the slightly higher speed achieved by Supermarine did not give a margin sufficient to justify a claim to the FAI.

Nevertheless, the S.5 increase of 24 mph over the fastest of the Gloster IV machines, both equipped with the same powerplant, was a vindication of Mitchell's monoplane approach. The S.5s were thereafter to be relegated to practice machines for the eleventh competition although, in the event, N219 was returned to Supermarine in July 1929 for a complete overhaul and the fitting of a new geared Lion engine. On the strength of which it competed in the competition of that year and came third, with a speed only 2 mph slower than the new Macchi M.52, with an engine developing over 100 hp more thrust.

* * * * *

In 1928, Vickers (Aviation) Limited acquired the Supermarine Company and its chairman, Robert McClean, set about reorganising many of the rather haphazard management practices of the Woolston company. Whilst friction was virtually inevitable, and the appointment of Barnes Wallis as Mitchell's assistant was particularly insensitive, it was an acknowledgement of the achievements of Mitchell's design team that it was retained as a separate entity at Woolston – although major decisions would have to be approved by McLean. Thus, in subsequent volumes of *Jane's*, descriptions of designs from Mitchell's team were kept separate from other Vickers products under the following title:

THE SUPERMARINE AVIATION WORKS, LTD
(DIVISION OF VICKERS (AVIATION) LTD)

Vickers had no doubt acknowledged the potential of the much smaller company as early as 1920 when its Commercial Amphibian came a very close second to their Viking; then came the Sea Eagle, the Seagull II, the Scarab and, pre-eminently, the Southampton. However, in view of Vickers' dominant position in the armaments industry, Mitchell's Schneider Trophy contributions must have been particularly noted and thus made Supermarine's by-no-means-expected design of the Spitfire that much more probable.

Mitchell's name now appeared among the list of Vickers Directors and the 1925 publicity description of him as 'one of the leading flying-boat and amphibian designers in the country' was now significantly expanded to:

> one of the leading flying-boat, amphibian *and high-speed seaplane* designers in the country. [my italics]

He was still only thirty-four years of age.

* * * * *

Enter Rolls-Royce

Once the excitement of the 1927 Schneider competition win had subsided, the design and cost realities of competing for the Schneider Trophy were now even more clear to all potential entrants, especially as only two of the six competing aircraft had finished the course this last time. Thus it was that the Royal Aero Club, which would organise the next competition, now supported the change to biannual competitions, starting in 1929, and the FIA, at its January meeting in 1928, concurred. The rules of the competition were amended to the effect that a permanent holder of the Trophy would have had to gain three victories, not within five years as before, but in the course of successive contests.

Britain would certainly welcome a longer lead time in order to improve upon the performance of the S.5, in time for the defence of the trophy which would take place in front of a home crowd: the failure of the S.5 to significantly improve upon the World Speed record of the Macchi M.52 allowed for no complacency and suggested that the present Supermarine machines or engines might not be capable of much further direct development. But for a redesign, further Government support would be vital – as Mitchell himself had said in his 1927 lecture to the Royal Aeronautical Society:

> After the failure of the British team to win the race in America in 1925, it was brought home to all interested that our machines were a long way inferior to the American machines and that if we wished to hold our own again in this important field of aviation we should have to treat the matter much more seriously. Furthermore, it became obvious to all that machines could no longer be entered for these races by private enterprise. It is true that the Air Ministry had loaned machines for the race but very little opportunity had been given for research and experimental work and the engine designers had been working independently.

Perhaps Mitchell's words reflected a fear that the Treasury would not support the cost of another competition entry as word had probably got around that Churchill, then Chancellor of the Exchequer, had taken some persuading to agree funds in 1927 and that Sir Hugh Trenchard, the Chief of the Air Staff, was not at all keen on his RAF getting mixed up with air racing. But once again events favoured Mitchell as Churchill, no doubt because of the resounding success over Mussolini's aircraft in Venice, gave the new project his full blessing, asking Sir Samuel Hoare how much money he needed. Trenchard then not only accepted the inevitable but took a personal interest in the arrangements and in the selection of the next High Speed team.

With this backing, the Air Ministry's support was extended, in the direction that Mitchell wished for: allowing more time for the design of special airframes, racing engines and fuel, and for the mating of the engines to the propellers and to the shape of the airframe or vice versa; and also more time for consultation between engine and aircraft designer.

One particular development in the competition, particularly as far as British engine designers were concerned, was the emergence of a Rolls-Royce racing engine. Britain and Mitchell had been fortunate that their winning Napier engine of 1927 had not been a new and, therefore, possibly unreliable design but had been in continuous development since before its use in his 1922 Schneider Trophy winner; it had been producing 450 hp in 1919 and this had actually been increased to over 900 hp for 928 lbs weight by 1927. It is also noteworthy that, during this time, no Schneider aircraft powered by these Napier Lions failed to complete the course because of engine problems. Nevertheless, the question had to be asked whether this remarkable engine was now reaching the end of its development potential, especially as Napier had not developed a supercharger.

By 1927, Napier were producing about 50 Lion engines per month but had not gained Air Ministry funding for the development of their next engine, the 60.3 litre Cub. Meanwhile, in 1924, Rolls-Royce had begun an engine to rival the Curtiss D-12 which had powered the 1923 and 1925 Schneider Trophy winners. The resultant Kestrel then went on to power a whole range of classic RAF Hawker biplanes: the Hart, Fury, Audax, Hind, Demon, et al.

In 1921, Arthur Rowledge, who had been responsible for the Lion engine, had joined Rolls-Royce and been influential in recruiting James Ellor from the Royal Aircraft Establishment to work on supercharging. His current concern was the proposed Buzzard; it was larger than the Kestrel and developing 825 hp but it had a tendency to crack its cylinder head after only a short run. Mitchell's only direct experience of this company's engines at this time, 1928, was in connection with their Eagle in his earlier slow-flying aircraft and he apparently asked Major G. P. Bulman, the Air Ministry official responsible for the development of aero engines, for his views.

Bulman later reported:

> The Rolls-Royce engineers had discussed separately with Reggie Mitchell, the Supermarine designer, and myself, the possibilities latent in this engine, given intensive effort for a short life, and were enthusiastic in their hopes. To Mitchell it would mean a considerable rehash of his S.5 to accommodate the bigger and heavier engine, with its extra cooling and heavier fuel load. To me it meant a desperate gamble to back something virtually untried, entirely contrary to my habit, and to commit the Air Ministry and the nation to a gigantic bet, instead of playing safe by putting all one's money on the well tried faithful Lion.
>
> Mitchell and I met together alone three times over a short period to resolve the problem ...

Commander James Bird of Supermarine and Bulman accordingly called on Henry Royce then living in semi-retirement at his West Wittering home on the Sussex coast. They pressed on him the matter of national prestige as well as the eventual benefit to the British aircraft industry and Royce agreed that the company should take up the challenge of developing a special racing engine; it was also agreed that, as it was already October 1928, the partially developed 36.7 litre Buzzard engine would have to be the basis for the project. According to Bulman, Rowledge, Ernest Hives, the head of the Experimental Shop, and A. C. Lovesey, his assistant went down to West Wittering:

> It was a bright authumn morning and Royce suggested a stroll along the beach; as they walked he pointed out the local places of interest. But Royce who walked with a stick, was a semi-invalid and he soon tired.
>
> 'Let's find a sheltered spot,' he said, 'and have a talk'. Seated on the sand dunes against a groyne, Royce sketched the rough outline of a racing engine in the sand with his stick. Each man was asked his opinion in turn, the sand was raked over and adjustments made.

It would have to have a different crankcase and supercharger to conform to the sort of shape that Mitchell was likely to develop out of the S.5 design and so it was separately designated the 'R' engine, with a hoped-for output of 1,800 hp.

There followed another example of how Mitchell's career was at times influenced by events outside his control. Bulman reported back to his chief, Sir John Higgins, the Air Member for Supply and Research, who immediately asked the Managing Director of Rolls-Royce to call and agree the decision in principle. Bulman then described how the meeting seemed destined to dash hopes of a formidable powerplant for Mitchell's next design:

> Claude Johnson, who had built up the Firm's worldwide reputation on Royce's technical brilliance and vision, a man of striking personality,

had died in 1926, and his successor [his brother, Basil] it was who came
to see Josh and myself, alone. To our utter amazement he begged to be
excused from our commission. Racing and all its aspects were things,
he said, strictly to be avoided by his firm. Its reputation for sheer quality
and perfection must not be smirched by sordid competition of this sort.
To participate unwillingly, and quite possibly fail, would be a calamity
for the firm with the loss of its prestige. And so on, in dreary defeatism.

Bulman then described how, knowing that the firm's engineers were only too
keen to take up the challenge, he came out with 'a single word, unprintable in
polite context and essentially masculine'. Luckily, Sir John was well aware of
the desire of 'his masters' to fend off attempts by rival countries to win the next
Schneider Trophy competition:

> Higgins turned and looked at me for a long second, and then in a steely
> voice of real Air Marshall caliber said to our guest, 'Mr [Johnson] …
> I order your firm to take on this job. We have complete faith in your
> technical team. The necessary arrangement will be made between our
> respective staffs. Good afternoon.' As our disconsolate and vanquished
> visitor closed the door behind him Josh said to me, 'Thank you for sum-
> ming up the discussion so succinctly,' and gave a huge chuckle as I shot
> out of his office to telephone the glad tidings to Rowledge in Derby.

The 'desperate gamble to back something virtually untried', including
designing a supercharger to give boost at very low level, contrary to its usual
high altitude operation. And, of course, Mitchell was also gambling his future
reputation in high speed flight. As Bulman said, it was 'mutually agreed that we
should back the Rolls project, largely – and for my part wholly – on the faith that
I had in the Derby team.'

* * * * *

Meanwhile, it had become clear that Italy was just as determined to win the
Trophy next time and why it had supported the move to biannual contests: taking
note of the previous determined and thorough British campaign which had led
to their 1927 success, it ordered new machines from Macchi, Fiat, Piaggio, and
Savoia-Marchetti as well as engines from Isotta-Fraschini and Fiat:

- the Fiat C.29 proved to have a layout similar to the Macchi but with a
  mere 22 foot wingspan for it was to be powered by the lightest engine of
  its power in the world, the 1000 hp Fiat AS.5;
- the Macchi M.67 emerged as a further development of the M.39
  and M.52 – after all, the latter had set the new world record which
  Supermarine had been unable to better and was now to rely on an Isotta-
  Fraschini engine with three banks of six cylinders, expected to deliver
  1800 hp. Because of its extra weight, there was now to be no sweepback
  of the wings and, as with the 1927 British aircraft, float strut bracing was
  now replaced with wires.

The other two Italian designs were very different from the layouts now coming to
be expected for the Schneider Trophy contest:

- the Savoia-Marchetti S.65 was, admittedly, a floatplane but it had twin
  booms supporting the tail unit, thus allowing for two engines to be
  mounted in a central nacelle. The pilot was placed in between the tandem arrangement and the tractor/propeller combination of two 1000 hp
  Isotta-Fraschini engines promised formidable power. It would also present no torque complications on take off, something that, before the advent of variable pitch propellers, was becoming even more of a problem;
- the Piaggio P.7 represented a quite revolutionary departure from all other
  Schneider Trophy designs as it had neither floats nor any conventional

flying-boat hull. Instead, it had a watertight fuselage with a marine propeller to get it up on hydrofoils, at which point a conventional airscrew was to be engaged for take-off. It also had a cantilever, elliptical wing (which closely foreshadowed that of the Spitfire – *see* p. 245).

A German, an American and two French entries were projected but none of these materialised, leaving the contest as before, between the Italians and the British. This time, the British entries were ordered only from Supermarine and Gloster. The design of the latter company, designated Mark VI, now finally embodied the mono-floatplane approach. This change of philosophy was particularly influenced by the fact that the pilot's less than satisfactory forward view in the Mark IV would have been even worse if a top wing had had to be placed sufficiently far forward to support a new and heavier engine. This engine showed that Napier was still not done and their, now supercharged, Lion was developing 1320 hp. It also gave some insurance against the possible failure of the brand new Rolls-Royce engine employed by Supermarine. The wings were an interesting shape: like the Short Crusader of 1927, the new Gloster had wings which could not be exactly described as elliptical but which had a thin, high speed, section at the roots, widening out to a thicker section towards the tips for low speed lateral control (*see* drawings on p. 245).

S.6 and Squadron Leader Orlebar

Mitchell was content to rely on the rightness of his previous design for the significantly more powerful engine and so his main design effort was now directed towards the realisation of an all-metal aircraft larger than the S.5 in order to accommodate the projected bigger and heavier engine: the 930 lbs. of the 1927 Lion was to be replaced by an engine weighing 1530 lbs. First configuration drawings were sent to Mitchell on 3 July 1928 and so he was able to influence the shape of the cam covers so that they would conform to the streamlines he was developing for his new machine – as Bulman reported: 'the cylinder blocks and valve gear casing were trimmed down and externally reshaped to come within the frontal area and line of the new Supermarine S6 designed by Mitchell'. An eventual 1900 hp was achieved by August – a power increase of 211% over the previous Lion engine for what turned out to be a loaded aircraft weight increase of 78%. Mitchell was reported to have said: 'Go steady with your horse-power' – alluding no doubt to having to accommodate such a massive power increase as well as the cooling problems that would be encountered.

And so Mitchell had to consider the airframe implications of the change from the Lion engine with a 24 litre capacity to the proposed Rolls-Royce R of 36.7 litres. The most obvious change to the S.5 shape was the different cowling necessitated by the 'V' shape of the new engine. Its extra weight also involved placing the cockpit further back and the increased fuel consumption meant that both floats had to be used for the tanks – the new engine was going to consume nearly 2.5 gallons per minute. As the empty weight of the S.6, at 4,471 lb, was 1791 lb heavier than that of the S.5, the wingspan was slightly increased, giving an additional area of 30 square feet and the front float struts had to be attached further forward on the fuselage to support the combined effects of a longer and heavier engine. Supermarine drew attention to an advantage of this change of position: 'In place of the cantilever engine-mounting used on the S.5, the front float struts have been moved forward to provide a substantial saving in weight'.

Solving the constructional and loading problems in itself had justified the new 'S.6' designation but these matters were relatively straightforward compared with contending with the heat generated by the new engine. The extra plumbing for the fuel transfer from two floats was as nothing compared with that required for cooling the engine oil. The channels for oil cooling running along the sides of the S.5 were now increased and new ones added to the underside of the fuselage as well, with the collection point now being the inside of the fin. Their efficiency had additionally to be increased by devising some method of ensuring maximum contact of the oil with the outer surfaces of the system which was exposed to the slipstream. Mitchell's Chief Metallurgist, Arthur Black, came up with a method described in the following Supermarine publicity:

> By a new form of internal construction, the oil-coolers have an increased efficiency of about 40 per cent ... A large number of sloping gutters are arranged along the sides of the fin, so that the oil, after being sprayed from the pipe at the top of the fin, is made to trickle down the gutters and over the internal structure, thereby ensuring that the greatest possible amount of oil is in contact with the metal all the time.
>
> A similar purpose is served by the oil-coolers along the sides and belly of the fuselage. Those along the sides take the oil to the fin and those along the belly return it to the engine. These coolers are shallow channels of tinned steel attached to the sides of the fuselage. They owe much of their efficiency to a number of tongues of copper foil athwart the flow of oil. These are soldered to the sides of the cooler and project at right angles into the stream of oil. They are staggered in such a way that the flow of the oil is not seriously impeded.

Nevertheless, Greig had found that the position of the oil pipes, running along the outside of a very narrow fuselage, 'turned the inside of the cockpit into something approaching an extremely hot Turkish bath' with the oil temperature gauge reading 'around 136 degrees centigrade'.

Other aspects of the design reflected constructional changes beginning to take place in the aircraft industry and Supermarine were thus anxious to point out that their move to metal construction was not just with respect to the framework of their machine but that it placed them in the forefront of the use of load-bearing external skinning – particularly in respect of the wings: instead of being plywood-covered with the radiator panels externally attached, as with the S.5, the wing structure was now covered by the panels alone. Supermarine publicity draws attention to both stressed skinning and to the use of new materials:

> Unlike the S.5, which had wooden wings covered with plywood, over which were placed the wing radiators, the S.6 has metal wings, and the radiators, which [now] consist of two thicknesses of duralumin with water spaces between, are made as a wing covering to take torsional loads ... This method saves a considerable amount of weight over previous practice. The fuselage is all of metal and the skin takes practically all the stresses. The front portion of the fuselage acts as an engine-bearer and the [laminated] skin in this region takes all the engine loads.

Mitchell had never less than an extremely practical approach to his designs but there was usually some concession to aesthetics. It would seem that the inputs from wind tunnel data, the substitution of the larger 'V' engine and its enormous demands, had somehow now given the S.6 an unusually stark appearance, compared with the rival Gloster VI, described by one newspaper as 'more the conception of an artist who can make and create his own lines by the stroke of a brush, than the work of a designer who is bound by the principles of engineering and the comparative inelasticity of metal and timber'.

Despite the gap of two years between competitions that had now been agreed on, the scheduled start of the eleventh contest was less than six weeks away before Mitchell's uncompromising and complex airframe could be tested in the air and, even then, flying time was limited. The main reason for this delay was to be found at Rolls-Royce. The new engine had been first sketched out in the previous October and, by May, it had reached 1545 hp. During this time, the

Mitchell attending the completion of the S.6 engine installation.

company progressively redesigned and substituted more robust parts as they began to fail until, on 27 July, the new engine passed the magic one hour mark at full throttle and supercharger boost. A few days later, with the competition scheduled to begin on 6 September, the blending of a special fuel by Rodwell Banks, produced an engine run of 100 minutes and 1850 hp.

The end of testing was much to the relief of the good citizens of Derby, as the tests had also required the simultaneous running of three 600 hp Kestrel aero engines. These engines drove fans to cool the crankcase of the R engine, to disperse fumes in the test house, and to simulate 400 mph airflow conditions in flight – designing a supercharger for low level operation was a unique development at that time and involved a forward-facing air intake which converted air-speed energy into pressure energy.

These were the days before modern health and safety regulations and Rod-well Banks, describing the din of a total of four aero-engines within the test sheds, wrote that 'reverberation from walls and roof is such that at certain engine speeds one cannot keep still: the whole body seems in a state of high frequency vibration. One shouts at the top of one's voice but cannot even feel the vibration of the vocal chords'. People living up to fifteen miles away, on the outskirts of Nottingham reported still being able to hear the engine runs and the ears of the Rolls-Royce workers were plugged with cotton wool. They were also well supplied with milk to counteract the notoriously laxative effect of breathing in the Castrol engine oil, ejected out of the open exhaust ports of the R engine and deposited on the walls of the test cell: *Flight* reported that one early run consumed oil at the rate of 112 gallons per hour and that the state of the test shed inside was 'a wonder to behold'.

Meanwhile, a new High Speed Flight had been formed in the February of 1928. Greig, who had been posted in after the death of Kinkead, had recommended members of his Hendon aerobatic team: Fl. Lts G. H. Stainforth and R. D. H. Waghorn and Fl. Off. R. L. R. Atcherley, the last being long remembered for his aerobatic display in a Gloster Gamecock at the 1926 RAF Display at Hendon. (He also recommended Fl. Lt J. N. Boothman, of whom we shall hear later, but he was on an overseas posting). He then prepared to hand over command to a Squadron Leader, A. H. Orlebar, with the new title: 'Officer

Commanding the High Speed Flight'; he was also to command the Flight for the following contest. It was to prove a wise choice, as Bulman later said: 'He was the inspiration and father of the whole outfit, yet always modest, approachable and self-effacing. He himself flew each aircraft before he would allow any of his team to take over, and personally tried out every modification and adjustment made.' Here was another example of good fortune aiding a Mitchell design.

Sqd. Ldr A. H. Orlebar and Mitchell in front of S.6, 1929.

As before, none of these airmen had been trained as maritime pilots and so time was needed to convert onto seaplanes that were expected to be the fastest in the world. Webster, the 1927 winner, had narrated:

> I was test flying at Martlesham Heath when I was sent off to Felixstowe, which wasn't far away, to join the High Speed Flight. I had never flown a seaplane but took to it like a duck to water! We were flying everything at Martlesham – the Inflexible, the Vixen, a Gamecock, the Hinaidi and a Sidestrand – so we tended to take everything in our stride. I had over one hundred and twenty types in my log book. We had a Flycatcher on floats for practice flying and my first flight in that was on 17 February; then we moved on to the old Bamel and the Gloster IIIs. I did quite a lot of flying in the Bamel. We used to pop back to Martlesham to do normal testing between practice flights. I also had three flights in the Crusader at Felixstowe.

In the following year, speeds had increased significantly but Waghorn's experience was similar:

> We started with the Fairey III.D, and Flycatcher, and then worked through to the Gloster IV, the Gloster IV.A, and Gloster IV.B. I don't think any of us found any jump from a land plane to the ordinary seaplane as from the handling point of view; they are remarkably similar. However, our first flight in the Gloster IV saw a decided jump. The whine of the fast revving engine, the seemingly endless take-off with its attendant jolts and jars magnified in some extraordinary way; the difficulty of knowing what speed you are travelling at and the apparent magnification of any inaccuracy in flying, all helped in giving me, at any rate, a very vivid impression of my first flight in a high-speed seaplane.

Waghorn also described how they put their time to good use in the practice machines by concentrating on devising the best technique for cornering: to achieve a turn at a constant height, as was the previous practice, it was necessary to apply rudder to correct the tendency of the banking aircraft to yaw upward owing to aileron drag; this, in turn, increased resistance during the turn. It was decided, therefore, that any tendency to climb at the pylons would not be corrected too strongly. Additionally, two scientists from the Royal Aircraft Establishment were attached to the High Speed Flight and they installed instruments in the aircraft to monitor speed, acceleration, and climb in order to evolve the most efficient turning circle – a compromise between tight, high G sharp turns and loss of speed and wider sweeps which incurred less drag.

With the increased speed in the turns, pilots had now to get used to blacking out, as Atcherley recalled: 'I went "out" halfway round a turn at Calshot Castle, the sharpest of the four turns, and flew completely unconscious at about 500 feet halfway back to Cowes before regaining my senses. (*See* map on p. 167.) Even then, there was a very frightening lapse of seconds when one realised that one was flying and had been "out" but still could not see or move one's hands.' Not surprisingly, he admitted that 'it made me brood a bit'.

N247, the first of the two S.6s.

These preparations received a serious setback when the actual contest aircraft finally arrived, as it was found that the effects of the greatly increased torque from the new engine had not been fully anticipated. The S.6 revealed an exaggerated tendency to dig in the left float and describe circles in the water – Orlebar reported that the gyrations 'had rather shaken' Mitchell. Waghorn described the situation as follows:

> With the arrival of the S.6 our hopes had risen considerably only to be immediately lowered to the depths when Squadron Leader Orlebar started his initial tests in Southampton Water. The S.5 in her take-off had been so straight-forward that we had assumed that her elder brother would also prove himself equally docile while being broken in. [Indeed, the floats were no longer rigged asymmetrically] We were therefore very surprised to see the behaviour of the S.6 on her first test. The S.6 behaved much as a horse refusing a fence. She sat on her tail and it seemed as if no amount of coaxing would get her forward. Furthermore, she dug her left wing into the water and not content with so much mischief started a gigantic porpoising. Time and again the Squadron

Leader tried and, although he had overcome the porpoising, she still continued to dig her left wing in and to swing viciously to the left. To the rest of us in the Seacar [speed boat] alongside, it was a heart-rending although impressive sight. From the Seacar we had a close-up of the whole proceedings and a very good view it was, not that one could see much of the pilot and fuselage, as most of the time they were enclosed in a whirl of spray. After about half an hour of this we returned to Calshot in a rather dejected frame of mind, as it certainly had not been a good beginning.

One can easily, therefore, imagine the Chief Designer's feelings, seeing his aircraft quite unwilling to *fly*; and when Orlebar pointed out to Mitchell that the first machine's number – 247 – added up to 13, 'the poor chap replied with feeling that he had not designed that', so the Commanding Officer kept the matter to himself. D'Arcy Greig also tried without success but, later in the day when a little wind had got up, Orlebar finally succeeded in taking off. Waghorn's narrative continues with description of the solution:

The main trouble was the wing digging business and due, without doubt to the enormous torque effect of the slow revving engine and propeller. Mitchell's first move was, therefore, to shift nearly all the petrol into the starboard float and put in hand the immediate construction of a new and larger petrol tank for this float. The result of this was in the end satisfactory enough though there were a good many anxious trials before she got safely into the air. To start with, it was a peculiarly delicate task for Squadron Leader Orlebar. He was swinging, he knew, and his left wing wasn't very far from the water and still he couldn't tell how much owing to the mass of spray enveloping the fuselage. He found out subsequently that a lot of the initial resistance to any acceleration was in part due to the very smooth, almost oily state of the water on which the first taxiing trials took place …

The difference in behaviour in the S.6 when she passed from an oily to a rippled patch was most interesting … I was once watching Atcherley trying to take her off. The sea was oily and the machine obstinate. She never looked like getting on the step. Atcherley shouted to me that he was packing up. We had, however, noticed a patch of rippled water in the distance, and got him to try once more over on that particular bit. The result was magical, and he got off on the first attempt. The torque effect is greatest at slow revs and the trouble was that being at the peak of the power drag curve, the drag of the floats was just about counterbalancing the thrust. The nose of the machine coming out very high and the tail of the floats digging right into the water, set up a very high resistance. We had, therefore, to fit a faster revving propeller, with more power for the take-off. This also gave more power for the top speed; but we already had more power than specified and therefore more heat to dissipate than the original radiators were designed for. Hence it was going to be necessary to throttle down to keep the water cool. That very slight increase in wind, by about 4 mph, made the difference; whether it was chiefly the increased control given to the rudder or chiefly the surface of the water affecting the floats, I am not prepared to say – perhaps a combination of both. But certain it was that provided you kept the machine into, or slightly to the right of the wind, you could get her on to the step. If she once got to the left of the wind it was hopeless … Whilst discussing these difficulties it is perhaps easy to assume that the S.6 had a bad take-off. Actually, this was not the case; provided one got her into the wind and on her step she accelerated like the proverbial gun.

A solution to the torque problem at take-off would be found in later years with the invention of variable-pitch propellers but, in the meantime, a special technique was worked out whereby it was necessary to keep the stick well back, contrary to all basic flying instruction, in order to maximise lift at the extreme low end of the aircraft's airspeed; additionally, the take-off had to begin with any breeze kept on the left quarter: this allowed for a nice judgement of acceleration whilst being pulled by the propeller torque to face directly into wind by the time that lift-off speed was attained.

Once in the air, Orlebar was extremely impressed with the accuracy of Mitchell's forecast of the new machine's behaviour: 'He had told me about the possibility of the wing dropping when she first got in the air, and that is why I was prepared for it and shut the throttle momentarily'; he then found that the wing came up easily and, having touched down momentarily, was then able to climb away for the new machine's first air test: 'Mitch had said he hoped for a speed of something up to 340 mph, and I achieved an indicated 345'.

However, in the short time remaining before the contest was to begin, modifications had to be made. Having now achieved take-off, the cooling system was found to be inadequate and so extra radiator piping had to be created along the sides of the floats and, additionally, small wing-tip scoops were fitted. These scoops, facing forward and with exhaust ports at the wing roots, created an extra flow of ram air at a velocity of about 35 mph over the inner surface of the radiators to increase their efficiency – an unforeseen bonus for using the radiators as load-bearing skinning for the wing. Flying problems had also to be addressed: aileron stiffness due to friction had to be overcome, re-rigging had to counteract a tendency to fly left-wing low, and tail-heaviness had resulted in an elastic bungee being fitted to the control stick (the tailplane was not adjustable). But the S.6s were now coming up to expectations and testing could thereafter be mainly confined to assessments of different propellers, to fuel consumption and to perfecting flying techniques in the very short time that was now left before the competition was due to start.

* * * * *

The Italians had had to wait until August for their new machines and it was soon found that the Piaggio P.7 was a non-starter as it was impossible to achieve transition from water taxiing to flight. Also, one of the Fiat C.29s caught fire on its second flight, was repaired, and then stalled after a third attempt at a take-off and sank. At least the first of the Macchi M.67s was looking decidedly promising, reaching 362 mph, but then it too crashed at low level. This time, the pilot was killed. Visibility had been similar to that when Kinkead died and, additionally, the windscreen might have been fogged by exhaust fumes. As the second C.29 and the other two M.67s were not then ready for the competition, Italy requested a one month postponement on August 22 but the Royal Aero Club stuck to the rules and refused the next day; the FIA concurred.

It is interesting that when this request was turned down only one of the new British contenders had flown: Orlebar's first successful take-off was on August 10 and the Gloster VI and the second S.6 were not first tested until August 25, by which time the first S.6 had gone back to Supermarine for the fitting of extra radiation. Nor could adequate practice time on the new aircraft be certain – the engines could only be guaranteed for a high speed run of one hour and they were required to be taken back to Derby for overhaul after every five hours' running. The delays caused by the removal of time-expired engines and the fitting of overhauled power plants were compounded by the British weather. These highly specialised Schneider machines required good visibility, gentle winds, and short, choppy water without 'white horses'; too much of a swell would cause the noses of the floats to dig in and set up an eventually uncontrollable porpoising and, on the other hand, flat calms would prevent 'unsticking' as well as producing a mirror-like surface which would make it both difficult and dangerous to judge the aircraft's height above the water – the long, flat landing approach of over 100 mph that had had to be judged for the S.5 had not diminished.

As it turned out, the Gloster VIs were soon to be withdrawn as their engines could not be made to run properly. So the British had been taking something of a gamble by not allowing a postponement, although the first S.6 was proving an unproblematic machine in the air (when the over-riding requirements of high speed flight might well have created a machine that was difficult to fly – as was the case with the American Gee Bee Racers of the early 1930s, for example).

If one sets aside the peculiarities of the take-off procedures, the S.6 was a remarkably viceless aircraft to fly and also revealed that Mitchell had made improvements over the flying qualities of the previous, smaller, machine, even though the wing loading had risen from 28 to 40 lb/sq ft. As Waghorn testified:

> While flying, she gave me the feeling of great stability, and when not flying low, the slow revs of the engine gave me the impression that I remember I got when I flew a Horsley [bomber] after having just left the seat of a Gamecock [fighter]. On turns she was delightful. Perhaps she was a little heavier laterally than the S.5 and the Glosters, but then she was a much bigger and heavier machine. There was no noticeable torque effect against a left-hand turn which had been so tiring in the S.5, and, generally speaking, gave me a feeling of great trust and confidence, and I never had cause to change my opinion …
>
> The S.6 appeared to stall about 3 mph slower than the S.5, but air speed indicators are not infallible at such a speed. However, it can be taken that she stalled in the region of 95 and was certainly no faster than the S.5. She was extraordinarily stable at the stall. The S.5 would quiver at the stall and flick over either side at the slightest provocation. The S.6 showed no tendency to drop either wing, but would sink on an even keel. On one such occasion, while testing the stalling speed, I found the machine on an even keel sinking at about 87 mph. When one considers the behaviour of her elder brother at a similar speed it is all the more interesting, especially when you realise the extra top speed of the S.6.

The eleventh Schneider Trophy Contest, 7 September, 1929, at Calshot
To avoid Britain winning by a fly-over, Italy had decided not to withdraw and sent over their, as yet, untested aircraft – the last Fiat C.29, the Savoia-Marchetti S.65, and the two Macchi M.67s – as well as the Macchi M.52R world speed record holder and an M.52 practice machine. By this time, both the Supermarine aircraft were ready but, as the Gloster's problems could not be solved in time, it was decided to call up one of the 1927 S.5 machines – Worsley's N219, now fitted with a geared Lion engine.

Despite lack of flight testing, especially on the Italian side, the navigation tests on 6 September went well and all the aircraft were moored out for the water-tightness test. Some hours later, Mitchell had to be aroused from sleep in the Officers' Mess, after being up late superintending final preparations: it was discovered that one of the new planes, Atcherley's N248, was listing, with over two hours to go of the required six hours flotation test. Mitchell decided that it would hold out for about three hours and went back to rest. By the due time, the machine had a very pronounced list but was able to be beached and a leak repaired.

Meanwhile, a further problem was also discovered, this time to Waghorn's N247, when traces of white metal were found on a spark plug during the routine plug change; internal damage was strongly suspected and thus a change of the offending parts and cylinder block would be necessary. The Schneider Trophy rules now did not allow changing 'any major component' at this stage in the competition but, luckily, a substitution of parts was permissible. Whilst major overhauls would normally be carried out at Rolls-Royce with the engine removed from the aircraft, at Calshot, it was necessary to devise some means of offering the intact machine up to the replacement block and, as Orlebar later reported, 'poor Mitchell was hauled out again' to supervise the operation. Luckily, a number of Rolls-Royce mechanics had come down by coach to see the competition and they were rounded up by policemen from various hotels around Southampton. By working through the early hours of the morning, under the supervision of Lovesey (one of the men who had originally discussed the engine with Royce) they were able to make the change, especially assisted by a left-handed fitter who was able to knock out one particular gudgeon pin. It was also fortunate that, thanks to Rolls-Royce craftsmanship, the spare block fitted in all particulars, as all the R engine parts were hand made.

The effort had been entirely necessary as it was found that one piston head had almost melted through and its cylinder lining was badly scored: an engine failure, at the very least, would have been inevitable, on the next run. (The damage

An S.6 on its launching pontoon prior to the 1929 contest. (The air scoops mentioned above can be seen projecting under the starboard wing tip).

Crowds on Southsea beach for the 1929 Competition.

was attributed to unmixed fuel being drawn into the engine from the super-charger during slow running before take-off and thus washing lubricant from the cylinder walls. As a precautionary measure, thereafter, it was decided that no engine was to have long periods of slow running prior to the beginning of the contest.) When Orlebar arrived in the morning for the contest, he was not a little surprised to see a group of unknown and tired mechanics on site but he thought it best not to distract Waghorn by telling him what he learned of the work which had only been finished by 6 a.m.

Among the spectators of the competition was the Prince of Wales, who had been flown round the course in a Southampton flying boat, and Lady Astor, who was seen by the Chief of the Air Staff, Sir Hugh Trenchard, talking to Lawrence of Arabia. (Lawrence, having enlisted as Aircraftsman T. E. Shaw, was now act-ing as secretary to the Wing Commander in charge of race organisation. 'Keep your eye on that damned fellow' Trenchard told D'Arcy Greig: Lawrence had embarrassed the RAF, still a relatively fledgling service, by joining the 'other ranks' after his earlier charismatic desert operations. According to Atcherly, he had told Lawrence that 'if I ever hear of you in the Press again Lawrence, I'll sack you'.) The new Labour Prime Minister, Ramsay MacDonald, and the Prince of Wales watched the race from aboard the aircraft carrier HMS *Argus* and other notable spectators were Mitchell's mother, father and brothers and sisters; this was the first Schneider competition to be held in England since his early days with Supermarine and so Mitchell ensured that his family were also given VIP treatment.

Schneider Trophy rules still restricted entries to three aircraft per country and it was announced that they were to fly in the following order: Waghorn in his S.6, Warrant Officer Dal Molin in the M.52, Greig in the S.5, Lieutenant Remo Cadringher in one of the new M.67s, Atcherley in the second S.6, and Lieutenant Geovanni Monti on the second M.67. There was to be a gap of 15 minutes between the take-off times of each competitor – the Schneider event was never organised as a race but as a competition to determine which of the aircraft had the best performance.

The British tactics for the new competition resulted from the fact that the new Gloster machine had had to be withdrawn and the older S.5 substituted and because the unknown M.67s were due to set off after the first British machine. Consideration had also to be paid to the compromise that had had to be worked out with reference to the fuel consumption and engine temperature of the British entries: cooling was so critical that a temperature of 95 degrees had not to be exceeded although the necessary throttling back did allow a nicely judged decrease in the weight of fuel carried. Thus it was decided that Waghorn would fly as fast as possible, consistent with keeping to a safe engine temperature; that Atcherley, in the second S.6 would risk a higher temperature if the performances of the two preceding Italians made it necessary to go faster than Waghorn; and that Greig, in the slower S.5, would provide additional backup.

The Italians' new engines had been refusing to run smoothly at full throttle so it had been decided to leave the new M.67s to the last in the hope that the first two British pilots would overstrain their new engines or run out of fuel for fear of being overtaken – thus allowing the Italians to avoid pushing their own relatively untried engines unnecessarily. In response, Orlebar arranged for Atcherley to delay his start for almost all of the 15 minute gap allowed between competitors so that, if Cadringher went off on time, the British would have nearly half an hour in which to assess the speed of the M.67 and to adjust the performance of the second S.6 accordingly (a similar tactic to that adopted in 1926 by Ferrarin).

Scoreboard showing the results – before the disqualification of Atcherley (No. 8 on board).

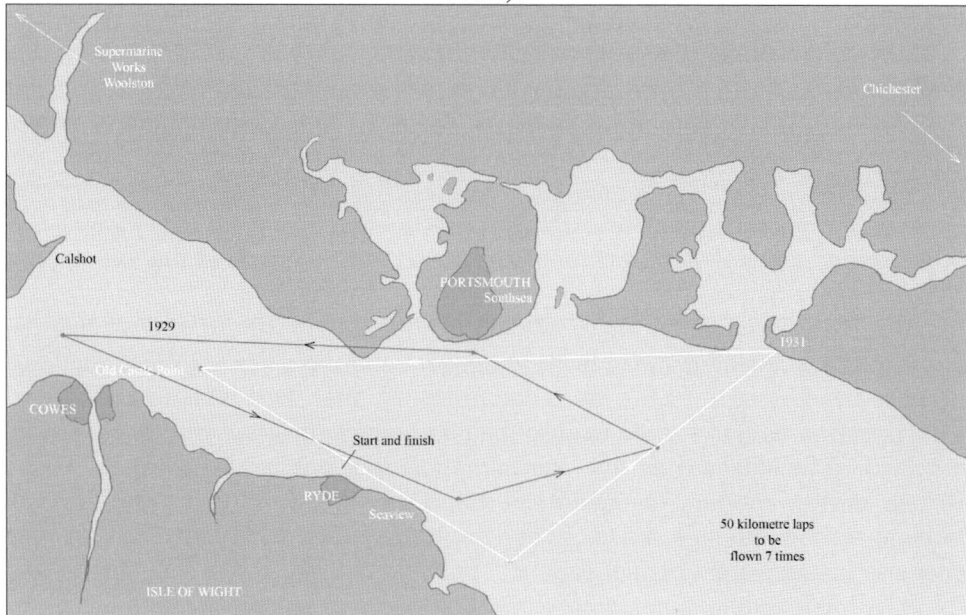

Schneider Trophy Contest Courses: Calshot, 1929 and 1931.

But things did not work out as planned, especially for the Italians. Waghorn began with a disappointing first lap of 324 mph owing to a somewhat erratic flight path as shipping had made it hard for him to pick up a sight of the second pylon. It was, nevertheless, no surprise that he was seen to be faster than Dal Molin, who was timed at 286 mph; this was at least 2 mph faster than Greig whose S.5, surprisingly, was not being overtaken by Cadringher on the first of the new M.67s. But then the Italian (on only his second flight in the machine) retired on lap two as it transpired that he had been nearly blinded by fumes from his exhaust on the windscreen and half suffocated. (The course was a left-hand circuit and the exhaust ports of his central bank of cylinders were on the left-hand side.) It was now necessary for Atcherley to fly faster than Waghorn to be sure of seeing off the last M.67 but he too had a visual problem:

> I thought she was never going to 'unstick', and if it had not so obviously been an occasion of 'now or never' I would have throttled back in an effort to damp the vicious porpoising that had begun ... I believe my take-off took one minute forty seconds ... and I became literally soaked to the waist as she careered blindly through the water in a cloud of spray ... my goggles became opaque with brine from the spray and I found I could no longer wipe them clean with my left glove. I knocked them up from my eyes but the slipstream took charge and snapped them from the strap behind my helmet. Although I carried a spare pair round my neck, I found I could not get these up to my eyes with only one hand to spare.

Flying at over 300 mph and without goggles, he had to tuck down as far as possible behind the windscreen and press on. As a result, he came near to killing not only himself but also Commander Alan Goodfellow the observer at the first turn. He and an Italian official had climbed to the top of the 30 foot pylon which was mounted on an old destroyer between Seaview and Chichester Harbour and he reported that 'as he came rapidly nearer we realised that he was heading straight for the top of the pylon ... At the very last moment he saw us and swerved sharply, passing not more than a wing span inside the pylon.' The moment the incident was over Goodfellow described how the Italian observer 'danced an excited jig on top of our somewhat perilous perch and shouted down to his fellow Italian observer on the deck, "Eliminato, eliminato." He probably doesn't know to this day how near he came to getting my foot in his backside.'

Atcherley carried on and in the process established the fastest lap of the contest at 332 mph. This speed was far better than that of the last Italian, Monti, who was also suffering from fumes in the cockpit; and then his misery was added to by a serious leak in the cooling system which sprayed back steam and nearly boiling water. Fortunately, he managed to get down safely – also on his second lap. Waghorn, meanwhile, had run out of fuel within sight of the finishing line but, like Webster before him in 1927, he had miscalculated the number of laps and had come down on an extra one.

And so, despite the expenditure by both nations on eleven new machines, the 1929 Schneider Trophy was won by one of the only two to complete the course. Waghorn, in the new S.6, was first with an average speed of 328.63 mph – 20 mph more than the current (straight line) Absolute World Speed Record held by a Macchi 52bis; Dal Molin, now using a flatter parabolic cornering technique, was second in the 1927 M.52, at an average speed of 284.2 mph; and Greig, also in an aircraft from the previous contest, averaged 282.11 mph. Atcherley, in the second new S.6 had flown at an average speed of 325.44 mph but his disqualification denied Mitchell the satisfaction of having his two aircraft coming first and second as well as being the only machines designed in 1929 to complete the course. On the other hand, such were the vicissitudes of the Schneider Trophy that the contest might have been won by Italy but for the fortuitous presence of the Rolls-Royce engineers who had worked overnight to get Waghorn's engine ready.

Atcherley was compensated for not being placed in the competition by achieving the World Closed Circuit Speed Records for 50 km. and 100 km. at 332.49 and 331.75 mph. respectively, on his sixth and seventh laps, despite having his goggles removed by the slipstream. Extensive press coverage and eulogy was sustained by a competition to establish a new World Absolute Air Speed Record between the S.6 and the Gloster VI, whose engine problems had now been overcome. The latter achieved 336.3 mph three days after the Schneider contest but Orlebar, as the CO of the High Speed Flight, fittingly took the record with a speed of 355.8 mph. The existing Italian record of 318.62 was further exceeded two days later, on 12 September, when the S.6 reached 357.7

mph. N247 had been the chosen aircraft as it had slightly larger float radiators and, unlike N248, did not have the slight drag penalty of needing a little right rudder to keep it straight.

The S.5 flown by Fl. Lt D'Arcy Greig.

Mitchell and Henry Royce.

As a final comment on to 1929 Contest, a continuation of Waghorn's account of flying the new S.6 gives a fascinating insight into the problems of negotiating the course made unfamiliar by the sudden assemblage of spectators' boats and of the very limited cockpit view from low down behind the engine:

The day was unique, a deep blue sky of a type rarely seen in this country coupled with an amazingly good visibility. At the time it was blowing 10 miles an hour, and all was bustle on the tarmac … At about seven minutes to two my engine was started by Lovesey, the Rolls expert, and was run by him for barely two minutes [to avoid another piston problem – *see* pp. 163–4]. I then climbed in and made myself as comfortable as possible. At two minutes to two I was lowered into the water and started to take-off immediately.

... once off the water I made my way towards Old Castle Point, and then turned left and dived over the starting line at about 350 miles an hour. The pylons were mounted on destroyers and stood out quite well, provided they were not anchored against a background of shipping. One could not get a view directly ahead and I had to pick up the correct line largely while turning the previous pylon. On the long legs we picked our course mainly by landmarks or shipping we had passed over. As an example, the Seaview turn was anchored, say, half a mile from the shore. By plotting our radius of turn on the chart, and from previous practice, we knew that we should have to have the coast, say, five hundred yards on our right. By aiming to do this we would arrive in approximately the correct position; when within about 200 yards off the pylon we could see it, so the actual turn itself was gauged with the pylon in view.

The first lap was naturally the most difficult, because we were not used to the various groups of shipping, which afterwards helped so much on our course keeping. As an example, while passing the Seaview turn on my first lap, I looked for the Chichester turn ship and picked out the only isolated vessel in that area. I made for it, and while still some little way from it, saw the pylon away on my left. I had been quite unable to see it as it had had a background of shipping immediately behind it. The ship which I had mistaken for the turn ship was, in fact, an oil tanker, and should not have been allowed to stray where it had. Atcherley actually turned round it. My own detour cost me six miles an hour, and this is the reason my first lap speed was only 324. From the Chichester turn I could see the Southsea pylon while still turning and had no difficulty at all in passing it, the esplanade on my right being also a great help. Next I came to what was the most difficult leg of the course – that from Southsea to Cowes – as there was no land and practically no shipping to guide one on approaching the turn. To make matters more interesting for the competitors, someone had conveniently parked a Flotilla of Destroyers immediately behind the pylons; hence the amazing turns of some of the Italians embracing all the Destroyers ... Once round the Cowes turn the course was plain sailing again, there being plenty of shipping and the shore of the Isle of Wight to help one.

I had completed several laps, everything was going beautifully – never a miss from the engine, and the machine handling perfectly – when I noticed the Italian Macchi diving towards the starting line just as I was coming up to the Cowes turn; at the Seaview turn I couldn't see him at all; at the Chichester turn I saw him a speck in front, and at the Southsea turn I saw him disappearing over Alverstoke. This time much nearer, and I was obviously overtaking him rapidly, the question was – could I overtake him on the straight before the Cowes turn, or just after? I hoped for the latter, for if I should catch him before the turn I should not be able to see him.

However, it planned out as I hoped, for on rounding the Cowes pylon I saw him just coming out of his turn a few hundred yards in front. I decided to pass him on the inside and swung about a hundred yards to the left to clear him. I passed him about half way down the straight.

By now I had completed five laps and everything was going just as it should. The air in the cockpit was very hot, but owing to a stream of fresh air from a ventilating pipe over my face I wasn't too uncomfortable. An attempt to rest my knees on the sides of the fuselage was abruptly stopped when I discovered that they were, to all intents and purposes, 'red-hot', a slight exaggeration, perhaps, but that is what it felt like, and through my slacks, too! I was flying at about 150 to 200 feet, as I found at that height I got the best view of the course, and it was sufficiently low to be able to keep level. I had been running all the time somewhat below full throttle, as owing to the unexpected increase in power and consequent petrol consumption of the engine, she would not last the course with the petrol we were able to safely carry. The rate that petrol can be poured out of a two-gallon tin will give some idea of the rate the engine was consuming its petrol during the race. I had therefore been told on no account to use full throttle as I shouldn't finish the course; imagine, then my feelings when the engine momentarily cut right out and started missing badly just after I had finished what I imagined was my sixth lap. Would the Rolls engineers ever believe that I hadn't given full throttle? I began to gain height and continued round the course with the engine spluttering and only taking about half throttle. I climbed as much as possible in the hope that should she run right out, I could perhaps glide the remaining distance over the line. I was incidentally getting a very fine 'bird's eye view' of the entire course, but under the circumstances was not impressed. I got to the Cowes turn, and while banking, the engine cut out completely, and I was forced to land off Old Castle Point – only a few miles short of the finish. I leave my feelings to your imagination.

It was twenty minutes later that I learnt I had done an extra lap, and I also realised how deadly accurate had been Lovesay's estimation of the petrol consumption [*the amount had been reduced when it had been decided to run slightly throttled back*].

Enter Lady Houston

By March, 1930, the Air Ministry was aware that government backing was being proposed for French and Italian entries for the next Schneider Trophy contest in 1931. With the possibility of a third win, and therefore the outright capture of the trophy in front of a home crowd, Supermarine and Rolls-Royce began discussions and, in March 1930, wrote to the Air Ministry predicting an increase of 25 mph on the Schneider course, assuming that the S.6s would be loaned back for uprating and that they would be piloted again by High Speed Flight pilots.

At the victory ceremony on the evening of the 1929 win, the Prime Minister had said that Britain would accept any challenge for 1931 that might be forthcoming; however, a cooler appraisal of the cost of a fresh competition, given the pressing problems of a worsening economic situation, resulted in an official statement shortly afterwards in which it was announced that the original aims of Jacques Schneider were no longer being fulfilled and that sufficient data about high speed flight had been accumulated from the previous competitions. The government did not, however, wish to discourage participation in future events 'on the basis of private enterprise'.

The United States had come to such a decision in 1926 and the failure of so many of the specially designed aircraft in 1929 could have done little to help the pro-Schneider lobby. It had been, indeed, fortunate for Mitchell's career that the long life of this international seaplane contest had continued thus far but, when the government had just placed an order for the Hawker Fury as the RAF front-line interceptor, it might be forgiven for questioning whether esoteric *floatplanes* and their competitions were the most obvious or economic path for the development of the nation's military aircraft.

On the other hand, as *Flight* said, 'The importance of the spur which a fixed definite date of such a contest provides cannot easily be overrated' and Mitchell, obviously with a more vested interest in competition work, joined in. In a 1929 article, 'Racing Seaplanes and their Influence on Design', he wrote:

> During the last ten years there has been an almost constant increase in speed of our racing types. To maintain this steady increase very definite progress has been essential year by year. It has been necessary to increase the aerodynamic efficiency and the power-to-weight ratios of our machines; to reduce the consumption and the frontal areas of our engines; to devise new methods of construction; and to develop the use of new materials.

And he put his finger on the essence of the matter (and, as it turned out, on the importance of the Schneider events to the Battle of Britain and the Spitfire) when referring to engine design:

> It is quite safe to say that the engine used in this year's winning S.6 machine ... would have taken at least three times as long to produce under normal processes of development had it not been for the spur of international competition. *There is little doubt that this intensive engine development will have a very pronounced effect on our aircraft during the next few years* [my italics].

More immediately, the winning of the Schneider Trophy outright was tantalisingly close and Mitchell was beginning to look towards a top speed of 400 mph for his next Schneider machine.

Such hopes were dashed by the announcement on 15 January, 1931, that 'no assistance can be given by His Majesty's Government either direct or indirect, whether by the loan of pilots or material, the organisation of the race, or the policing of the course, or in any other way'. It would be hard to imagine a more comprehensive rejection, covering, as it did, all past ways in which the government had helped as well as 'any other' that could be imagined. The response, especially in aviation circles was outrage. The Society of British Aircraft Constructors sent a circular letter to every member of Parliament saying that 'the British victories in the last two contests have given to British aviation and technique, both in aircraft and engines, a prestige in the minds of foreign buyers of aircraft that probably could not have been attained in any other way' and the prominent aviation writer, C. G. Grey made the typically waspish response that 'a Government that will give £80,000 to subsidise a lot of squalling foreigners at Covent Garden and will refuse £80,000 to win the world's greatest advertisement for British aircraft is unworthy of the Nation'.

Also, the *Daily Mail* discovered that the Under-Secretary of State for Air, Fred Montague, was a member of the Magicians' Circle and claimed that 'the disappearance of the Schneider Trophy appears to be one of his most amazing feats'. The Stoke-on-Trent *Evening Sentinel* quoted an interview with Mitchell which reflected the general unhappiness within the aircraft industry, and, particularly, within Supermarine: 'British aircraft today are unquestionably superior to any other aircraft in the world … But if we drop our research work now and allow things to drift, in a year or two's time we may have lost that position'. Even Sir Samuel Hoare, by now the Secretary of State for India, wrote to the *Times* saying, 'Now when every other industry is passing through a period of unprecedented depression, the export of our aircraft to foreign countries, already to be valued in many millions, is steadily rising. This change I mainly assign to the reputation that we have won for ourselves in foreign markets, and that we should not have won to the same degree without the resounding victories in the Schneider Trophy.' The government then made a slight concession by undertaking to help with the provision of service pilots and facilities if 'a definite undertaking is given immediately that the necessary funds will be made available from private sources'.

Whilst the Royal Aero Club had received promises of financial support totalling £22,000, it was clearly unlikely to underwrite the £100,000 they had estimated as the cost of supporting the production of two new and improved machines and the necessary engines to support them. Fortunately, at this point a formidable and extremely wealthy lady steps onto the stage – Lady Houston. It would appear that a prominent flier of the time, and President of the Royal Aeronautical Society since 1927, Colonel the Master of Sempill, who knew Lady Houston well and was one of her favourite Britons, had much to do with engaging her well known championship of matters British. In the event, 'to pre-vent the Socialist Government being spoilsports', she promised £100,000 (based on average working wage figures, about £5 million in 2000) to sponsor Britain's entry – and, incidentally, to try to embarrass the Labour Prime Minister, as her jingoistic press release clearly revealed:

> I am utterly weary of the lie-down-and-kick-me attitude of the Socialist Government. To plead poverty as a reason for objecting to England entering a race against teams supplied by nations much less wealthy than our own is a very poor excuse. To down anything that extols and glorifies the wonderful spirit that even a Labour Government cannot

knock out of we British seems their chief aim. It is down with the Navy, down with the Army, down with the Air Force, down with our supremacy in India – but up with Ghandi, up with strikes which every honest workman detests, the ultimate aim of which is to bring about revolution and ruin and beggary of all in the kingdom. Everyone will soon have to prostrate before every foreigner and cry 'Forgive me for living'. That is why I have guaranteed the money necessary to give England the chance of winning and retaining for ever the Schneider Trophy. I live for England and want to see England always on top.

Lady Houston subsequently felt it necessary to cable the Press Association: 'I have received a telegram saying the Government has insisted on a banker's guarantee being given for the £100,000 promised. This is the sort of insult only a Labour Government could be guilty of, but I am instructing my bankers to do this'. It was unfortunate for the government that this former small part actress (see Appendix Six) was now in a position to embarrass them so effectively and Geoffrey Salmond, soon to become Chief of the Air Staff, kept her in touch with the preparations for the contest. Despite the eccentricity of her receiving Salmond recumbent beneath a Union flag, it was thanks to her that Supermarine and Rolls-Royce were guaranteed the chance for Mitchell to crown his high speed designs with a third consecutive winning machine – perhaps the new aircraft should have had a similar union flag image on its fin, as did the S.4.

S.6B – Schneider Trophy outright winner

By the time that the political points had been scored and the necessary money allocated, there was less than a year left for all the work required in time for the competition in the coming September. Because of this and the finite funds available, the British hopes would have to be limited to uprating the previous Rolls-Royce engine and modifying the S.6 design somewhat to handle the increased power; as a result, the aircraft was to be designated 'S.6B'.

Development of the engine was assisted by interim work that had taken place in connection with Sir Henry Segrave's water speed record boat, *Miss England III*, and also by silent funding: Major Bulman had obtained £25,000 of Lady Houston's gift to cover the engine development for the contest, to which he (later) confessed allocating a further £52,000 from his Air Ministry Engine Vote, under the heading of 'general development work' for the Service.

Far less silent was the actual development work that had to take place; Derby had once more to put up with the noise of engine testing, accompanied by the three Kestrel engines driving fans. The Mayor had to make appeals to the patriotism of its citizens as the tests ran from April the 1st to the 12th of August before the uprated engine could run for a full hour at 3,200 revolutions per minute and full power (exactly one month before the scheduled contest date). The contribution of Rolls-Royce to the winning of the Schneider Trophy can be encapsulated in some statistics: the Buzzard engine, from which the Schneider Trophy 'R' engine was developed, produced 825 hp and this was increased to 1900 hp for the 1929 contest engine – and then to 2350 hp two years later – but, whilst the power had now increased by 24%, the weight had only risen by 7%.

In the course of this development the consumption of oil had again risen – to a rate of 50 gallons per hour – but this was subsequently brought down to a more manageable 14 gallons per hour. Mitchell could accommodate the requisite amount of oil within his still slender fuselage and fin but the increase in power by 450 hp at an additional 300 rpm presented him with even more heat to dissipate and an expected increase in the tendency of the aircraft to swing to the left as the greater torque pushed the opposing float even further down on that side.

New floats were provided with additional radiator surfaces right down to

Supermarine S.6B (1931)

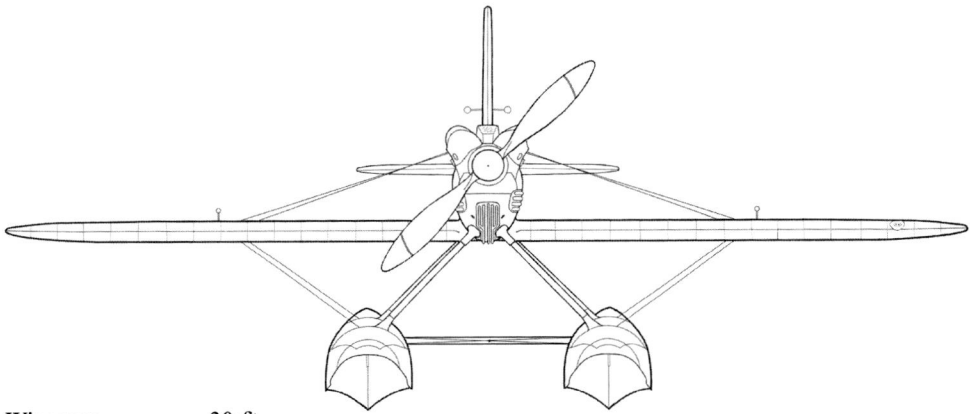

Wingspan	30 ft
Wing area	135 sq ft
Loaded weight	6,086 lb
Maximum speed	407.5 mph

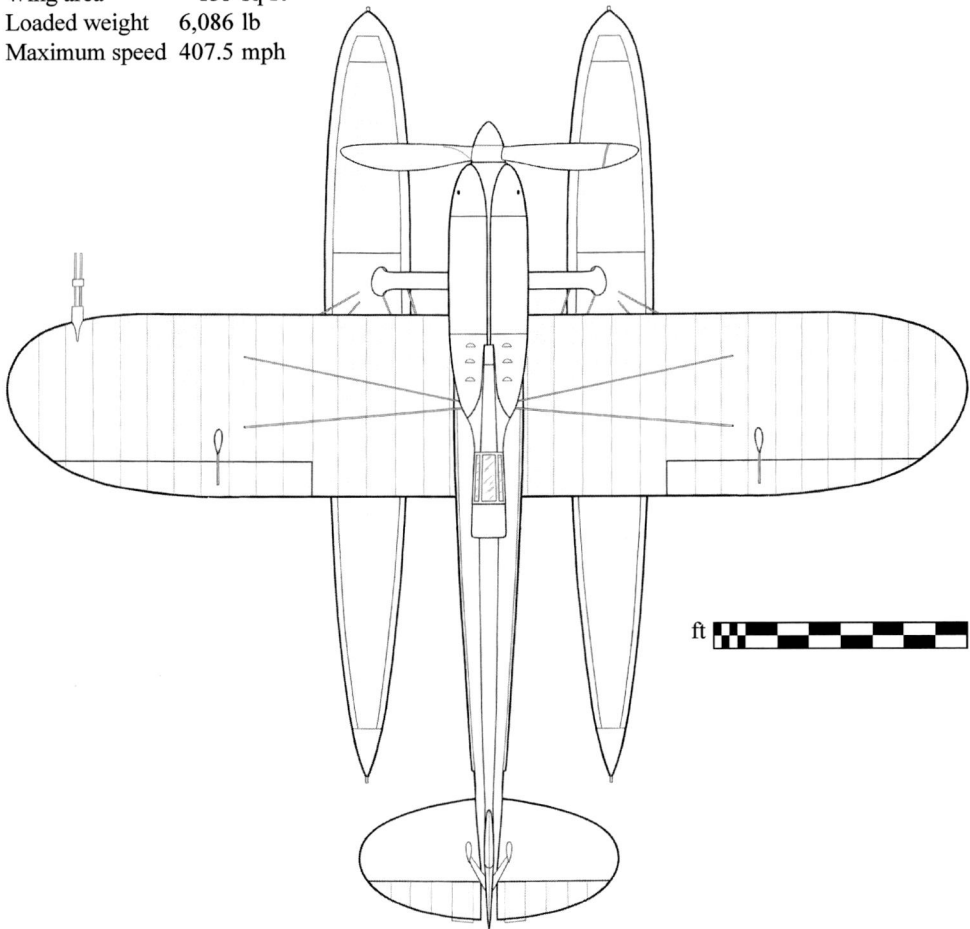

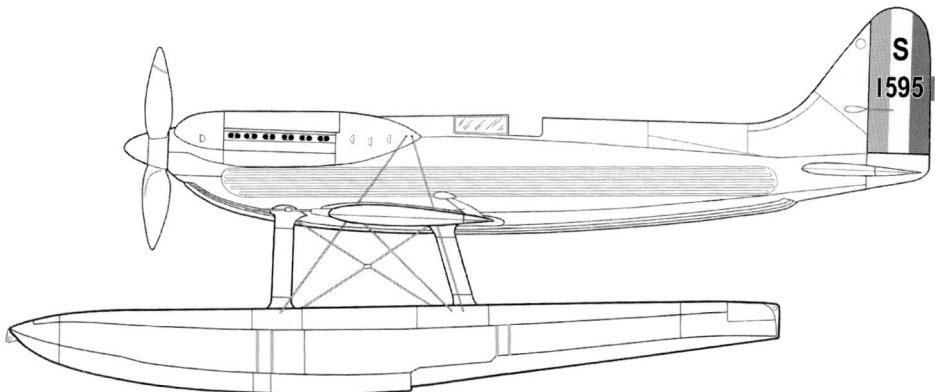

ft

S
1595

their chines so that now no less than 470 sq ft of the 948 sq ft of available surface area was being used for cooling purposes; additionally, experimentation was undertaken to improve the efficiency of the wing tip air scoops that had been used on the S.6 to cool the inside surfaces of the wing radiators. There were now what Supermarine publicity estimated at 'something like 40,000 btu' to be dissipated every minute – the equivalent of 234 modern fan heaters running at full power and so it is understandable why Mitchell, in a radio broadcast after the competition, described the S.6B as a 'flying radiator'.

The need to enlarge the cooling area was at least facilitated by the otherwise unwelcome need to increase the size of the floats, as the increase in the fuel consumption of the new engine required their capacity to be enlarged to accommodate even more fuel, as did a modification to the competition rules. This change was designed to allow the whole contest to be held in one day and so avoid the possibility of bad weather on the navigability test day causing a postponement of the following speed competition: the aircraft were now required to take off and land immediately prior to the start of the race proper instead of the navigability and seaworthiness tests being carried out with minimal fuel the day before.

The floats would now have to withstand landings with the increased weight of oil needed for the new engine and fuel used for the new navigation requirement, as well as the fuller load needed for the actual circuit flying. Extensive wind tunnel testing at the National Physical Laboratory and tank testing at Vickers, nevertheless, led to a narrower float design, albeit of increased length, which reduced drag significantly as well as giving increased fore-and-aft stability on the water. Additionally, the side plating was extended by ¾ in. below the chines and as far as the step to improve their running and to decrease spray. The elevators no longer had a vee cut-out close to the fuselage but were now made to operate with minimum clearance in order to reduce turbulence at this point.

The two S.6Bs built with these various features were given the serials S1595 and S1596 and the two 1929 machines, N247 and N248, were equipped with the new engines and, as such, they were redesignated S.6As. As a result, the only noticeable difference between the S.6A and the S.6B was that the former had its original floats, which were two feet shorter than those of 1931. In anticipation of there being four Supermarine racers, Flying Officer L. S. Snaith was added to the High Speed Flight which had now consisted of Flight Lieutenants J. N. Boothman, E. J. L. Hope, F. W. Long, and G. H. Stainforth (Atcherley, Greig and Waghorn had by now been posted away).

Flying with the new machines began when the first of the S.6As, N247, arrived on May the 20th and almost at once more alterations to the basic S.6 design were found to be necessary, the most potentially serious being that of the unpredictable phenomenon of flutter from control surfaces at high speed. This revealed itself by an alarming oscillation of the rudder during an early high speed run which caused the buckling of rear fuselage plates, stress cracks around some of the rivet holes and stretched control wires. (For Mitchell's reaction, *see* Chapter One, p. 7.)

As there was little time available for fundamental investigation or possible redesign before the contest was due to take place, Mitchell adopted the expedient of placing streamlined weights on forward projecting brackets attached to both sides of the rudder – in order to place the centre of gravity of these surfaces at their hinge lines and so to dampen any oscillations which might occur. For good measure, these bob weights were also added to the ailerons and the last bay of the fuselage was strengthened. All the aircraft were so modified. Orlebar explained the problem of flutter and the solution as follows:

The weight of the movable control surfaces was all behind the hinge, and they, therefore, had a tendency to lag behind any movement, caused by vibration, of the fixed surfaces. Having lagged, they would want to flick over, and this tended to increase the whole movement, so causing the serious flapping which develops into flutter. This accentuating of the movement is avoided if the whole dead weight of the control is equally in front and behind the hinge, when the control only tends to conform to any movement of the fixed surface. Therefore, since the hinge position could not be altered [in the time available], it was necessary to fit horns on to the rudder and ailerons carrying heavy streamlined lead weights well forward of the hinge in order to adjust the balance.

The first of the S.6Bs at Calshot.

Weights were also needed in response to pilots' reports of some instability on take-off and during turns: Mitchell decided that this problem was due to the centre of gravity being too far back and so he had about 25 lb of lead placed in the nose of each float and reduced the amount of oil which was, again, carried all the way back into the fin. However, Orlebar had also reported nose-heaviness during level flight but the proposal to add some backward pressure on the stick by fitting a bungee chord was not favoured by the pilots. Mitchell therefore 'produced a splendid gadget to cure the trouble' – he fitted metal strips, nine inches long and one inch wide, to the trailing edge of the elevators and had them bent downwards by about one degree, thereby utilising the slipstream at high speed to deflect the elevator upwards slightly and to prevent any load on the stick. This principle of trimming had been established by Anton Flettner during World War One and was evident in the servo rudders on the D.H.10 and the Short Singapore I in 1926 but Orlebar might not have come across any examples of the trim tab approach – being regarded as an advanced feature with the Boeing 247, which first flew two years later, in 1933. The CO was certainly impressed by the immediate effect of Mitchell's modification: 'I was able to take my hand and feet off the controls at about 330 mph and the machine carried on straight ahead perfectly happily. It was an extraordinarily good shot to get her so exactly right the first time.'

A third problem to emerge was more worrying – the first new S.6B, which had arrived on July 21st, would not get into the air at all as it gyrated in 'a very good imitation of a kitten chasing its tail', as Orlebar said. During the course of trying to overcome these rotations, S1595 damaged a wing radiator by fouling a barge and had to go back from Calshot to Supermarine for repair. In the

meantime, its smaller 8 ft. 8 in. diameter propeller was fitted to N247 which
then also obliged by simply gyrating on (and in) the water, thus pointing to the
problem: the slipstream from this sized propeller would appear to be creating
a side pressure that the rudder was unable to counteract. The wider thrust
'cylinder' from a larger propeller might therefore take some of this pressure off
the fin and rudder. A 1929 propeller which was 9 ft 6 in. in diameter was fitted to
S1595 when it was returned on July 29th and the new machine then took off with
no more difficulty than was usual (*see* below).

'Mitch' with (l. to r.) Fl. Lts Snaith, Stainforth and Hope and Lt Brinton.

S1595, moored off the Woolston slipway, after the fitting of bob weights.

Three of the six airscrews ordered from Fairey Reed for the S.6B had been finished to the smaller size and had therefore to be abandoned but, as an expedient, it was possible to beat out the ends of the remaining blanks to produce 9 ft 1½ in. diameter propellers whilst the 1929 sized ones were also reused. With the S.6, a reduction in the pitch of the propeller had assisted take-offs; now, an increase in propeller diameter had the desired effect but, again, the effect in flight was an increase in fuel consumption and a higher engine temperature. Again, Mitchell had to accept a slightly lower airspeed than his design was capable of.

Meanwhile, the uprated competition engines were prone to cutting out because of choked fuel filters. This was found to be the result of the fuel mixture (25% benzole, 74.78% aviation petrol and 0.22% tetraethyl lead) causing the excess compound used to seal the joints in the fuel system to come adrift. Mitchell's response was both practical and blunt: 'You'll just have to bloody-well fly them until all this stuff comes out'.

Fl. Lt Long returning from a test run.

Whilst trim problems had been overcome, propellers matched to engines, and fuel lines were being cleared of excess sealant, take-offs were still far from being, literally, a 'racing certainty'. The problem had been described by Waghorn in 1929:

> Owing to the slow revs of engine and propeller, coupled with the great power and consequent torque effect, the first thing that happened on opening up the engine was that the left wing tried to dig itself into the water. This almost submerged the left float, and the drag so produced swung the machine rapidly to the left, making her quite uncontrollable; the more the machine swung to the left of the wind the more rapid did the swing become, until centrifugal force became greater than the drag of the left float, and she would suddenly throw her right wing down rather violently making it essential to shut off the engine. With a fairly fresh wind and full load it is advisable to take-off directly into wind, and with that end in view we found it essential to point the machine about 70 degrees to the right of the wind and to have right rudder on from the start. The machine then runs along with its left wing a few inches from the water across wind, but not swinging. Having got her therefore running across the wind at 40–50 mph, one is now confronted with what is really the trickiest part of the proceedings, and that is to get her into the wind without letting her swing right round, which she will want to

do; once left rudder is applied the machine will accelerate rapidly; provided you have not put on too much rudder she should reach her hump speed by the time she is directly into the wind. At this point she assumes a new position on the water – very much lower in front – and accelerates rapidly up to taking-off speed. She seems to leave the water at about 100 miles an hour.

With the much increased power of the new engines, this problem now called for even greater vigilance and judgement. There was also a tendency still for the noses of the floats to dig in, causing the aircraft to porpoise, and this had to be resisted by holding the control column right back, contrary to all the training pilots are given for normal take-offs. Snaith described how these procedures would cause the S.6 to quite suddenly 'leap off the water and into the air at a pronounced angle and in a partially stalled condition, virtually hanging on its propeller ... The whole manœuvre was complicated because many a time we had to take off blind, our goggles being misted up or covered with water'.

S1596 taking off amidst the usual spray.

Setting aside any speculations about these pilots' belief in their own immortality, one must undoubtedly admire their skill and 'press on' attitude, sitting behind a 2350 hp, engine, keeping the head well down to shield the goggles from the worst of the spray until the aircraft got on to the step, and perhaps swinging the nose of the machine slightly from side to side to get a view ahead; all this in order to make the nice calculation of aiming to the right of the wind in order to be pulled by the propeller torque straight into wind by the time take-off speed was reached.

As mentioned above, it was also necessary to be alert to any development of porpoising as the aircraft came onto the step and developed a flatter running angle; when flying speed was being reached, it was then imperative to hold the stick unnaturally well back in order to get well clear of the water and to keep it there – but bearing in mind that, with marginal lift, it would be very easy to stall, fall back and possibly crash. If the nose attempted to come up too high, the stick could be eased forward slightly.

Blacking out at very low altitudes during the high speed turns had also to be contended with and, at the other end of a flight, there was the problem of landing, possibly when there was a swell running, with the nose of the aircraft increasingly impeding forward vision as the angle of attack was increased to maintain lift as the airspeed dropped; a long flat approach was necessary, touching down on the heels of the floats. (Is it adequately appreciated that the British success in the Battle of Britain and later owes much to the highly skilful but dangerous developmental work undertaken by the High Speed pilots on these racers?) One thinks particularly of the first group of High Speed pilots who had experienced the initial dramatic leap in the performance of their aircraft: Schofield had reported how he had had to learn to land the Gloster at over 130 mph – only about 20 mph slower than the top speed of current RAF fighters – and feel his way on to the the surface 'without seeing it at all'. He found the Supermarine monoplane design slightly better but he soon discovered that it was not possible to help matters by looking down the side of the machine because of the 'searing wave of flame' coming back from the exhausts.

Supermarine publicity gave particularly detailed and interesting accounts of some of the design and structural considerations:

The problem of supplying enough cooling surface water and oil was one that presented the greatest difficulty.

In the 1929 Schneider Trophy Contest the two S.6s were flown with the engine slightly throttled down, because there was not enough radiator surface on the machine to cool the engine when running at full throttle. In that Contest the Rolls-Royce R engine gave an output of 1,900 hp, and for this year's Contest the power output of the new R engine was increased to 2,350.

In the S.6B the entire upper and lower surfaces of the wings and the upper surfaces of the floats constitute water radiators.

Some idea of the difficulties confronted by the designers can be gained if one realises that to keep the engine running at normal temperatures something like 40,000 BT units of heat must be dissipated each minute from the water and oil cooling surfaces, which is equivalent to approximately 1,000 hp in heat loss from these surfaces.

The new floats have greatly increased aero and hydrodynamic qualities, and the starboard float carries considerably more fuel than the port float, the differences in load balancing the tremendous turning moment of the engine, particularly during the take-off. Full engine torque has the effect of transferring a load of approximately 500 lbs from one float to the other.

The construction of the new floats was complicated by the necessity for fitting water-cooling surfaces on the whole of the upper surfaces. When filled with water from the engine at a temperature near boiling point, the radiators expand nearly half an inch, and to prevent buckling of the outer skin, the designer had to incorporate an ingenious elastic framework to take up this expansion. It was also found essential to insulate the fuel tanks from the water-cooling surface to prevent evaporation of the petrol.

Like the floats of the S.6, they carry all the petrol in steel tanks which are built as part of the floats. The fuel is forced by engine-driven pumps to a small pressure-tank in the fuselage which feeds direct to the engine. On steeply-banked turns the sudden application of centrifugal loads, equal to 5 or 6 G, prevents the pumps from working, and the small pressure tank carries just enough fuel to keep the engine running during each turn. Immediately the turn is concluded the pumps begin to operate again and the pressure-tank is replenished.

Whilst Schofield seems to have made more of the new landing problems than, for instance, Greig, it is notable that the British team had now begun to have its accidents. Mitchell had been required to produce the fastest aircraft in the world but he had watch them fly virtually at sea level, often in hazy conditions, and taking-off and landing without the aid of flaps or variable pitch propellers, amongst busy shipping lanes. After Kinkead's death in 1928, Flight Lieutenant. E. J. Linton Hope had virtually written off one of the S.6A machines. A piece of the cowling from N248 worked loose in flight and, as he was managing a successful emergency landing, the wash from a passing ship caused the sensitive machine to cartwheel and sink in fifty feet of water. The pilot survived but was withdrawn from the team because of a punctured eardrum. (He was the son of the influential hull designer mentioned in Chapter Two; as Group Captain E. J. Linton Hope, AFC, he was killed in action in August, 1941.)

He was replaced by Lieutenant G. L. Brinton. On 18 August, on his first take-off in N247, the second back-up machine, he apparently seemed not to get the take-off technique right; at a height of about ten feet, it would appear that he had pushed the stick forward and the machine hit the water, bounced back at a sharper angle of attack, hit the water again, and then, from about thirty feet, dived in – like the pilot of the Sea Lion III before him. D'Arcy Greig explained the circumstance as follows: 'If the machine developed fore and aft pitching motions, known as porpoising, during any stage of take-off, there was only one sure remedy – close the throttle, slow down and taxi back for another attempt. Any other form of remedial action or an attempt to carry on, was only courting disaster, for in these circumstances the amplitude of the pitching invariably increased until the machine was eventually thrown into the air without flying speed. If this happened, a crash was almost a certainty.' Brinton's inevitable crash was described by Penrose: the machine 'bounced, dropped back, bounced again, stalled its starboard wing, tilted steeply and struck the water hard. The floats were torn off … and young Brinton was jammed in the fuselage and drowned'. *Time* magazine reported that

> a young lieutenant last week climbed into the cockpit of the Supermarine S.6, which won the Schneider Cup two years ago. He was Lt Gerald Lewis Brinton, 26, youngest member of the British Schneider Cup team. It was his first flight in the S.6. The plane slid along the surface of the Solent until it was going about 200 mph. It cleared the water for a second and then dropped back to it. A tower of spray shot up. The S.6 bounced 40 feet in the air and then plunged down into the Solent, nose first. When Lt Brinton's fellow officers reached the ship in a speedboat, it had risen again, upside down, with wings and tail torn off. The wreckage was towed ashore and the dead body of Lt Brinton removed from the tail of the fuselage, where the shock had wedged it.

The writer of this slightly dramatised account did not elaborate on the fate of Brinton's body: it was first assumed to have been lost at sea and only later was it found jammed into the rear of the fuselage. It is not recorded how the actual discovery of the body was received by Mitchell but, in view of his well-known concern for the pilots of his machines, an explanation of the need to cut into the damaged machine for its recovery must have required considerable tact.

At about the same time, a French and an Italian pilot were killed while practising with their teams. The French entries were to be, essentially, uprated versions of the 1927 machines which did not compete that year and which were suffering from stability problems because their wing areas had now been reduced in order to increase their competitiveness. One machine was considerably damaged in a landing accident and, on 30 July, another was completely destroyed, killing its pilot. Meanwhile, Macchi were developing their M.67 layout into a new machine which was also to kill one of its pilots. Prior to this, on 18 January, the Italians had responded to the disappointment of the last competition with an attempt on the World Speed Record. Unfortunately, the chosen Savoia S.65 plunged into Lake Garda, killing its pilot, Dal Molin, who had come second in 1929. It was suspected that he had been overcome by fumes.

The main feature of the new Macchi was, like that of the S.6B, the incorporation of a more powerful power plant into an existing airframe but the Italian design also featured contra-rotating propellers. This bold approach to the elimination of torque problems on take-off was to be achieved by, essentially, bolting two Fiat engines in tandem and by driving separate propellers via individual reduction gearing to coaxial shafts. Extra speed would also be gained by not having to rig the wings slightly out of true and by not needing such large floats to counteract torque. Since the demise of the S.65, Italy had to pin all its hopes on this new Macchi M.72 which, with the 2,500 hp now to be available to it for only 2050 lb dry weight, was looking extremely promising. The surface cooling arrangements were very similar to those of the S.6s, except that the float struts had a very broad chord which could also be utilised for heat dissipation.

Not surprisingly, the revolutionary engine was plagued with problems, especially carburation, and during a fly-past to demonstrate its erratic behaviour to those on the ground, Monti, another 1927 contest pilot, crashed and died in unsolved circumstances, but probably a stall. This accident took place on 2 August, at which time the French modified planes were still not ready and the remaining pilots were also in considerable need of experience of high-speed flight in contest machines.

As a result of such accidents and other setbacks to the French and Italian teams, a joint request for a postponement of at least six months was received on 3 September by the Royal Aero Club and this presented a difficult decision. By this time, it was felt that all the significant problems with the S.6Bs had been solved and Hope's S.6A had been salvaged and was well on the way to being restored to flying condition (it flew again on 6 September). On the other hand, bad weather had prevented any practice in the competition machines since August the 26th and so the risk of being barely prepared to compete on the due date of September the 12th had to be weighed against the hope of winning by a possible flyover which would immediately ensure permanent possession of the Schneider Trophy for Britain. It had also to be borne in mind that finance for future contests was extremely uncertain and there was, additionally, a very strong obligation to ensure that Lady Houston's generosity was not wasted. These considerations were all evident in the response of Harold Perrin, the secretary, who wrote: '… my Committee has decided with regret that it is impossible to accede to your request. We took into consideration that only nine days remain before the appointed date, that the elaborate preparations are virtually complete, and that very large expenditure has been incurred by all concerned, including many local authorities and private interests [i.e. Lady Houston whose £100,000 had, obviously, now been spent].'

As a result, the Air Ministry was informed on 5 September that neither France nor Italy would be able to compete. (Italy attempted to upstage what looked like an inevitable win for Britain by going for a new world-speed record in the M.72 which, when it functioned properly, promised to approach the magic 400 mph mark. However, after a few successful runs, it was opened up again and then flew into rising ground on far side of Lake Garda, killing yet another pilot. A study of the remains later suggested that the engine had exploded.)

The Final Schneider Trophy Contest, 13 September, 1931, at Calshot
Despite the withdrawal of the opposition, there was a last minute panic in the Supermarine camp which Maj. Bulman, the Air Ministry official, reported as particularly affecting Mitchell, already over-anxious and overworked:

> The night before the Contest, nerves were taut … Mitchell and I walked down to the sheds on Calshot Spit from the mess to see the engines doing their final run … Jimmy Ellor of Rolls tore up to us quite distraught to announce that the wing radiators were stone cold, 'the system wasn't working' … The effect on poor Reggie was appalling. Half-crying he spluttered, that 'This was the end', he'd been a fool to go ahead with the wing-cooling idiocy, we'd be the laughing stock of the world, he wished he were dead. For a moment I thought he was going to throw himself into the sea. I grabbed him by the arm and yelled, 'Hold on you ass. Jimmy has gone berserk. It's all right I tell you. Everything is fine'. By the time we had got down to the Hard where the S.6s were gleaming under the floodlights, the engines ticking over and then gradually

opened up. No panic. Radiators all fine! What had persuaded Jimmy that all was not well I never found out. But it nearly killed Reg.

And so, in the end, the only postponement of the twelfth contest was a not unfamiliar delay of one day owing to bad weather. The appointed day was squally with rain and the sea was running with an unpromising lop, but initially the Clerk of the Course, Col. W. A. Bristow, was not for cancelling. Bulman has described how Orlebar once again displayed his leadership:

> Orly gave me a terrific wink as they climbed in [the Fairey floatplane hack] and I guessed his intent. They came back twenty minutes later. The rotund and normally rather pompous Clerk looked grey and was very wet. 'Quite impossible,' he said. 'Out of the question. Race postponed.' Orly had persuaded him of his perfectly correct decision by a superb measure of piloting, skimming over the wave tops and occasionally dipping his float tips into them to produce a shower over the cockpit.

In contrast, the following day was almost perfect with visibility of over ten miles. In view of the flyover situation, it was decided that the first S.6B, S1595, was to complete the course without putting undue strain on the engine or airframe – especially as Mitchell's valiant efforts to prevent engine overheating could not quite cope with prolonged full throttle flying. If this attempt were to fail, then the repaired S.6A, N248, would aim to finish the course and therefore to win the Trophy outright; with less radiator area on its shorter floats, it had to be flown slower than the others, also to avoid engine overheating, but it still ought to win the trophy at a very respectable speed, if called upon. The second S.6B, S1596, would be available to make trebly sure of a win but it was hoped that it could be retained instead to give the crowds the additional thrill of seeing the setting of a new World Speed Record. Orlebar gave the senior pilot, Stainforth, first choice and he opted for the proposed attempt on the speed record; the next most senior man, Boothman, then opted to fly first in the competition itself and, hopefully, to have the honour of winning the trophy.

And so, just before 1 p.m. on the 13th of September, Fl. Lt Boothman taxied out in S1595, the machine never having flown in practice for longer than 27 minutes. Nor had it been considered wise to practice the landing, with all the necessary fuel for the 350 kms of the competition course and for the required preliminary landing, as the 1931 rules required. Nevertheless, he took off without any apparent difficulty, landed at about 160 mph without mishap, and then took off again after a period of less than two minutes; he then flew the prescribed seven laps, all within about four mph of each other, slightly throttled back, taking the turns wide and with a gentle bank and finished with an average speed of 340.08 mph – just over 11 mph faster than Waghorn in 1929. Then, as if to em-

Plaque to Lady Houston at Calshot.

phasise the superiority of the Rolls-Royce/Supermarine partnership and also to post a more impressive speed, Stainforth (who in 1929 had unluckily been chosen to fly the Gloster which was withdrawn) took out the other S.6B a little later and proceeded to capture the World Absolute Air Speed Record at 379.05 mph.

Lady Houston had come over in her steam yacht, *Liberty*, to watch *her* machines flying and two days later gave a celebratory lunch on board, which was attended by Mitchell and his wife and by the High Speed Flight. Cozens gives the following anecdote relating to Lady Houston: 'she was afforded the rare privilege of mooring her yacht *Liberty* on the RAF buoys inside Calshot … In the evenings the *Liberty* had a string of electric lights from her bowsprit to the mastheads and down to the stern, and this seemed to add just the right touch to the celebration of victory.'

The Air Ministry then set about disbanding the High Speed Flight and restoring the Calshot base to its normal flying boat duties (the resident squadron having been relocated to Stranraer). However, Rolls-Royce particularly wanted to have produced the first aero engine to exceed the new magic mark of 400 mph. Mitchell had indicated the same in an interview with the Southampton *Daily Echo*, after the competition, when he said that 'with a specially tuned up engine, I am very hopeful we may get very near to an average speed of 400 mph, which is our ambition'.

This crowning success of his S.6B was achieved after the intercession of Sir Henry Royce (who had been knighted in 1930 for his services to the aircraft industry). For this special sprint machine, Mitchell had the wing-tip air scoops removed and a specially prepared engine was supplied – with a new fuel mixture of 30% benzole, 60% methanol and 10% acetone plus a 5cc/gal. tetraethyl-lead solution, which required the wearing of goggles when filling up the fuel tanks. To absorb the increased power of the new engine, a 9 ft 6 in. propeller was fitted.

On 16 September, during a test flight, Flight Lieutenant Stainforth lost control of S1596 when the heel of his shoe jammed under the rudder bar during the landing and so another S.6 went under the waves. After delays caused by bad weather, Stainforth squeezed into the cockpit of S1595, fitted with the special sprint engine that had fortunately not yet been installed in S1596, and the required four runs were photographically measured. There was some concern that bad light and a low evening sun might prevent confirmation and that a rerun, which would necessitate the engine going back to Derby for inspection, might not be allowed in view of the continued disruption of normal RAF duties at Calshot – it was now thirteen days since Stainforth's ducking. But eventually, at 4 a.m., the results were telephoned through and Mitchell was informed; he was 'too sleepy to be more than mildly enthusiastic' that the World Absolute Speed Record had just been raised by nearly 30 mph to 407.5 mph. (From February to September 1932, the outright land, water and air speed records were all held by Rolls-Royce R powered machines.)

* * * * *

By way of a postscript to these last successes, a review of the top speeds and different records of Mitchell's Schneider trophy racers indicates the rate of aircraft and engine development spearheaded by Mitchell in one formative decade:

1922 Sea Lion II	129.66 mph (1st World Record for Maritime Aircraft)
1923 Sea Lion III	157.17 mph
1925 S.4	226.75 mph (World Speed Record for Seaplanes)
1928 S.5	319.57 mph
1931 S.6B	407.5 mph (World Absolute Air Speed Record)

These entries can also be seen to chart Mitchell's gradual rise to public notice: the performances of the S.4 and S.5 revealed that the Sea Lion II performance was not just a flash in the pan and being elected a Fellow of the Royal Aeronautical Society in 1929 was an acknowledgement by his fellow professionals of his contributions to advanced aviation technology.

In an article in the Aeronautical Supplement to *The Aeroplane*, Mitchell gave a remarkable set of figures which showed how his engineering skill and attention to the detail of cooling his Schneider racers contributed to their success:

Radiator type	Externally corrugated brass	Externally flat brass	Externally flat dural	Externally flat dural with 12% internal cooling
Aircraft:	Curtiss CR-3	S.5	S.6	S.6B
Weight of radiator per hp dissipated	300	410	75	67
hp dissipated per unit of cooling surface	50	66	81	92
Resistance per hp dissipated	15	0	0	0

An invitation to give a talk on the BBC in 1932 indicated recognition by the wider public of his technical achievements; and in this broadcast he explained in layman's terms the broad design problems that 'the designer' had had to over-come and expressed his admiration for 'the great courage and great skill' of the pilots of the High Speed Flight. It was a typically self-effacing speech despite his appearing in the New Year's Honours List of that year. In view of Supermarine's current lack of success in supplying aircraft to Imperial Airways (*see* p. 187 ff.), Mitchell might have been permitted a wry smile on reading the official Honours List letter from King George V, who was entitled 'of the British Dominions beyond the Seas, Emperor of India and Sovereign of the Most Excellent Order of the British Empire'. Mitchell particularly disliked having to wear bows on his Court shoes but it was surely impossible for one who had designed so many aircraft for the British armed forces not to accept becoming a Commander of the Order of the British Empire. He was still only thirty-six years old.

* * * * *

As Mitchell always gave full credit to others in his speeches, there is no reason to suppose that, in this chapter concerning his notable Schneider Trophy successes, he would begrudge a final word about the pilots concerned. Various instances of these pilots' skill and courage are to be found in the preceding pages, including flying and alighting at unprecedented speeds, in machines with extremely limited vision, and at altitudes that gave little margins for error and little possibility of survival if things went badly wrong.

Undoubtedly, other pilots would have accepted the challenges of flying beyond the boundaries of previous experience but it was those mentioned above to whom credit must go; if these pilots had not successfully flown their, frankly, dangerous aircraft, the Spitfire might not have been ready in time for the Battle of Britain. One notes at least D'Arcy Greig's dedication in *My Golden Flying Years* 'to all those involved with the Schneider Trophy races that helped so much in the development of the Spitfire in later years'.

Supermarine publicity poster.

A Walrus awaiting retrieval by warship. (One crew member is well outboard on a wing to prevent aircraft swinging into ship; the other is preparing to attach the hook of the ship's crane to the aircraft centre-section – with a safety line to prevent a slip into the aircraft's propeller.)

— *Chapter Seven* —

1930 to 1933
Mitchell's
Last Seaplanes

Although the three consecutive Schneider wins had made the standing of Supermarine in the aviation world an enviable one, success in the world of the larger military and civil flying boats was far less secure and far more subject to the direction (or misdirection) of the Air Ministry. The eventual outcome was that, despite the Far East Cruise of the Southampton IIs, the excellent squadron service of this type, and its continuous development, Supermarine was not able to capitalise on these successes in the civil aviation field. In the mainly flying-boat operations of the developing Imperial Airways Empire routes, Shorts were to sweep the board. An early warning sign was that, at almost the identical time of the Far East cruise, Sir Alan Cobham had used the prototype of the Short Singapore, not a Supermarine aircraft, for his 20,000 mile Africa survey. This expedition, which must rank as one of the greatest of early aerial survey flights, went clockwise via Malta, the Nile and the Great Lakes, round the coastline via Durban and Cape Town and returned along the western seaboard between 17 November 1927, and 31 May 1928. Also in 1928, the Short Calcutta replaced the Supermarine Swan on the Channel Isles service.

In view of Great Britain's strong maritime concerns and its vast Empire across the oceans, it is not surprising that other aircraft companies had wished to enter the flying-boat field that was the especial concern of Supermarine. Shorts, especially, put up with many setbacks after their move from floatplanes to flying boats. This development commenced with the Cromarty, begun in 1918 and finally completed in 1921. This aircraft did not secure a government contract nor did their revolutionary Silver Streak landplane of 1920, the first British aeroplane to be built entirely of metal and with the first all-metal stressed skin fuselage. However, the Air Ministry was far-sighted enough (at least in respect of hull design) to see the advantages of such metal structures after the water soakage problems of the World War One Felixstowe flying boat. Accordingly, they contracted Shorts to make a metal hull for a Felixstowe 5 superstructure at the very time that the first orders for the wooden Supermarine Southampton I were being contemplated. The resultant Short S.2 first flew in 1924 but was soon written off when it stalled into rough water from about 30 feet. At least the strength of its hull was demonstrated by the fact that the hull was virtually undamaged and remained watertight.

There then followed Ministry orders for both the N218 Southampton, whose metal hull was the prototype for the Southampton II, and for the all-metal Short S.5 Singapore I. Again, Shorts did not get a production order but it was this latter aircraft which was loaned to Cobham by the Air Council for his portentous Africa survey flight. Just before this, on the 12th of August, 1927, the Singapore had joined a Southampton in a 9,400 mile Baltic cruise, accompanied by aircraft from two other companies which had by now entered the flying boat field.

Montage of publicity from rival flying boat manufacturers.

One machine was the Saunders A.3 Valkyrie which, like the Short Singapore, had a wingspan about twenty feet greater than the Supermarine aircraft. However, its all-wooden construction, completed at the time of the change to metal-hulled Southamptons, was an important factor in no production orders being issued for this aircraft. But it did mark the employment of three engines, as did the other participant in the Baltic trip – the 95 foot wingspan, metal-hulled Blackburn Iris II. It was faster than the Southampton II, could carry a heavier load and had a tail gunner's position – not standard in flying boats before 1930, and only appearing in a production Supermarine machine with the Stranraer of 1934. Ominously for Mitchell's company, it was chosen as the flagship of the Baltic cruise and accordingly carried Sir Samuel Hoare, in his second term as Secretary of State for Air, on his visit to the Aero Exhibition in Copenhagen.

Thus, after the production and development of Supermarine's Southamptons had peaked by 1928, there was a significant falling off in their flying-boat activity until the appearance of the Scapa in 1932 (*see* below). Meanwhile, in 1928, Shorts developed a civil version of the Singapore, the Calcutta, and later, the Kent, which supplied the flying-boat components of the Imperial Airways fleet and, in the military field, their Rangoon of 1930 gave way to their Singapore III, of which thirty-three were built.

The Solent/Nanok design had been an early Supermarine response to the rival three-engined types beginning to appear but this order was cancelled – perhaps partly as a result of comparison with the performance of the new British types on the Baltic cruise. After this cancellation and in view of the new standards of the Iris II, the Air Ministry encouraged Supermarine to 'stretch' the Southampton in the search for larger, more powerful types to compete for the eventual replacement of the standard Southampton equipment. The result was the Saunders-Roe A.17 and the Southampton X.

The first order, N251, was to be an aircraft which mated a Southampton superstructure to a hull which Saunders wanted to develop as a successor to the wooden Valkyrie type. The new metal hull used corrugated panels for the outer skin instead of stringers and was assembled at the Cowes factory now owned by Saunders-Roe Ltd. N251 was given the newly formed company's type number A.14 and features no more in the Supermarine story – except perhaps for its leading to the Saunders-Roe A.27 London which later competed successfully for orders against Mitchell's Stranraer in 1934.

The second order, N252, produced an aircraft which, despite being design-nated a Mark X version of the Southampton, was quite unlike earlier marks. In fact, it was very similar in appearance to the next Saunders-Roe flying boat, the A.7: both were sesquiplanes and had three engines and twin fins. In particular, Mitchell virtually abandoned the Southampton approach to hull design and util-ised the external horizontal corrugations of the A.14 hull, thus avoiding the need for internal stringers and notched ribs. He also specified stainless steel sheeting below the waterline.

The Mark X notably had a position for a rear gunner in the tail, a first for Supermarine, although it did not become a production machine. Other departures from the Southampton II were the shapes of the tail surfaces and of the floats which were larger, in order to accommodate extra fuel – something learned from the Schneider Trophy machines, perhaps. The superstructure was also a depart-ure, literally, in that its construction was to be the responsibility of the parent firm at Weybridge.

Flight testing began in March, 1930, no longer by the long-serving Biard but by Capt. J. 'Mutt' Summers, Vickers' chief test pilot – an early consequence of the Vickers takeover. The Mark X turned out to be over 300 lbs above its

estimated weight, no doubt largely because of the complicated engine mountings
(Vickers) and an underestimate of the weight of the stainless steel plating (Supermarine). Thus, not surprisingly, it performed below expectations: the estimated
15,000 ft ceiling was found to have been extremely optimistic and its maximum
speed to be 15 mph below estimate. Its original Armstrong Siddeley Jaguar VIC
engines were then exchanged for other engine combinations and, with Bristol
Jupiter XFBMs, attained a ceiling of 11,800 feet and a top speed within 5 mph of
its specification.

Supermarine publicity montage for the Southampton X.

There were also changes to the strutting, the engines were given Townend
drag-reducing ring cowlings, smaller floats were employed and the flight deck
was enclosed. At 130 mph, it was now faster than the three-engined Saunders-
Roe A.7 Severn, the Short Rangoon and the Blackburn Iris III. But signs were
not good for Supermarine: the Short S.15, built for the Japanese Government in
the same year could achieve 136 mph and it was a Saunders A.7 Severn which
was sent, also in 1930, on a 6,530 mile proving flight to the Middle East and

back; nor was it accompanied by the Southampton X but by the faster and more powerful Short Singapore II whose four engines gave it a top speed of 140 mph.

Perhaps because both the Supermarine and the Vickers contributions to the Southampton X came out overweight and in view of earlier tensions in the Supermarine design office mentioned in the Introduction, no further joint projects were initiated; Alan Clifton's report that Supermarine's next, individual, attempt to improve on the Southampton, the Scapa, came out 'bang on the weight target' might therefore be seen to contain a certain amount of self-satisfaction. And Webb was no doubt reflecting general company morale at the time when he noted that 'since Vickers took over in 1928 the only successful aircraft produced by us had been the S.6 Schneider seaplanes'; he also wrote that 'we naturally considered that we built rather better aircraft than Vickers and the idea of being bossed about by them did not appeal to us at all. A rather crude joke went round the workshops in the form of a question: "Why are we like a bunch of choirboys? Because we're being buggered by Vickers!"'

By the end of 1929, Supermarine had several flying-boat projects under consideration: a Southampton replacement, a civil version called the Sea Hawk, an Air Yacht – all with three engines, a four-engined civil project, and the six-engined 'Giant'. Of these, only the Southampton replacement and the Air Yacht were completed.

AIR YACHT – the Dornier Influence

The aircraft which later became known as the Air Yacht began as Air Ministry specification 4/27, calling for an armed reconnaissance flying boat, larger than the Seamew. The response was originally drawn up, between the 1927 and 1929 Schneider Trophy activity, as a biplane and with the previous Southampton X provision of bow, midships and rear gunner positions; however, in late 1927, Mitchell produced a monoplane design which featured, for the first time, sponsons instead of wingtip floats –the first indication of the influence of Dornier thinking and practice (*see* p. 193 ff.):

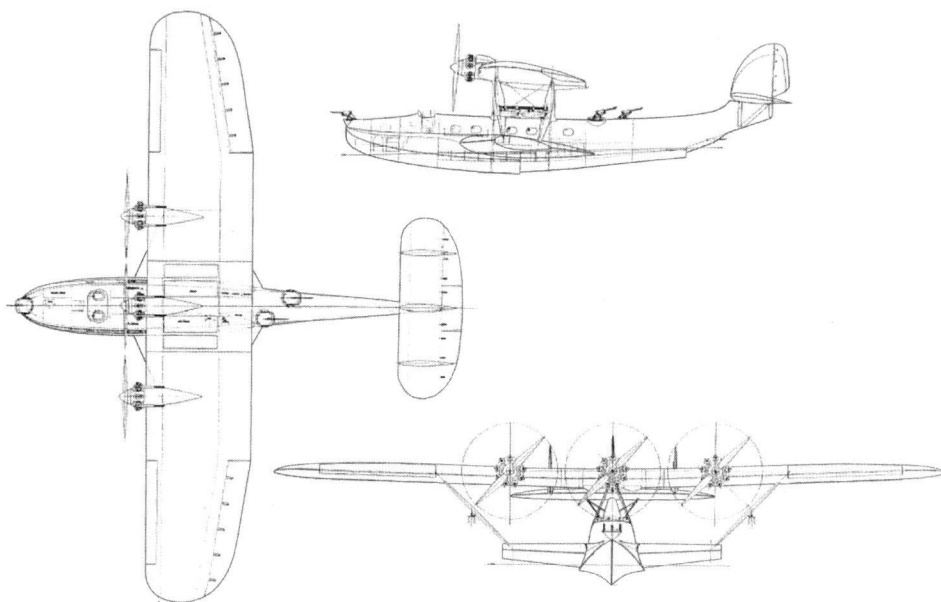

By 1930, the actual machine which emerged showed a monoplane of extremely utilitarian appearance but with an engineering structure of considerable aerodynamic cleanness. The sesquiplane compromise between the monoplane

Supermarine AIR YACHT (1930)

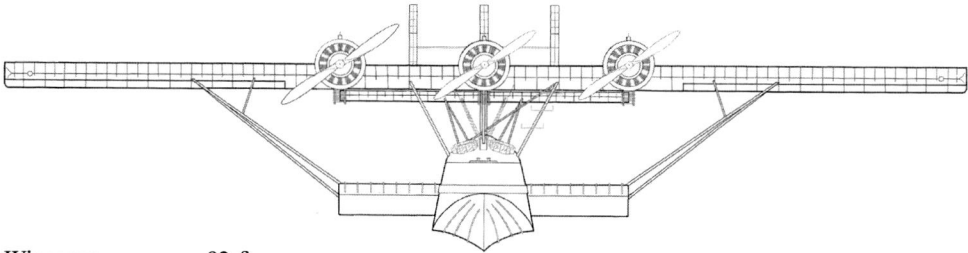

Wingspan	92 ft
Wing area	1,472 sq ft
Loaded weight	23,348 lb
Maximum speed	118 mph

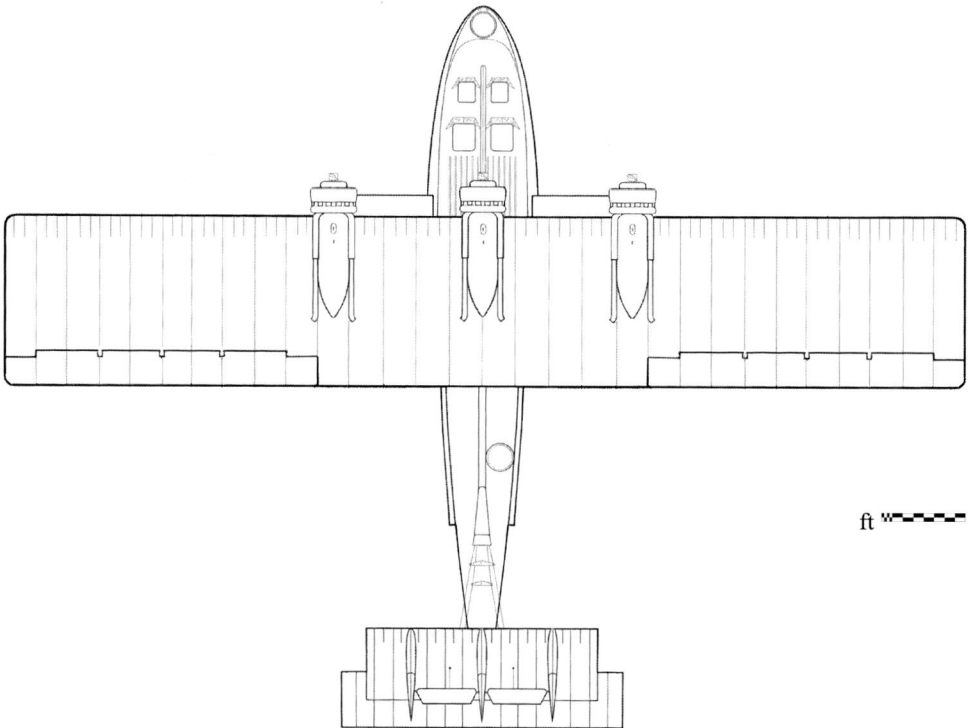

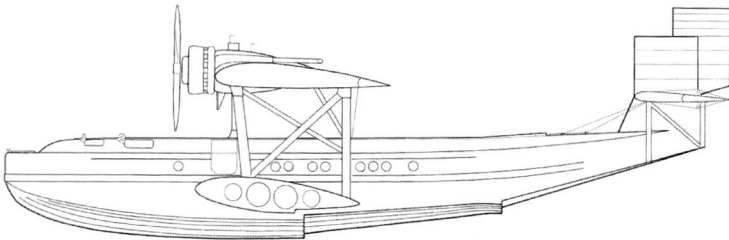

ft

and biplane formulae, which the Southampton X represented, was gone and, instead, there appeared an all metal monoplane with a wing-span of 92 feet and powered by three engines which were now faired into the wings – as with all Mitchell's later multi-engined flying boats. In the present case, their thrust lines were above the top surface of the wing to maximise water clearance for the propellers.

Air Yacht with original 'V' struts and Armstrong Siddeley Jaguar VI engines.

Mitchell had designed monoplanes before this, particularly the Schneider racers, but this was the first large Supermarine flying boat that was not a biplane. The new aircraft, however, looked more to the earlier Sparrow II than to the Schneider machines as it had a plank-shaped parasol wing with sloping V struts supporting the wing about two thirds out from the centre-line. The hull, on the other hand, was of the corrugated flat-sided type like the contemporary Southampton X and Saunders-Roe A.14. Southampton triple fins were retained but, in keeping with the other lines of the machine, they were extremely angular. The cabane of struts supporting the wing was laterally braced with typical Mitchell elegance: they were not cross-braced but had a minimalist pair of struts extending from the right-hand side of the fuselage to the strut position under the left wing.

Supermarine's move to all metal aircraft with stainless steel fittings was by now well established but, like most of contemporary British flying boats, the new design still had fabric covered flying surfaces. It had, however, one feature which made it stand out from all other Supermarine aircraft and its contemporaries and this was the employment of sponsons attached to the lower sides of the hull instead of the customary wing-tip floats. The final result was a completely different design from that of all previous Supermarine flying boats and, had it not been for the different mounting of the engines, the similarity with the Dornier Wal series of flying boats would have been, to say the least, uncanny (*see* overleaf).

Whilst it was a visit of the later Do X version of the Wal which made a great impression in British aviation circles (*see* p. 198), Mitchell's new design made its first flight on 30 February before the Mk.X flying boat visited Britain and was being designed only slightly later than was the German machine. But one of the earlier Wal machines had visited Southampton in 1925 and two years later a Lion engined version had been tested at Calshot and so it is more relevant to look towards the earlier Wal series, developed between 1922 and the middle 1930s, as the main inspiration of the British machine. The earlier Wals had two engines and carried up to ten passengers, whilst their development, the Super Wal, had four engines and could carry nineteen passengers at a cruising speed of 115 mph. Mitchell's aircraft was expected to have had a performance somewhat similar to the latter machine. For the early flights it was powered by Armstrong Siddeley geared Jaguar VIs with Townend drag-reducing cowling rings; these engines de-livered 490 hp each and Biard regarded the resultant top speed as 'quite fast' (he was still testing for Supermarine as the Air Yacht flew a month before the South-ampton X, with Mutt Summers, although being designed later).

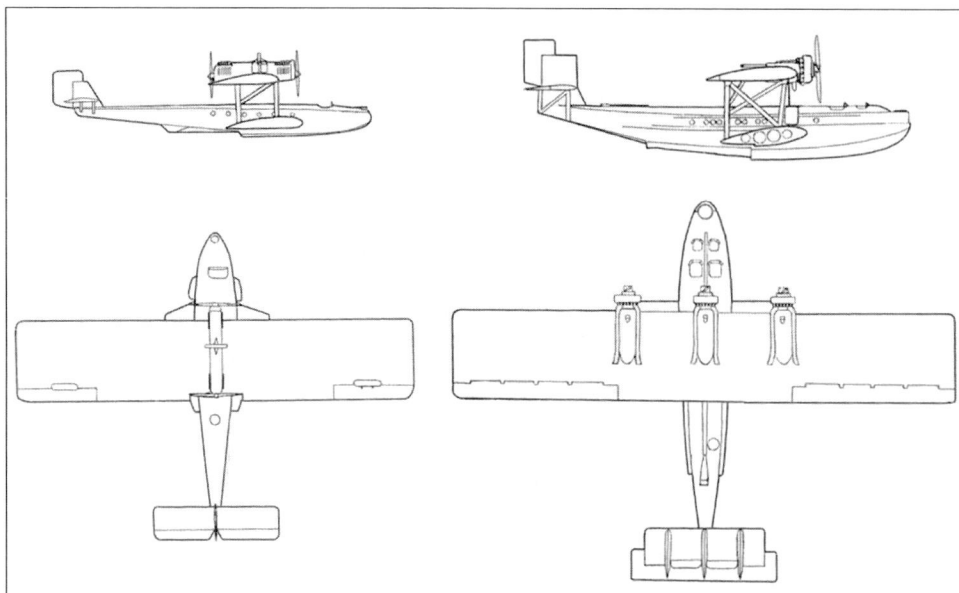

Dornier Wal.   *Air Yacht.*

On 8 May, 1931, Supermarine's flying boat was flown to the Marine Aircraft Establishment at Felixstowe but there was no evidence of any urgency in the testing programme of Mitchell's proposed armed reconnaissance machine: it was reported to have only flown nine times by the end of July and to have spent 570 hours at its moorings. Perhaps the Air Ministry civil servants were reflecting the views of MPs who had even queried the cost to the British taxpayers of the visit of the Do X; more probably, their attitude was influenced by the Government's 1928 planning assumption that no major war seemed likely for the foreseeable future.

At Felixstowe, its general flying qualities were found to be good but that, with one of the three engines throttled back, it could not maintain height and that it was unstable with one engine out (*see* details of its crash, p. 197). The Air Ministry continued to support the Supermarine design by, at least, paying for repairs when the starboard sponson failed in fairly rough seas and the other one showed signs of similar structural problems – perhaps giving the lie to the usually held view that the powers that be were entirely hostile to the monoplane concept. Nevertheless, it has to be admitted that the aircraft did not give a dra-matic argument for going over to monoplanes: its maximum speed was well below the 130 mph that a contemporary biplane, the Southampton X, eventually achieved with only slightly more powerful engines.

The Supermarine aircraft was accordingly re-engined with 525 hp Arm-strong-Siddeley Panthers and was then found to be capable of 117 mph; however, it was still not possible to maintain height with any significant payload with one engine throttled back. Penrose, whilst reporting design problems, did compare its design favourably against two more traditionally-built flying boats: 'unfortunately the sponsons suffered battering by waves and even on calm water gave inferior take-off compared with the usual chined British hull' and he added that there were problems with aileron snatch; but then he went on to say that 'assessed as an engineering structure of considerable aerodynamic cleanness, the Air Yacht was a big step forward compared with the established three-engined Iris biplane, of which four were in the course of delivering to the RAF, or the Calcutta-derived Short Rangoon prototype due to fly in the summer.'

Another view of the Air Yacht, in its original form.

Air Yacht 'on the step'.

By 1931, Supermarine had not only replaced the original V struts from the spon-sons by a firmer bracing of N struts but had also begun to consider insuring their investment by seeking civil registration, in the expectation of fulfilling an order from the Hon. A. E. Guinness for a replacement for his 'Solent' Air Yacht. As *Flight* reported, 'Mr Guinness is a man who dislikes publicity, and so his use of a flying-boat "air yacht" during the past two years has been much less heard of than could have been desired.' It was explained that 'a few people directly inter-ested in aviation have known for several months that Mr Guinness had placed an order with Supermarine's for a new flying boat, to be used by him as others use their surface yachts; but a promise of secrecy was extracted from those of us who knew of this machine, and not until now could this be revealed'.

Registered G-AASE and now known as the Air Yacht, its boxy hull provided very suitable dimensions for the passenger cabin, which Supermarine quoted as a generous 35 feet in length, 6 feet 6 inches in height and 8 feet in width. It was luxuriously appointed with owner's cabin complete with bed, bath and toilet, a galley with full cooking facilities beneath the wing, and additional wash basins, toilet and comfortable lounge with settees and sideboards in a separate cabin for five other passengers. There was rather more basic accommodation for the crew in the nose, with seats for two on either side of a gangway and with a folding seat in between for a mechanic; the rear cockpit was available for the use of up to four passengers and berths for crew members were situated behind these cockpits. Electric lighting was fitted throughout the interior – which Biard described as 'one of the most luxurious that anyone had then seen' and 'fitted out in glass and silver, with deep-pile rugs underfoot ... the chairs were deeply sprung, the cabins softly lighted ... The interior was roomy, with plenty of height and elbow-room' and the temperature could even be regulated by a blown air system. Long exhaust pipes were fitted above the wings to carry the engine fumes well away from the passengers.

By June, 1931, the total cost of the Air Yacht had risen into the region of £52,000 but, in the event, it was never sold to Guinness, who bought a Saunders-Roe aircraft instead. Webb probably reflected company gossip at the time when he gave the following downbeat assessment of this episode in Supermarine's history:

> At this time we were having trouble with the Air Yacht which was well down on performance and so we fitted Panther engines of 525 hp in place of the Jaguar VI engines of 490 h.p. I think this increase of hp was to a large extent negatived by the addition of several large struts between the sponsons and mainplanes ... Henri Biard and, later, Tommy Rose, were about the only pilots who could do anything with it ...
>
> My impression was that R.J. who had always been more of a practical engineer than a technician had allowed himself to be lured by some of his bright boys into following other people's ideas instead of his own.

Another Supermarine employee, Harry Griffiths, also gave a negative report:

> It had a very long take-off run and there was always doubts as to whether it would leave the water at all with a full load of passengers, stores and fuel.
>
> Refuelling in those days was done with hand pumps from barrels taken out on a barge. There is a story (unconfirmed but, knowing the man, possibly true) that Biard, the test pilot, refused to attempt a full load take-off and 'went through the motions' of filling with fuel by pumping from a number of barrels, some of which were empty.

Had the Air Ministry been more actively interested in the future of monoplanes, Mitchell may well have been able to do a great deal more with this aircraft but, as with his other parasol winged design, Sparrow II, the Air Yacht was taken for storage to the Hythe flying-boat hangar.

Eventually, rescue came in the formidable form of the American, Mrs June Jewell James – as reported in *Flight*: 'on October 1932 the Air Yacht was bought by a Mrs J. J. James, of Kenya Park, Rownhams (Southampton) – a keen motorboat and flying enthusiast – has acquired the Supermarine "luxury air yacht" G–AASE ... On October 11, the *Windward III*, as the air yacht is called, piloted by Capt. H. C. Biard, and with Mrs James on board, left Southampton on a cruise "somewhere around the Mediterranean and North Africa".'

Mrs James had been shown over the aircraft by the caretaker of the Hythe base and, as a result, negotiated the purchase of the Air Yacht from Supermarine. She then became so impatient to have the use of her new purchase that she insisted on starting some days before a prearranged departure date and Biard, who had been seconded to her by Supermarine, has supplied a description of the one-and-only cruise attempted in the aircraft as well as the problems of accommodating a very rich and self-willed owner.

Biard's account of the Cherbourg Episode, en route to the Mediterranean:

I put the machine down according to orders in Cherbourg harbour, and took a look at the weather. The clouds were gathering blackly over the Atlantic ... I had a hard look at the barometer, and it had fallen a good deal even in the couple of hours we had taken to cover the 150 miles from Southampton. Cherbourg harbour gives very little protection, and I strongly advised the owner to let us go on to a more sheltered spot, but she said we would stay where we were. Moreover, she said that the party would sleep on board the air yacht.

There were seven people aboard, and I felt a good deal of responsibility at the time. Judging by the look of the sky and the sea, which was getting dirtier every minute, there was going to be something really unusual in the way of a gale before morning; and a monoplane with an enormous wingspan is about the nastiest thing in the world to try to keep peaceful at moorings in a real storm. About ten o'clock that evening, while I was watching the barometer falling perceptibly minute by minute, a smart motor launch came chuffing alongside and hailed us. It bore a message from the Admiral of the Port – a message that sounded uncommonly like a command – saying that our position was very dangerous, that there was going to be a devil of a storm, and that no responsibility whatever could be taken for us unless the passengers came instantly ashore. Our owner sent back a message saying that she had no intention whatever of quitting the air yacht ...

The storm seemed years in coming, and when it did come, it just arrived without the slightest warning. One moment we were rocking gently in the long swell; next, a cyclone had struck us like a giant fist, and the air yacht was leaping, squealing at her cables, throbbing as if stricken unto death, bouncing from wave crest to wave crest, and every second trying to dip first one and then the other wing tip under the waves, which had become mountains high all in a moment.

I kept her nose into the wind, to lessen the strain, and I sat at my controls doing what I could – attempting to keep the wings more or less level and the nose from swinging round. Within two minutes from the time the first squall struck us, all the passengers were clamouring in the passage and in my control cabin ... The wind yelled, the waves thudded on the hull like gigantic hammers, and presently the gale became so bad that the wind under the wings made the air yacht try to fly, and actually did lift her time after time, two or three yards off the water and into the air, until her mooring ropes jerked her down again with a dreadful wallop on to the rearing waves ...

It was nearly three hours after the first squall struck us – three of the longest, most horrible hours I have ever spent in any aircraft – that the pilot tug suddenly loomed up alongside us in the screaming wind, and lowered a lifeboat to take us aboard. Our machine was still frantically bumping up into the air and down again on to the sea, and we were all more or less dead of bruises and exposure. But even then the owner was undaunted. She had a favourite little dog aboard with her, and she wanted a lifebelt to tie to him before she transferred him to the lifeboat ...

The time was not one for niceties of behaviour, and we had no lifebelt to spare, so I lifted the owner into the boat when a suitable moment came, flung the dog in as well, hurled in the rest of the passengers one after the other without a single casualty or miss, and finally jumped myself. I certainly expected, when I went down next morning to look, to find the harbour swept clear of all signs of our air yacht. But, to my surprise, she was still there, practically undamaged ...

Despite some exaggerations, his (ghosted?) account is of interest to social historians as well as to those interested in the ability of another Mitchell design to withstand what were, clearly, extremely unfriendly conditions. After the near-drowning of the owner and her companions during a severe storm in Cherbourg Harbour, Biard then flew the Air Yacht down to Naples whence Mrs James proceeded to obtain audiences with both the Pope and Mussolini. Having flown to France to collect Mrs James, who had then gone on to Paris, Biard had to hand over the Air Yacht at Naples to a relief pilot as his stomach muscles, which had been torn in the S.4 crash, needed surgery. Unfortunately, this pilot, Fl. Lt Thomas Rose, although a very experienced pilot (as Webb had indicated, above), lost the power of one engine on take off and stalled into the sea in the vicinity of Capri on 25 January, 1933. The owner suffered a broken leg but otherwise there were no serious injuries sustained; the aircraft was too badly damaged to be worth salvaging.

* * * * *

Thus ended the Air Yacht. By this time, any hopes of a military role for it were well past yet this unique Supermarine aircraft did look forward to the Saunders-Roe A.33, eight years later. This aircraft, another 90 foot parasol monoplane with similar 'N' struts from its sponsons, was built in 1938 to the same specification as the Short Sunderland, but the old porpoising problem caused structural failure of the mainplane on the first high speed taxiing test and it was not proceeded with. Had Mitchell lived long enough and had the Air Ministry generally shown a more single-minded faith in the future of monoplane flying boats, one wonders if Mitchell's last flying boat, the Stranraer, would have been a Supermarine Air Yacht type equivalent to the American Catalina which equipped twenty-one RAF and RCAF squadrons during World War Two. Incidentally, the predecessor of the Catalina was the Consolidated Commodore, with a parasol fabric covered wing like the Air Yacht and with about the same span; it appeared in the same year as Mitchell's machine but was far less clean, aerodynamically, with well over thirty supporting struts.

Because of the design activity around the Air Yacht, the Southampton X, and the S.6, as well as all the later Southampton developments, it is not surprising that, in the hiatus between the Seagull III of 1926 and the Seagull V of 1933, Saunders-Roe were able to successfully enter the smaller flying-boat field which had been the almost exclusive province of Supermarine ever since 1918. The company's first product, the A.17 Cutty Sark of 1929, was a metal-hulled amphibian which employed a semi-retracting undercarriage similar to the well-tried Supermarine type; and it was in one of these aircraft that the Prince of Wales arrived to witness Supermarine's triumph in the 1929 Schneider Trophy competition. A scaled-up version, the A.19 Cloud, followed a year later and seventeen were ordered by the the Air Ministry for RAF pilot and navigator training. And it was an aircraft of this type which was sold to Guinness instead of the Supermarine Air Yacht. A Cutty Sark also offered a Channel Isles service during the summer of 1929 and later appeared on the first Isle of Man service (piloted by Tommy Rose, mentioned above).

TYPE 179 – 'Giant'
It might be recalled how Biard had been impressed by the standard of the Supermarine accommodation in the Swan and the Solent in the 1920s and even greater luxury was available a decade later when the Short C class flying boats operated the Imperial Airways routes. But, during this period, the Dornier company had been developing their series of monoplane flying boats, with an ever increasing payload of passengers and baggage. At the time that their Mark X arrived at Calshot in November, 1930, for a two week stay, this version was powered by twelve 610 hp engines and with a crew of ten; it was capable of carrying seventy passengers in seven luxurious and roomy cabins. The Westland test pilot, Harald Penrose, described it as 'a humbling sight' and narrated how, like every other designer and pilot, he tried to find an excuse for visiting Calshot.

Biard managed an invitation to handle the controls, as did the Master of Sempill, who left an account of what was obviously a very memorable experience of the 157 ft span aircraft:

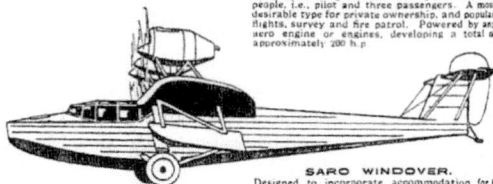
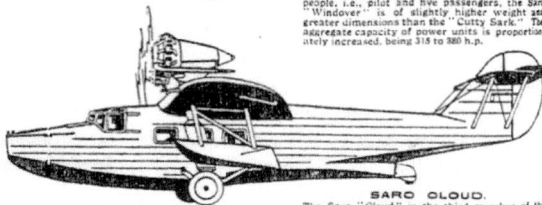

Publicity for Amphibians from rival manufacturer, Saunders-Roe.

Just aft of the control room is the entrance to the main spars [of the wings] and passageway to the engines so that the mechanics can make adjustments if necessary. Throughout the flight an inspector is able to check every mechanism and fuel connection from stem to stern and tip to tip. By the time you have explored the whole ship, climbing up and down ladders and watching what is going on, you are glad to have a rest in the luxurious seats of the huge multi-partitioned cabin, and there you will find that the well-equipped galley is by no means an ornament.

Not entirely to be outdone, on 18 May, 1929, the Air Ministry sent to Supermarine specification R20/28, for a forty-seat civil flying boat. Mitchell's response turned out to be a considerable departure from the large British biplane flying boats mentioned above as it was first projected as a high-winged monoplane, with three fins, a relatively flat-sided fuselage and with bulbous floats attached to the underside of each wing.

Artist's drawing of Type 179 as originally proposed.

Six radial engines were to be mounted in tandem on pylons above the wing, Dornier X fashion, and, notably, it had provision for passenger seating in the leading edge of the very thick wing. Whilst the new design was essentially a development from the Air Yacht, the former plank-shaped wing was now to be replaced by the first appearance of Supermarine elliptical flying surfaces, both wing and tailplane; the proposed torsion-resisting 'D' section leading edge of the wing also looked forward to that of the Spitfire, as did the use of a single spar – although in this case it was to be six feet in depth.

Had Mitchell's design been completed, its size would certainly have put his company ahead of other large flying boat contenders: it was to have a wingspan sixty-five feet more than the contemporary six-engined Short Sarafand and nearly thirty feet more than that of the imposing Dornier X. About this time, there was another very large, seven-engined aircraft – the Russian K-7 designed by Constantin Kalinin – which, interestingly, also featured an early example of the elliptical wing. It should, however, be pointed out that, whilst this plane has always attracted the attention of air historians because of its size (and because it flew), Mitchell's projected machine would have had a wingspan greater by ten

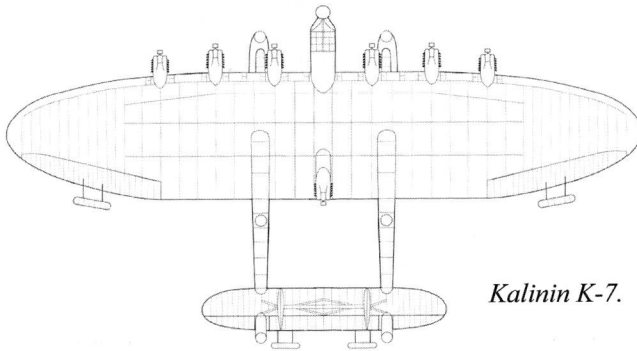

Kalinin K-7.

feet. Perhaps, more importantly, it should also be noted that the K-7 did not precede the Supermarine design, as its construction began the same year.

With a span of 185 feet, the name 'Giant' was therefore appropriate and it would have represented a significant departure from the almost universal formula of braced, fabric-covered biplanes. And it would have been a considerable challenge to the traditional fifteen-seat Short Kent, which had been ordered for the Mediterranean section of the Imperial Airways UK-to-India route in 1931: the Air Ministry *Report on the Progress of Civil Aviation 1930*, referring to Type 179, speaks of a 'Mediterranean Flying Boat' being built 'with the idea of increasing length of flight stages on routes such as through the Mediterranean. Luxurious accommodation will be provided for forty passengers, while detachable bunks are being fitted in order that sleeping accommodation may be available for half that number'.

It was almost twelve months after the issue of the original specification that a contract was drawn up for one aircraft, costed out at the remarkably precise figure of £86,585. By then, a replacement of the rather untidy arrangement of three rows of forward facing Bristol Jupiter radial engines was being proposed. Eventually Rolls-Royce Buzzard steam-cooled engines were proposed, in a neater pylon mounting arrangement, above a wing that now had straight tapered leading and trailing edges. Two inner nacelles housed two engines apiece, driving fore and aft propellers, and outer nacelles contained a single engine each, driving a tractor propeller. Feasibility studies were also made for using the leading edge of the wing as a condenser for the steam-cooling system (discussed in the next chapter), a variation on the wing surface radiator system of the S.5/6s. As a consequence, the forward facing seating in the wing had to be abandoned. Mitchell had also decided upon the wisdom of returning to conventional wing-tip floats instead of the earlier high-drag arrangement.

Thus by the time the keel of the Giant was laid down, in 1932. most of the Dornier influence had disappeared, as a company publicity photograph revealed:

Model of proposed Type 179 Giant showing revised hull, engine mountings and floats.

When one considers the Supermarine predecessors, the angular Air Yacht or the traditional Seamew, the original drawings of the Giant showed a tentative move forward, whereas the model shown above reveals a very considerable advance in design: the upswept tail section, the nicely stream- lined engine nacelles and the forepart of the hull, now reveal Mitchell's thinking to be in advance of forthcoming larger American flying boats. For ex- ample, the Sikorsky S-40, of

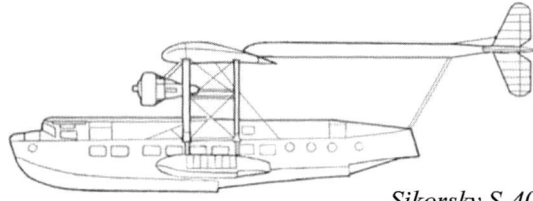

Sikorsky S-40.

the same year that the keel of the Giant was laid down, represented a traditional approach of struts and wires; also, the 'canoe' hull and the necessary twin booms

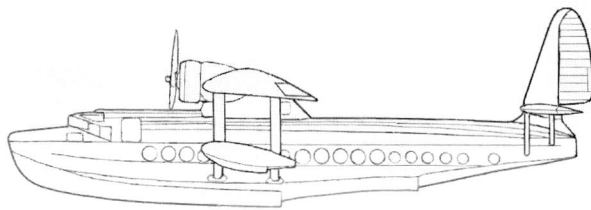

Sikorsky S-42.

for the tail section, which no doubt achieved a good weight/strength ratio, did not represent the way for- ward for later flying boat designs. The following Si- korsky S-42 had a tail unit integral with the main fu- selage and had lost most of its predecessor's struts and wires; coming a few years later than the proposed Giant, it had its engines neatly faired into the wing. It did, however, still retain wing and tailplane struts – and this in a machine that was to have a wingspan of 118 feet, compared with the proposed 185 foot span of Mitchell's projected cantilever structures.

Alas, early in 1932, the project was cancelled in view of the continued eco- nomic problems that faced the country: the Air Estimates for that year were £19,702,700, the lowest since the nadir of 1925). Joe Smith, the Chief Draughts- man, was left with the unpleasant task of laying off twenty of his drawing-office staff; and consternation was not limited to Supermarine, for questions were asked in Parliament – where the Under Secretary for Air justified the government's decision, claiming that over 70% of the estimated cost would be saved by cancel- lation.

Two years later, when Mitchell was invited to write something for the *Daily Mirror* about the Macpherson Robertson England–Australia Air Race, he was obviously still feeling raw about the Giant cancellation: 'As far as technique is concerned, British aviation is well to the front. Our Empire is so widely spread that fast aerial transport is perhaps the most vital necessity to our existence. Why are we so slow in the development of our big airliners?' A considerable amount of the Giant hull had been completed and extensive design work at the frontiers of current technology had been spread over nearly three years. Not surprisingly, the editor of *The Aeroplane*, C. G. Grey, saw this decision as nothing but short- termism and exclaimed:

> If this be the new Government's idea of economy, then God help England. A Chancellor who understood the difference between false economy and efficient expenditure and had sufficient intellect to keep in touch with the great developments of the day, of which air transport is perhaps the most important to the welfare of the Empire, would have realised that cancellation of the Supermarine is the falsest of false economy. To stop important experimental engineering, costing only a few tens of thousands, which when finished would show the way towards earning millions, is economy going mad.

Had the Giant been built, perhaps Mitchell's bomber (*see* Chapter Nine) might have been designed earlier and might even have been in the air when the critical need arose for such a weapon; on the other hand, it might be recorded in the Government's defence that the Germans had not felt justified in putting their huge Do X into quantity production and that one British aircraft of a somewhat similar type did actually fly. This was the Blackburn Sydney which in 1930 represented the first British flying boat in the heavyweight class; this monoplane with a metal-skinned wing, albeit braced and of only 100 foot span, could also have spearheaded the movement away from the traditional British fabric-covered biplane with an aircraft not dissimilar to the Air Yacht. The Air Ministry, however, did not place a production order. Shenstone later commented:

> One of the objects of the big six-engined flying boats was to compare biplane with monoplane. Shorts got the order for the biplane, which was not cancelled. This, the Short Sarafand, was completed, but nobody learned anything about monoplanes thereby. Even if the monoplane had been a failure when completed, it would have helped everyone's future designs, whereas the biplane was close to its end and in the view of some, had already outlived itself.

Another view of the Giant model, illustrating graphically what was lost to British aviation at the time (the Southampton in the background, left, makes the point).

It was thus fated that Mitchell would not be remembered, as might otherwise have been predicted, for his contribution to the main flowering of the Imperial Airways routes or for the creation of equivalents of the well-known wartime monoplane flying boats, the Sunderland and the Catalina.

Such setbacks might not seem too surprising when it is considered that only about a dozen new types per year were being dealt with at the main testing station at Martlesham Heath and this limited activity had to be shared amongst nearly the same number of main aircraft firms. Supermarine's employees, like many others, had had to take a reduction in wages as the price for retaining their jobs.

The chief factor that was to alter the fortunes of the industry, and the fate of Mitchell's last designs in particular, was the eventual collapse of the International Disarmament Conference and the final realisation that Germany's ambitions required an adequate response – especially in the matter of increasing Britain's air forces. A significant stage in the change was identified by *The Aeroplane*: 'The Air Debate of 8th March marked a turning point in the history of the RAF, for the House of Commons showed for the first time a proper appreciation of air power. No Debate on any Service Estimates has been so largely attended for many years'. There was also a proposal for a dramatic increase in the number of squadrons. It was now proposed to increase the existing 75 units to 116, with threefold expansion of the number of home-based aircraft – the 1923 proposed increase in the number of home defence squadrons to fifty-two had never been fully implemented.

Unfortunately, drift and indecision, influenced by strong pacifist sentiments and the depressed economic situation, were the most characteristic features of this period of government, despite Hitler's ominous withdrawal of Germany from the League of Nations and from the Geneva Disarmament Conference in 1933. The view of the Prime Minister, Stanley Baldwin that 'the bomber will always get through' did not encourage a strenuous policy of fighter development but nor, on the other hand, did it result in the creation of a retaliatory, modern bomber force which might deter enemy aggression.

In this climate of uncertainty, if not timidity, Mitchell was, however, encouraged to design two more traditional flying boats which, despite the obsolescence of their configuration, were ordered in some numbers and which both saw military action; he also designed the Walrus which was eventually ordered in large numbers and featured in innumerable wartime actions. The last of these came as a straight-line development from the early Supermarine pusher amphibian type, and the other two, the Scapa and the Stranraer, were in the tradition of the larger Southampton twin-engined biplane.

SCAPA

The first of the new designs, which was to become known as the Scapa, owed its origin to the last of the three experimental Southamptons, N253; it was ordered in 1928, to be fitted with a metal superstructure and with the new Rolls-Royce Kestrel engines. Whilst the experimental N251 design work had been largely to the benefit of the Saunders company and N252 had been the unsuccessful pairing with Vickers, this third order was to prove far more significant than just marking Supermarine's move to all metal structures. The new sleek Kestrel engines gave the N253 Southampton 10 mph more top speed and it took 14 minutes less to reach 5,000 feet than the standard Southamptons.

As the combined power of the Southampton's Lions had been 20 hp more than that of the original Kestrels, there was considerable attraction to the possibility of equipping the new design with the developed Kestrel III of 525 hp and also of staying with a two-engine formula. In the still relatively depressed economic situation, a more aerodynamically efficient version of the Southampton II with the cheapness of using only two of the new and more efficient engines seemed a good proposition and so in response to a R.20/31 specification for a general reconnaissance flying boat, Supermarine offered the last contracted Southampton, S1648, as the proposed Southampton IV prototype, at no extra cost to the Air Ministry.

Accordingly, the Southampton IV/Scapa has been regarded as, essentially, an 'improved Southampton', particularly as its hull planing geometry was closely based on that of the earlier aircraft. This similarity was very much a compliment to the intuitive designers of the Southampton hull of 1925, as the current use of tank testing did not suggest any real need to depart from the basic shape of its predecessor. Indeed, a wider beam behind the step, to discourage water striking the tail surfaces, was replaced by the older Southampton after-portion when actual take-offs revealed an unpleasant pitching when the rear step made contact with the water.

Supermarine SCAPA (1932)

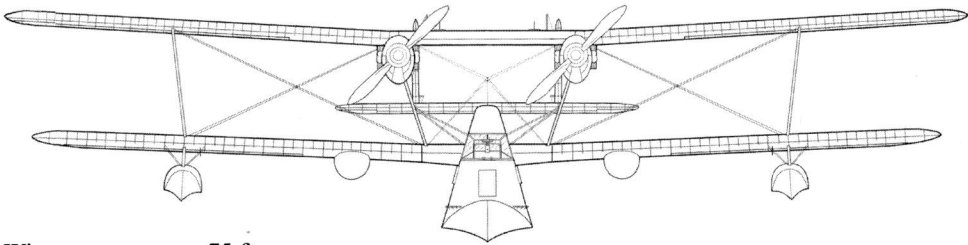

Wingspan	75 ft
Wing area	1,300 sq ft
Loaded weight	16,080 lb
Maximum speed	142 mph

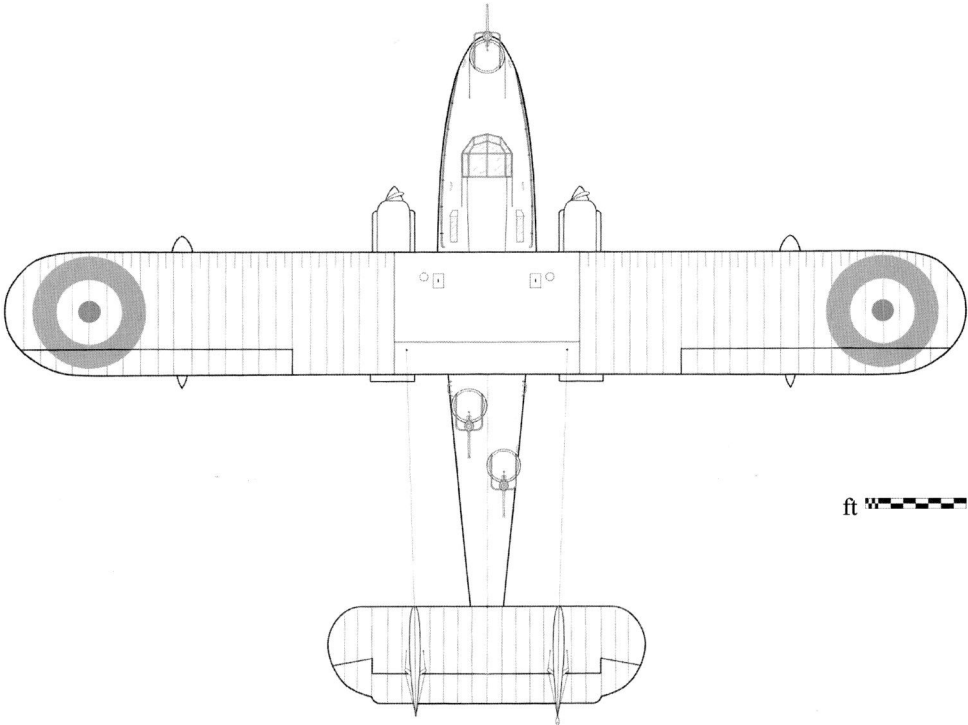

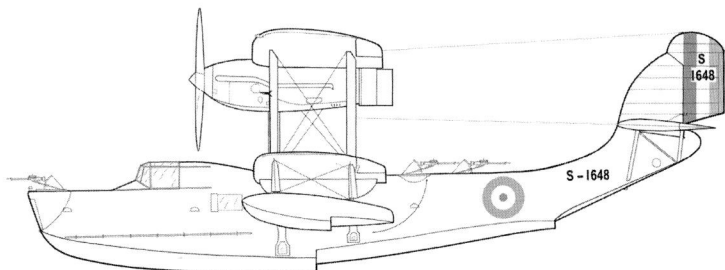

ft

S1648, the Scapa prototype.

Nevertheless, the eventual Scapa was, effectively, a new design. Finer water lines were developed by a change to the bow section of the new hull and the need to stretch the Southampton design resulted in a lengthened bow with a deepened forefoot which, with a flatter top-decking, gave more useable space within – as well as effectively altering the overall appearance of the previous Southampton hull. The new decking also allowed the now enclosed cockpit to merge better with the sides than had been possible with the 'Persian Gulf' Southamptons of No. 203 Squadron and the flatter coamings of the two midship gunners' cockpits would now offer less resistance to the airflow. The Air Yacht tradition of flat plates and rectangular sections (another reflection of the economic situation?) was still evident although the upswept tail section and other curvatures restored something of the elegance of the Southampton.

The superstructure offered even more evidence that the new prototype was very much more than a Mark IV version of its predecessor. The redesign meant that it was no longer necessary to sweep back the outer sections of the mainplanes as had been necessary with the changing service loads of later Southamptons. In fact, the outlines of all new flying surfaces differed significantly, with the tailplane being reminiscent of the Seamew and the fins looking not unlike those of the Southampton X or even the Swan. Two fins now replaced the triple-fin arrangement of the Air Yacht or standard Southampton and were well within the slipstreams of the engines, now positioned directly under the top wing.

This upward resiting of the engines was probably influenced by the water ingestion problems experienced by the Seamew and by Shenstone's concern for cleaning up drag-inducing features. Looking back at earlier designs, he had commented that 'the NACA work on optimum engine position had not been completed and many thought that engines in the leading edge would be bad'. Now, incorporation of the engines into the Scapa wing allowed Mitchell to dispense with the engine pylons of his previous inter-wing-engined designs and the large Warren bracing of the Southampton superstructure. His decision to return to wire bracing with the S.5 and S.6 must have been a factor here and also led to a single bay structure, even though the new machine's wingspan was to be 75 feet. Another aspect of the aerodynamic cleanup was the attaching of the lower wings directly to the fuselage via an elegant, slightly gull-wing, centre-section, unsupported below by struts. Shenstone, not surprisingly, considered the resulting aircraft 'perhaps the cleanest biplane flying boat ever built, with minimum struttage and clean nacelles faired into the wing'. He did not mention the very 'boxy' radiators which projected on either side to the rear of these nacelles, although their positioning would least compromise the overall lines of the new design.

Scapas under construction at the Woolston works. Foreground, right, shows an engine nacelle with radiator already attached to rear.

In view of all the new design features of the prototype Southampton IV, a new name for the type was justified far more than had been the case with the proposed Solent, which had been in essence a wooden-hulled Southampton. Nevertheless, no new name had materialised when 'Mutt' Summers took S1648 up for its first flight on 8 July, 1932. After numerous tests, it was delivered on 29 October to MAEE, Felixstowe, for further trials which included a maximum duration flight of 10 hours over the North Sea.

Scapa prototype over Southampton Water.

In the following May, the prototype was flown to the Kalafrana flying boat base, Malta, for overseas acceptance trials with No. 202 Squadron and these involved a long-distance flight to Gibraltar and back and a cruise to Port Sudan via Sollum, Aboukir, and Lake Timsah. It was also demonstrated to the Governor of Malta and, more importantly, to the Commander-in-Chief, Mediterranean, and the Air Officer Commanding, Mediterranean. On its return, the Scapa took part in the 1934 fly-past of 'the competition' at the Hendon RAF Display with, as Penrose reported, 'the clean Supermarine twin-engined Scapa leading, followed by the four-engined Short Singapore, triple-engined Blackburn Perth, the distinctive gull-winged Short Knuckleduster, the Saro R24/31 London and the three Saro Cloud trainers'.

The penultimate Scapa of the fifteen built.

The Air Ministry ordered twelve of the new Scapas, K4191 to K4202, and three replacements, K7304 to K7306, were ordered later. Fittingly, the first batch went to No. 202 Squadron whose pilots must have been particularly impressed by its performance during the acceptance trials, having been equipped since 1929 with Fairey III floatplanes. The Squadron soon undertook the almost traditional long-distance cruises with their new type, including a notable 9,000 mile return flight with two machines to Calabar, Nigeria, via Algiers, Gibraltar, and the Gambia. Additionally, the old No. 204 Squadron Southamptons were replaced by their Supermarine successors. After this, in 1937, No. 240 Squadron was re-formed at Calshot from the Seaplane Training Squadron C Flight and took over some of the No. 202 Squadron Scapas; they then were sent to Egypt during the Italy–Abyssinia conflict. And Scapas of No. 202 Squadron took part in anti-submarine patrols during the Spanish Civil War to protect neutral shipping. One Scapa was attached to No. 228 Squadron where it was involved with early radar experiments.

Scapa refuelling from a depot ship, Malta.

The Supermarine publicity for 1933 draws attention to its service testing, no doubt because actual performance data was now being withheld – an early indication of troubled times ahead; nevertheless, the information about structure, accommodation and equipment is, at least, quite full:

The 'Scapa' is the latest twin-engined reconnaissance flying boat, and was designed to replace the 'Southampton', which has been the standard RAF reconnaissance flying boat for nearly eight years. The 'Scapa' has been subjected to very severe and thorough testing both by the firm's personnel and by the RAF, at the Marine Aircraft Experimental Establishment, at Felixstowe. In these tests, the prototype maintained height with normal load and with one engine switched off. The 'Scapa' also showed that its top speed and ceiling are higher, the range greater and the take-off quicker than any other British flying boat with one, two, three or four engines [an oblique reference to the rival Blackburn, Saunders-Roe and Short aircraft].

WINGS. Unequal span three-bay biplane. Upper wing in three sections. Lower wing in four. Upper and lower centre-sections interconnected by two pairs of struts and form middle bay. Outer wing sections have one pair of parallel interplane struts each. Structure entirely of metal, with fabric covering. Spars and ribs of anodically-treated aluminium alloy, with stainless steel fittings. Leading-edge covered with duralumin sheet. Ailerons on all four wings.

HULL. Characteristic Supermarine two-stepped hull of anodically-treated aluminium alloy. Internal structure consists of a number of transverse frames with longitudinal stringers and flat sheet outer plating. Wing-tip floats of similar construction as hull.

TAIL UNIT. Monoplane type. Tailplane mounted on upturned end of hull and braced with 'N' struts on either side. Two cantilever fins and rudders mounted above tailplane. Balanced rudders. Aluminium alloy framework, with fabric covering.

POWER PLANT. Two Rolls-Royce 'Kestrel' III MS twelve cylinder Vee water-cooled engines, in monocoque nacelles, mounted directly to the under surface of the upper centre-section. Nacelles have quickly detachable cowling and folding working platforms. Large manholes give access to back of engines. Radiators at rear ends of nacelles. Fuel tanks (two) with total capacity of 460 Imp. gallons in centre-section. feed by engine pumps, but in event of pump failure, fuel supply maintained by gravity. Tanks have jettison valves. Oil tanks form leading edge of centre-section. Compressed air starter and alternate hand-turning gear.

ACCOMMODATION. Cockpit in nose for gunner observer. Scarff ring may be slid clear back to cockpit for mooring operations. All marine gear, bombsights and releases also located in this cockpit. Then follows enclosed cockpit for two side-by-side, with dual controls. Second pilot's controls detachable. Between pilot's cockpit and front spar frame is navigator's and engineer's compartment. Between spar frames on port side is wireless compartment. Aft of main planes are two staggered gunner's cockpits, with Scarff rings. Behind rear cockpit is a lavatory and stowage for collapsible dinghy, engine-ladder, maintenance platform and spare airscrew. Provision made in body of hull for cruising equipment, including cooking-stove, ice-chest, water-tanks, etc. Drogue stowed in trailing-edge of lower wing, near hull.

ARMAMENT. Three Scarff ring mountings and three Lewis guns. Five 97-round drums of ammunition for each gun. Provision made for 1,000 lbs of bombs.

[Reference to 'monocoque' engine nacelles indicates another influence of Schneider racing experience on commercial designs (*see* p. 157).]

* * * * *

Even though the Royal Air Force had not been very impressed with the Seagull II and although the Seamew had never proceeded beyond the prototype stage, Supermarine still had faith in their medium-sized reconnaissance amphibian formula. By now, British warships were being modified for catapult assisted take-offs and, in fact, as early as 1925 a Seagull II had been used at the Royal Aircraft Establishment to test the first British catapult for launching aeroplanes.

The Seagull V prototype on Southampton Water, originally numbered N-1 and later purchased by the Air Ministry as K4797.

Another view of the prototype – with wheels now fully retracted.

Thus it might reasonably be hoped that the poor deck landing characteristics of the early Seagulls might not be considered relevant to the ordering of an improved catapult version for capital ships. In fact, it was the Australian navy which came up with a requirement which led to a further Seagull development.

SEAGULL V

Compared with the Royal Navy, the Royal Australian Air Force had been operating their Seagulls more thoroughly, as has been described earlier (including a race between the machines at the annual regatta in Hervey Bay, north of Brisbane) and were thus more proactive in matters likely to be of continuing interest to Supermarine. By 1930, a broad specification had been sent around British aircraft firms for a reconnaissance amphibian which could be catapulted with full military load, which was also capable of stowage, and which could operate in six foot wave conditions. This requirement was such that C. G. Grey commented in *The Aeroplane* that 'The Australians want everything but the little black boy to cook the meals'.

In view of the limitations of the Seagull II/III, a replacement machine would have to be a complete redesign for reasons other than obsolescence (the earliest version of the type, the Seal, was first flown in 1921); in view of there having been a much modified Mk.IV, albeit not proceeded with, the eventual design was designated Seagull V. In fact, it might be fairly accurate to say that the only influence of the older type on the Seagull V was the basic layout of the last experimental Seagull II, N9644, a biplane which had reverted to a pusher air-cooled power unit. This 'parentage' was plain to be seen in the basic configuration of the new design but, otherwise, the move to metal structures, slab-sided fuselages, and the experience of the intervening years produced a quite distinct type within an older formula.

One important example of the redesign was the employment of a fully retracting undercarriage. Little encouragement had been given to such a feature in Air Ministry contracts and Supermarine had merely had to devise mechanisms which raised the wheels out of the water to facilitate waterborne taxiing but not out of the aircraft slipstream for flight; now, at last, the Seagull V hydraulics – modified later for the Spitfire – equipped a British military aircraft with drag-reducing full wheel retracts for the first time. Alan Clifton has recalled that he persuaded Mitchell to retract the wheels into the wing by saying 'We shall have to do it eventually, why not now?'

Seagull V showing simplified hull construction.

A more obvious aspect of the redesign of the Seagull II/III predecessor was the hull. It shared the more aggressive, slab-sided, features of the Scapa (and the later Stranraer) but had no upward sweep to the tail unit, as did the Seamew and these other two types. The result of this particular return to the Seagull and Scarab configuration was a more utilitarian appearance to the hull, accentuated by Supermarine's first move to one-step planing. Additionally, hull construction was simplified by the taking out of a licence on a Sunders patent – whereby the complex curves of the typical flying boat hull bottom were replaced by a flat surfaced 'V' below a horizontal 'bench' which terminated at the chine [*see* photo above]. The move to metal hulls also led to direct attachment of the lower wings to the hull and in this respect the new design was similar to the Scapa and the Stranraer, although in this case there was no elegant lower centre section.

Also, the wing upper centre section had a less than tidy trailing edge, as it had to be cut back for clearance of the pusher propeller (two two-bladers bolted

Supermarine WALRUS/SEAGULL V (1933)

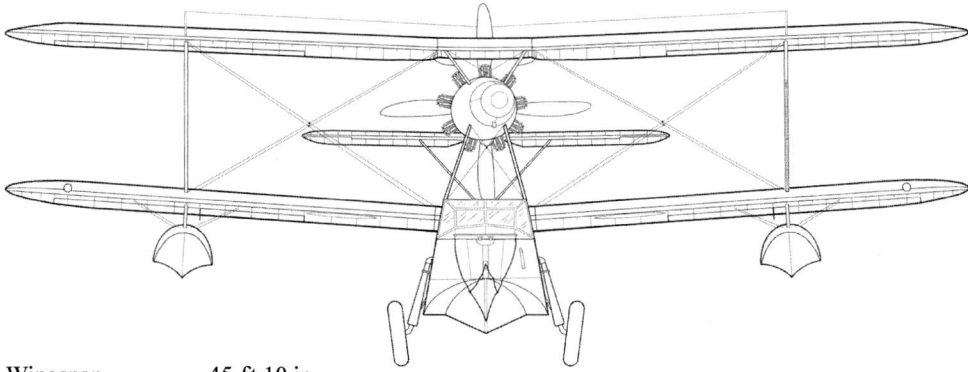

Wingspan	45 ft 10 in.
Wing area	610 sq ft
Loaded weight	7,200 lb
Maximum speed	135 mph

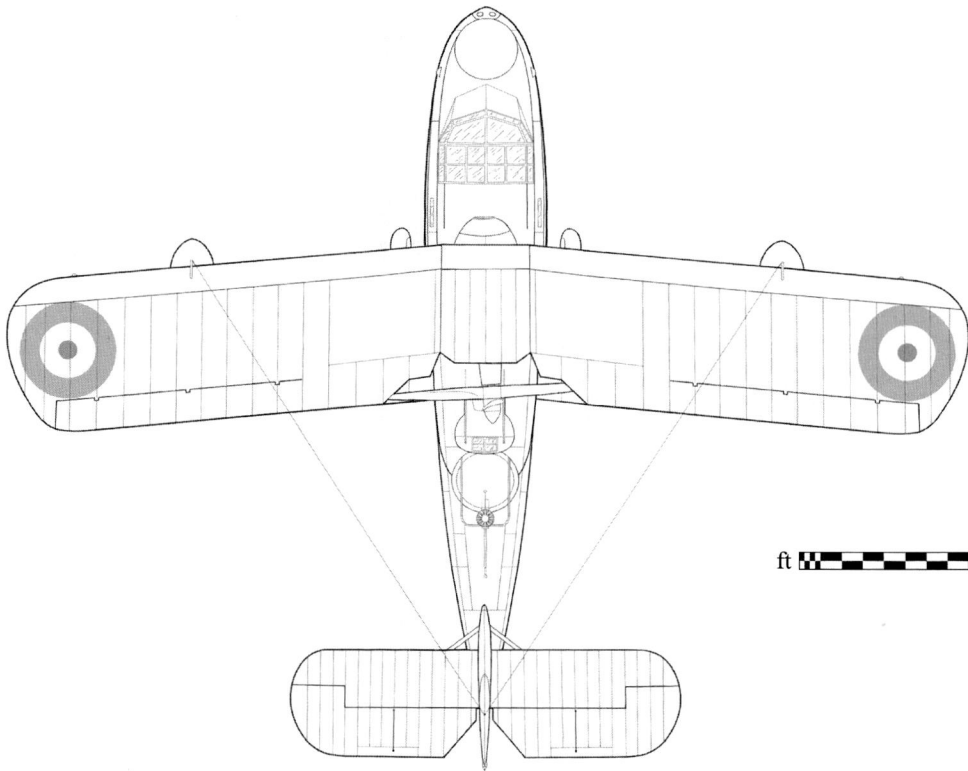

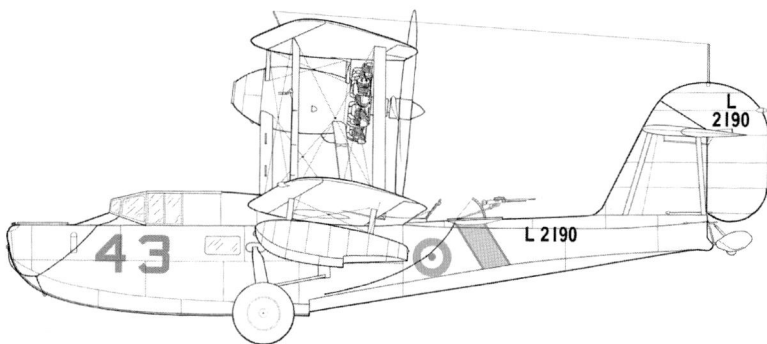

ft

43

L
2190

L 2190

together) and additionally cut back for the folding-wing arrangement. However, the lower wing had lost the large cut-outs of earlier designs as a result of the neater device of hinging the inner portion of the wing behind the rear spar: the inner sections of the wing could thus be folded away in order to clear the hull sides when the wings were stowed. The wings were supported by single bay struts similar to the recent Seamew and Scapa designs but the single engine nacelle had the more traditional Supermarine position between the wings. This nacelle also contributed to the 'minimalist' appearance of the Seagull V by being offset a few degrees to counteract the corkscrew pressure of the propeller slip-stream on one side of the fin.

By June, 1933, the prototype Seagull was complete and could be seen to be no beauty. Its, to say the least, functional appearance, allied with its very trad-itional configuration, appeared to have won it no friends when seen by the Air Ministry Director of Technical Development. Those at Supermarine with long memories of the inability of the company to win orders for their naval amphibian fighter could hardly have been encouraged by his comment to Clifton: 'Very interesting; but of course we have no requirement for anything like this'. Perhaps this reaction had some bearing on the test pilot's performance at the second SBAC Show at Hendon. *The Aeroplane* nicely described the event:

> This boat made its maiden flight on 21 June, five days before its first public appearance, but Mr Summers proved its qualities by throwing it about in a most carefree manner. Of its performance little is known but there can be little doubt about its amiability and general handiness in the air and on the ground. One must be prepared to see all sorts of aeroplanes looping and rolling with abandon nowadays, but somehow one has, up to now, looked to the flying boat to preserve that Victorian dignity which one associates with crinolines, side whiskers, bell-bottom trousers and metal hulls. The Seagull V destroyed all one's illusions.

Henry Knowler, Chief Designer at Saunders Roe who watched the display in the company of Mitchell, reported Mitchell's understandable surprise and anxiety at the low level antics of the five-day old prototype. 'He looped the bloody thing,' Mitchell kept repeating to everyone he met. (He had obviously not heard that a disgruntled American pilot had once done the same over Killingholme in a 95 foot Felixstowe flying boat. Nor did he live to see Pickering's testing of production Walruses (the later British Seagull V), as described by Peter Weston, a Supermarine apprentice in 1938: 'At the finish of the test he would fly very low over the river, by low I mean about 300 feet or so, in front of the flight shed and loop the Walrus to signal that it had passed the test; if not, he would just land and return up the slip and it would be worked upon. If you haven't seen a Walrus looped, you haven't lived.')

The combination of traditional and new features was continued in the mixture of metal and wood construction employed in the new amphibian. This composite structure is well documented in the following company publicity whilst, again, performance details are withheld:

The 'Seagull V' is the latest single-engined amphibian designed by the Supermarine Company specially for fleet-spotting work from aircraft carriers. The boat is therefore very compactly built and has folding wings.

    The addition of detachable dual controls widens the scope of the machine and the enclosed accommodation, in conjunction with the pusher engine, makes the cabin very quiet and comfortable for fleet-spotting, photography, wireless communication, etc. Catapulting points are provided.

TYPE. Single-engined fleet-spotter amphibian.

WINGS. Equal span single-bay biplane. Small centre-section carried on engine-mounting struts. Outer wings fold round rear spar hinges on centre-section and hull. One pair of parallel interplane struts on either side. Wing structure consists of two stainless steel spars, with tubular flanges and corrugated webs, and a subsidiary structure of spruce and three-ply. Plywood leading-edge and fabric covering. Inset ailerons on all four wings.

HULL. Flat-sided single-step hull, of anodically-treated aluminium alloy. Normal Supermarine system of construction. Wing-tip floats of similar construction.

TAIL UNIT. Monoplane type. Tailplane carried on top of fin built integral with hull. Tailplane and elevators built of steel spars and wooden ribs, with fabric covering. Rudder of wood, with fabric covering.

UNDERCARRIAGE. Retractable type. Each unit consists of an oleo leg and radius-rod hinged to the side of the hull. In raised position, wheels are housed in recesses in underside of lower wings. Lifting gear partly compensated and operated manually by hydraulic mechanism. Wheel brakes.

POWER PLANT. One Bristol 'Pegasus' II L2P nine-cylinder radial air-cooled engine, driving pusher airscrew. Monococque nacelle, with manhole to give access to back of engine. Two fuel tanks (each 75 gallons) in upper wings, with gravity feed to engine. Oil tank in nose of nacelle. Hand inertia starter.

ACCOMMODATION. Bow cockpit, with Scarff ring and stowage for marine gear. Enclosed cockpit, with pilot on left side. Detachable controls to right seat. Between pilots' seats and front spar frame is navigator's compartment. Between spar frames wireless compartment. Aft of wings is aft-gunner's cockpit, with special gun mounting.

DIMENSIONS, WEIGHTS AND PERFORMANCE. No data available.

* * * * *

It was at this time that Mitchell consulted his doctor just before taking a holiday in the August of 1933. His moments of irascibility that were well known to his staff were, rightly, attributed to the high standards he set but there was also another explanation that was not known to them. He had not been feeling well for some time and cancer of the bowel was diagnosed. An operation was performed almost immediately and Mitchell then went to Bournemouth to convalesce.

* * * * *

Whilst the Chief Designer was away from work until the end of the year, development work at Supermarine had to go on: the end of the line for the Southampton had been reached after the last batch for the RAF was ordered in 1931, the Southampton X remained in prototype form, the Type 179 Giant had been cancelled, the lone Air Yacht was no more, and the Seagull V had yet to be proven for catapult launching if it were to satisfy the Royal Australian Air Force. And although orders had been received for the Scapa, the Air Ministry had already issued another specification which was unlikely to be matched by a simple development of this new type – suggesting that the Scapa, which had been Supermarine's first production order since the Southamptons, had only been seen as a stopgap.

This latest specification, R.24/31, was for another general purpose coastal patrol flying boat, of robust and simple construction with low maintenance costs, but it had to be capable of carrying a 1,000 lb greater load for the same 1,000 mile range of the Scapa and to be able to maintain height on one engine with 60% of fuel on board. Thus, an enlarged and substantially altered version of the Scapa had to be projected and this was submitted, alongside one from Saunders-Roe, their A.27. Unfortunately for Supermarine, only the latter was accepted and, known as the London, was later ordered to replace the Southamptons and Scapas of Nos 201 and 202 Squadrons.

To further increase Supermarine's uncertainties, the Short Singapore III with a maximum speed of 145 mph was being ordered to replace other Scapas with Nos 204 and 240 Squadrons. However, the Short machines had about the same speed as that estimated for the new Supermarine aircraft and were powered by twice as many engines. The Singapore was, admittedly, larger than the Scapa and, therefore, needed more power for the same speed; and the even larger Short Sarafand was only a few mph faster with six engines. Thus, given an economic situation in which orders for these larger flying boats were likely to be kept to the minimum, it seemed a distinct possibility that a performance from Mitchell's smaller, twin-engined R.24/31 project, if significantly better than that of the Saunders-Roe London, might still stand a chance of winning Air Ministry contracts, given the growing calls for British rearmament.

Another reason for anticipating orders for the proposed new design was not simply based on the good performance figures that the Scapa had returned but on Mitchell's having come to believe in the virtue of employing a thin wing – for aircraft other than the Schneider Trophy racers – contrary to the generally perceived wisdom of the day. The eventual outcome was a flying boat that out-classed all its contemporaries of similar configuration. It was to be called the Stranraer.

STRANRAER

The engines chosen initially to pull the new machine's thinner aerofoil through the air and to give it the required one-engine performance required by the Air Ministry specification were 820 hp Bristol Pegasus IIIMs, providing a combined 590 hp more than the Kestrels of the Scapa. The two engines were to be mounted with the same thrust line as the Kestrels and in streamlined fairings but, being air-cooled radials, did not incur such extra weight and drag penalties of the Scapa radiators which had been fixed to the sides of the engine nacelles; long-chord Townend drag-reducing rings surrounded the cylinder heads and their oil coolers formed part of the top centre-section leading edge. Against these improvements, there was the additional twelve percent increase in wing area of the new machine; this extra drag and weight was added to by the two-bay strut arrangement required to support the extra ten feet of wingspan that was needed to meet the new load-carrying requirements of the R.24/31 specification.

The extra depth of the hull allowed the top of the enclosed cockpit to form a continuous line with the single midships gunner's cockpit, which was now placed in the centre of the hull top. Also, for the first time, Supermarine had built a larger service aircraft which made it possible to install the second rear gunner, more sensibly, in a faired-in cockpit in the tail. This had been proposed for the unsuccessful Vickers/Southampton X prototype, with its wingspan of 79 feet, and so presented little difficulty for the new 85 footer.

Supermarine STRANRAER (1934)

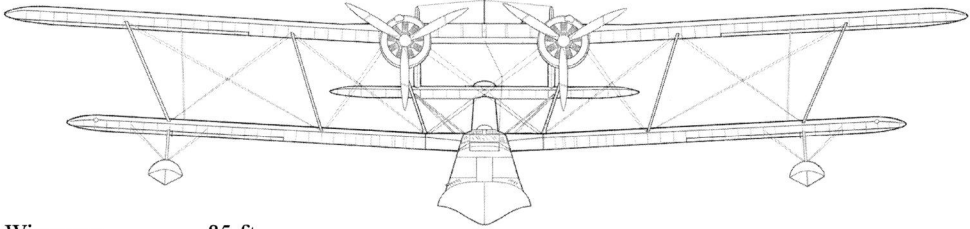

Wingspan	85 ft
Wing area	1,475 sq ft
Loaded weight	19,000 lb
Maximum speed	165 mph

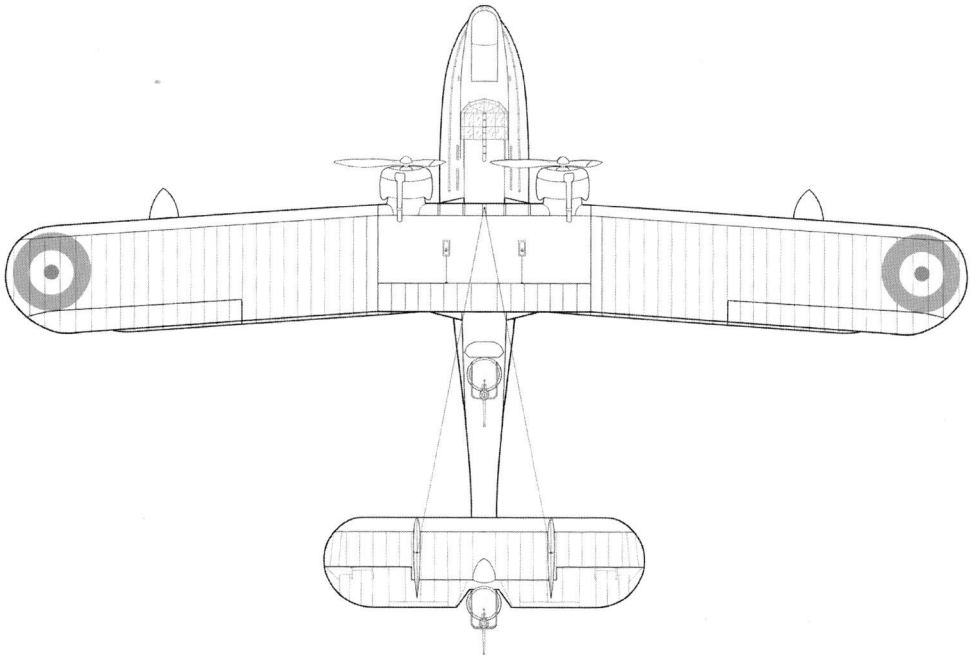

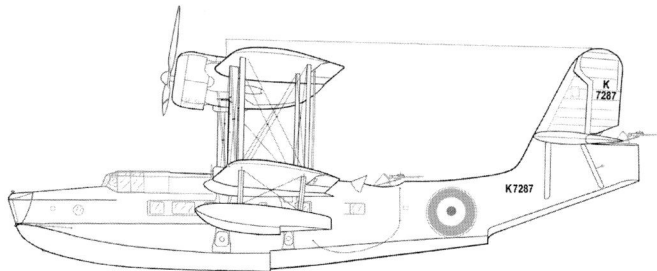

K3973 the prototype Stranraer.

This repositioning of the rear gunner and his armament, coupled with the shorter length of the radial engines in front of the centre of gravity resulted, from the outset, in the need for a sharper sweepback for the wings than was common in Mitchell designs. One other obvious departure from the Scapa was the fin and rudder arrangement. Two fins were employed, as before, but they were now closer together, in line with the thrust lines of the engines. As a result, the rudders were not increased in the same proportion as the other surfaces, with a consequent saving in weight and drag. The leading edges of the fins were now straight, matching the swept-back wings and the somewhat angular appearance of the hull; and they were a pleasing continuation of the line of the upswept tail pylon.

The Air Ministry requirement for a flying boat 'of robust and simple construction with low maintenance costs' resulted in the stainless steel fittings being given anodic treatment to inhibit corrosion and in the extensive use of Alclad, a new composite duralumin plate coated with pure aluminium on each side, again to counter corrosion. Its 'general purpose' character was evidenced by the fitting of carriers below the inner sections of the lower wings for up to four 250 lb bombs or extra fuel tanks; and the flatter fuselage section between the lower wing joins was even more convenient than that of the Scapa for transporting supplies, such as a spare engine.

At the time that the designing and the constructing of the new prototype were well under way, the Australian Seagull V order was still to be concluded, as was the Scapa flying boat contract. Nevertheless, the new prototype, K3973, was ready to be test flown by Summers on 27 July, 1934 and delivered in very short time to MAEE, Felixstowe for service assessment. The performance of the aircraft was such that an order for seventeen aircraft, K7287 to K7303, was placed with Supermarine by the following year. The standard service machine was fitted with the more powerful 920 hp Pegasus X engines and could outperform all contemporary flying boats of its class. It had a maximum speed of 165 mph, making it the fastest biplane flying boat to enter RAF service, yet had a stalling speed of only 51 mph. Its maximum ceiling was 20,000 feet and it could climb to the first 10,000 feet at a thousand feet per minute. As it was necessary to withhold these performance details because of the international situation, the company had to be content with the by-no-means despairing comment that the aircraft 'passed all its tests brilliantly' and went on to claim that 'the outstanding feature of this flying boat is that the performance obtained during a series of extended service trials, whether in respect of speed, climb, ceiling or take-off, is unequalled by any other British flying boat. All the specification requirements were exceeded by large margins'.

The Supermarine publicity concentrated particularly on the metallurgy of the aircraft:

TYPE. Twin-engined long-distance reconnaissance and bombing flying boat

WINGS. Unequal span biplane. Upper centre-section carried above hull by splayed-out struts supporting engine nacelles, lower ends of which are attached to lower wing-stubs. Each outer wing bay has two sets of slightly splayed-out parallel interplane struts. Wing structure of 'Alclad' with important fittings made of stainless steel, the whole being covered with fabric. Ailerons on all four wings.

HULL. Typical Supermarine two-step hull, made entirely of 'Alclad', except for the principal fittings, which are of stainless steel. Structure consists of transverse frames and internal longitudinals, the whole covered with smooth 'Alclad' plating.

TAIL UNIT. Monoplane type, with twin fins and rudders. 'Alclad' framework with fabric covering. All movable surfaces balanced. Trimming tabs in rudders and elevators.

POWER PLANT. Two Bristol 'Pegasus X' nine-cylinder radial air-cooled engines, mounted in 'Alclad' monocoque nacelles immediately below the upper centre-section. Townend rings. Openings in nacelles give access to all parts of engine requiring periodical inspection. Large manholes give access to rear of engines. Two 'Alclad' fuel tanks (250 Imp. gallons = 1,136.7 litres each) in upper centre-section, with pump-assisted gravity feed. Jettison valves fitted. Two oil tanks (26 Imp. gallons = 118.2 litres each) form leading-edge of centre-section. Oil coolers incorporated with the tanks. Hand and electric starting.

ACCOMMODATION. In the bow is a bombing and gunnery station. A hinged watertight door is provided in the nose for bomb-sighting. The gun-mounting is arranged to slide aft, clear of the cockpit, for mooring operations. The marine equipment is stowed in a compartment adjacent to the cockpit. Behind this is the pilot's enclosed compartment, with side-by-side seating for two, with dual controls. Immediately aft of the pilot's cockpit and forward of the front spar frame is the accommodation for the navigator, and between the spar-frames is the wireless operator's and engineer's position. Aft of the wings is the midships gun position and a further gun position is located in the extreme tail. A lavatory is provided in the rear portion of the hull. For cruising, provision is made for sleeping quarters, food and water storage, cooking, etc. Special equipment includes a collapsible dinghy, engine-changing derrick, engine ladder, maintenance platform, spare airscrew, etc. Folding drogues are stowed aft of the midship gun position. Provision can be made for the transport of a torpedo or a spare engine on the lower centre-section.

The Stranraer prototype on its beaching gear.

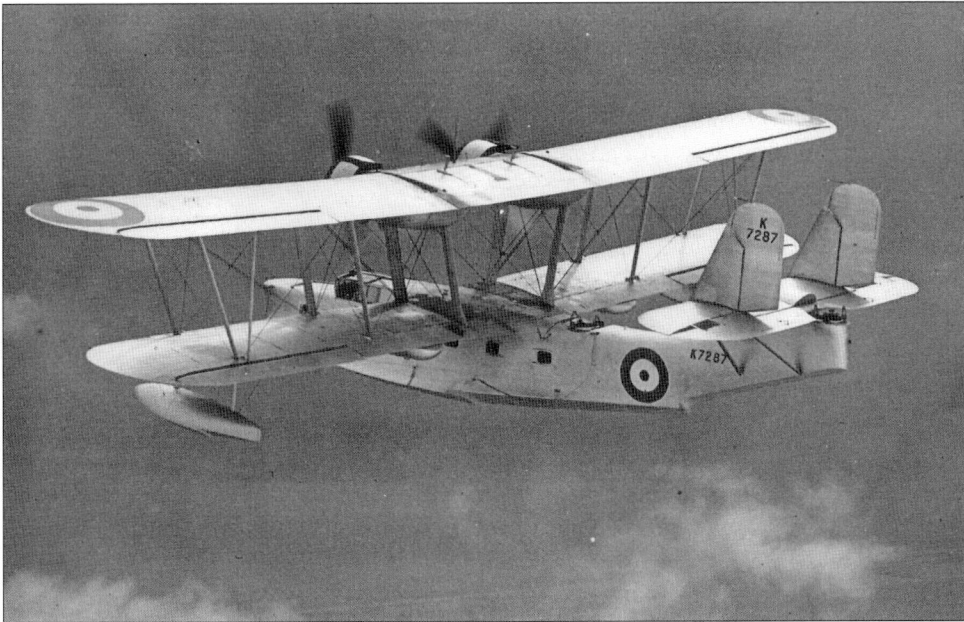

K7287, the first production Stranraer.

As with the Scapa, this aircraft had become quite distinct from the earlier machine from which it was developed and so a new name was chosen; the machine entered service as the Stranraer. It must have been gratifying for Supermarine to see the Saunders-Roe London flying boat replaced by their new aircraft with Nos 201 and 240 Squadrons and to see another rival company's aircraft, the Singapore III, superseded by the Stranraer with No. 209 Squadron. Other machines replaced the Scapas with No. 228 Squadron and so the total of Stranraers ordered from Supermarine, including the prototype, came to eighteen.

The pleasing new lines of the Stranraer prototype as seen from below.

WALRUS

Whilst the Stranraer was being built, the Seagull V was undergoing modifications and trials. 'Mut' Summers criticised the rigidity of the undercarriage and the lack of steering capacity on the ground and these deficiencies had had to be put right; and by this time it had been discovered that another make of aircraft had been allocated its N-1 number, so the one-and-only Supermarine prototype now became N-2. and went to the MAEE on 29 July, just over six weeks after its maiden flight. Evaluation tests then lasted until the end of October, after which the Seagull went to the Royal Aircraft Establishment at Farnborough for the catapulting trials required by the Australian Government. By this time, the empty weight of the aircraft had gone up to 5,016 lb and Mitchell, having returned to work at the end of 1933 after his operation for cancer, ordered an aerodynamic cleanup as parasitic drag had appreciably reduced the performance of the heavier machine.

Despite the lengthy testing, including successful catapult trials, no British order was expected; Webb, now in the business manager's department, quoted an Air Ministry letter, saying 'that we do not envisage any role for an aircraft of this type with HM forces'. He also related how serving officers at the nearby flying boat base at Calshot could also see no use for it: one of them asked 'What are you people doing wasting our time on a machine like that – it will be shot out of the skies by the fighters?' So Webb pointed out that there would be no fighters with enough range to shoot anything down in mid-ocean and that, catapulted from a cruiser or battleship, it would be the eyes of the fleet.

The second production Seagull V on HMAS Sidney.

A2-2, being catapulted from HMAS Sidney.

The prototype was then taken back to home waters for the continuation of trials at Sheerness and in the Solent until May when it was returned to Super-marine for the fitting of redesigned wing-tip floats for improved buoyancy, the removal of the wheel brakes for lightness and for an improved layout of the observer's compartment. Further fleet operation trials continued, including 'sea state' landings in 30-knot winds and six-foot waves off the Kyles of Bute and underway recovery onto a warship making up to 13 knots through rough water. In these trials the Royal Navy had been acting as 'programme manager' for the Australians and, as a result, their government finally ordered 24 production Seagull Vs, A2-1 to A2-24.

Thus it would appear that the future of Mitchell's design might rest solely with the Australian Government's requirement, but movements were afoot nearer to home, as reported by Caspar John:

> To the late Rear Admiral Maitland W. S. Boucher, DSO, Royal Navy [at that time serving in the Naval Air Division], goes the initiative for the introduction to the Fleet Air Arm of this somewhat improbable looking, yet highly successful flying machine.
>
> He said to me one day in late 1933, 'I've just been to Supermarines. I've seen a small amphibian. It looks handy, tough and versatile … something the Navy needs. I want you to put it through its service trials. Off you go' With a Supermarine Southampton flying boat course at Cal-shot and some tests at Felixstowe intervening, off I went to Woolston to collect Seagull V N-2 early in 1934.

Caspar John, son of the artist Augustus John and later Admiral of the Fleet, then took the Seagull to Gibraltar for rough weather take-offs and landings and for fleet co-operation exercises, after which it was described as 'the complete answer to our prayers'.

Thereafter, N-2 was purchased by the Air Ministry, renumbered K4797, and on New Year's Day, 1935, it was handed over to the Fleet Air Arm for a short series of official acceptance trials. As the prototype had first flown in June, 1933, it is clear that the Admiralty had needed some time to be convinced that open sea and catapult operation from their capital ships would work smoothly. No doubt the Australian initiative helped to overcome any doubters and an initial British order was now placed on 18 May for twelve aircraft. The first machine of the 1934 Australian order flew on 25 June 1935, and, after a further eleven had been completed, the Air Ministry took delivery of their own first batch. The rest of the Australian order was then completed, along with a second batch of eight for the Air Ministry, and the developing international situation then contributed to a much larger British order for twenty-eight.

A name was now to be chosen for the British machine, unlike those for Australia which retained the Seagull V appellation. In the past, Supermarine am-phibians had been favoured mainly with 'friendly' seabird names: Sea Eagle, Seagull, Sheldrake, Seamew and so on, whilst the name Sea Lion was a nod in the direction of the engine used. It is thus an interesting comment on this latest amphibian's 'somewhat improbable' appearance that the far less glamorous name 'Walrus' was now chosen.

Nevertheless, it was not only the first British aircraft to be catapulted with a full military load but it was also the first British-designed military aircraft with a retracting undercarriage. Thus it was that Admiral Sir Roger Backhouse received a ducking when the prototype Walrus, being used as his 'barge', was landed at sea with the undercarriage still down and turned turtle. After that, the Walrus was fitted with a horn to warn pilots of what was then a novel feature. The original batches of 48 aircraft ordered for the navy was increased dramatically in 1936 with the requirement for another 168 machines. Despite its initially very doubtful future and its backward-looking appearance, the Walrus then became the last and the most successful of all Mitchell's reconnaissance amphibians and the navy's standard wartime fleet spotter.

Alex Henshaw, better known for high speed flying and for testing the production Spitfires built at Castle Bromwich, has left the following affectionate memory of flying the Walrus (although some of his comments do not square with Supermarine publicity claims for pilot comfort quoted on p. 214):

> Most pilots looked upon the Walrus as an ugly duckling and I may have thought the same. There was, however, something endearing about it. I am not sure if it was the incongruity amidst the sleek fighters and that I felt sorry for it, or that it operated in an environment which appealed to me and that when the going got rough it did its job like a professional. Certainly it was one of the noisiest, coldest and most uncomfortable machines I have ever flown and I never seemed to be able to climb in or out of the cramped cockpit without leaving a piece of skin behind. Strong it certainly was, and it could be landed on grass with the wheels up without much damage. I never tried this but George [Pickering], on our first flight together, said, 'The Walrus is not for the absent-minded.' He then went on to tell me of pilots who had landed it on water with the wheels down, or landed on the tarmac wheels up – both with spectacular results. I always felt you could land it on a postage stamp or in a puddle of water when you got used to its rather strange ungainly ways. At first it reminded me of a large iron dustbin filled with empty soup tins: in rough water it seemed to float in about the same manner and with as much noise. Operating in calm weather was pleasant, orthodox and easy. In really rough seas, however, I can only describe the experience as a wrestling match blindfolded. The noise of the waves pounding over the foredeck, the hull hammering until it must surely cave in and the surging wind and the water cascading over the cockpit was all rather frightening. As you peered through a constant stream of water over the screen and opened the throttle the first bout of wrestling was on. If a sudden huge wave hit you before you were ready, you throttled back, took another breath and waited your opportunity to plunge in again. The trick was to judge your wave roll accurately and to watch out for the heavy foaming tops that sometimes accompanied them. Although I was nearly always cold when I started this exercise, by the time I had kicked the rudder hard to port and starboard a few dozen times, twisted, pushed and pulled the control column into my stomach, plunged through waves I felt sure would take us down to the ocean bed and then finally hung on to the prop as I literally lifted this clattering tin can into the air with the tail still clipping those furious waves below, I was in a bath of perspiration.

As this book has included a number of accounts of flying Mitchell's aircraft, one cannot resist a contribution from Ann Welch, a wartime ferry pilot:

> Further to my comment on sometimes being a bit vague as to what aircraft one was actually flying, because of the continuous chopping and changing of types – such as looking at the small print on the airspeed indicator to see if it was a Spitfire or a Seafire [i.e. calibrations in mph or knots]; one soon learnt little reminders, like if the rudder bar was under water you were in a Walrus. Not too much imagination is required here. Late one evening we had a Walrus for onferrying from Hamble to Lee. It had about 150 gallons of water sloshing about inside it, and the poor girl took off and sort of phugoided away into the dusk.

* * * * *

The eventual orders for the Walrus do not conceal the poorer showing of Super-marine's other staple product, the large flying boat: the Air Yacht did not live up to its original expectations and the cancellation of the Giant was a major blow to Supermarine's likely chances of becoming a leader in seaplane design. Whilst the Stranraer had the best range and endurance performance of all British Coastal Command aircraft until the advent of the Short Sunderland, it was the last of the biplane type and lack of urgent Air Ministry thinking about the needs of naval flying is evidenced by the fact that Short's Sunderland flying boat, the Stran-raer's successor, was the result of a civil order from Imperial Airways. The establishment in 1936 of Coastal Command might have been seen as a significant step in the right direction but, as the title of Andrew Hendrie's history of the service indicates, it was the 'Cinderella Service': rearmament was concentrated upon developing aircraft for Bomber Command and for Fighter Command whilst naval thinking was mainly concerned with its warship fleet.

Thus it was that Supermarine had had little encouragement to develop in any thoroughgoing way an advanced anti-submarine or reconnaisance aircraft to rival the Sunderland or the American imports, such as the Hudson or the Catalina. Ironically, however, Mitchell's last aircraft were to be a product of this unpromising situation: the Walrus was ordered in numbers because of concentration upon the warship fleet and, as the next chapter will show, the pressure to develop a land-based fighter, at the expense of dedicated naval aircraft, led to persistent support throughout the various teething troubles with the advanced technology of the Supermarine fighter project – support which was to produce the outstanding, if surprising, appearance of the company's non-marine fighter, the Spitfire.

Reginald Joseph Mitchell, CBE, in the garden of his house in Portswood, Southampton, following his 1931 Schneider Trophy success.

— *Chapter Eight* —

1934 to 1937
Mitchell's Last Landplanes – the Spitfires

The comprehensive winning of the Schneider Trophy in 1931 with a flyover might have been regarded by other nations as opportunist but it was the result of both Rolls-Royce and Supermarine being able to deliver, on the due date, new machinery that was perfectly reliable whilst being at the forefront of aviation technology. Indeed, in the last eight modern contests, twenty-two entries failed for mechanical reasons but, of the twenty-one aircraft which completed their event, one third were Supermarine machines. Only one, the early S.4 failed.

Thus, by the time that it became urgent to produce a new breed of British fighter, Mitchell's involvement in the design of a land-based single-seat fighter was not as unexpected as might have been the case earlier. In fact, Supermarine's engagement with the land-based fighter concept began almost at the same time as the last of his Schneider Trophy racers, when the Air Ministry issued specification F.7/30. This required a day and night fighter with a top speed of at least 195 mph at 15,000 feet, with the highest possible rate of climb, and armed with four machine guns – either all in the fuselage or two in the wings outside the propeller arc. (For full details, *see* overleaf.)

The fact that Mitchell did not succeed immediately in producing an effective response to this specification is partially because of trying to fulfill some of the Air Ministry requirements but his final achievement, with the Spitfire, can best be appreciated by looking first at the state of the British fighter provision in the 1920s and 1930s.

Specification F.7/30 and Contemporary British Fighters
Following the end of World War One, the huge surplus of military aircraft did nothing to encourage the development of new British state-of-the-art fighters, especially as the general mood of the country was for putting militarism to the back of the mind. Britain's pursuit of her interests and commitments in the Empire and the Middle East, it is true, led to new military actions but the employment of World War One types, unopposed against tribal uprisings, did not present a pressing need for newer machines. For example, the World War One Bristol F.2b fighter was used on the Turkey–Iraq border as late as 1925.

There was also the Government's 'ten year rule' assumption that no major war was likely to break out in the near future, thus giving plenty of time to speed up military development if and when danger was perceived. With such an optimistic view, it is not surprising that aircraft procurement procedures would also be leisurely (one notices that the 7/30 Specification is dated October 31 – perhaps also because of time taken up in agreeing compromises in its demands).

After an Air Ministry specification for a new type was issued, a design competition between interested manufacturers began; this was followed by an order for a prototype for one or more of the most promising proposals and this

The extracts below are from the Specification generally considered to be the most influential Requirement in the years before World War Two:

SPECIFICATION No F.7/30
1st October, 1931

1. General Requirements
    (a) The aircraft is to fulfil the duties of 'Single Seater Fighter' for day and night flying. A satisfactory fighting view is essential and designers should consider the advantages offered in this respect by low wing monoplane or pusher.*
        The main requirements for the aircraft are:
            (i)   Highest possible rate of climb
            (ii) Highest possible speed at 15,000 feet†
            (iii) Fighting view
            (iv) Manœuvrability
            (v)   Capability of easy and rapid production in quantity
            (vi) Ease of maintenance ...
    (b) The aircraft must have a good degree of positive stability about all axes in flight and trimming gear must be fitted so that the tail incidence can be adjusted in flight to ensure that the aircraft will fly horizontally at all speeds within the flying range, without requiring attention from the pilot ...
    (d) The aircraft must have a high degree of manœuvrability. It must answer all controls quickly and must not be tiring to fly. The control must be adequate to stop an incipient spin when the aircraft is stalled.
        An approved type of slot control, or other means which will ensure adequate lateral control and stability, at and below stalling speed, is to be embodied ...
2. Power Unit
    (a) Any approved British engine may be used ...
    (f) The airscrews preferably to be of metal construction, and is to be designed in accordance with the required performance of the aircraft as specified in paragraph 4 of this Specification ...
4. Contract Performance.
        The performance of the aircraft, as ascertained during the official type trials when carrying the total load specified in paragraph 3 [660 lb] and with an airscrew satisfying the requirements of paragraph 2 shall be:
            Horizontal speed at 15,000 ft not less than 195 mph†
            Alighting speed not to exceed 60 mph
            Service ceiling not less than 28,000 ft
            Time to 15,000 ft not more than 8½ mins
        The specified alighting speed must not be exceeded, but may be obtained by variable camber or equivalent devices provided that control and manœuvrability are not adversely affected ...
7. Disposition of Crew. Armament and Equipment
    (a) the Pilot's view is to conform as closely as possible to that obtained in 'pusher' aircraft. The following requirements indicate the ideal view which is considered to be necessary, and the aircraft should be designed to conform as closely to them as possible in practice.
        The pilot must have a clear view forward and upward for formation work and manœuvring, and particular care is needed to prevent his view of hostile aircraft being blanked out by top planes and centre sections ...
8. Arrangements for alighting and taking off ...
    (b) The aircraft is to be suitable for operation from small, rough-surfaced and enclosed aerodromes.

[*The implied alternative to a 'pusher' would be the more conventional 'tractor' configuration. Presumably, the specification assumed that, with a pusher arrangement, the pilot would be well ahead of the wing(s).

† Some accounts of F.7/30 erroneously give a specific speed to be aimed at.

The full text can conveniently be found in Alfred Price, The Spitfire Story.]

might lead to an order for a whole squadron of the type; after squadron evaluation, a larger order might then be forthcoming. C. G. Grey's estimation was that 'the time from issue of the specification till as many as three squadrons were equipped with the winning type was generally seven years'.

Air Ministry staff had obviously to justify their existence by requiring a modicum of new types but had no recent experience of aerial warfare to demand anything more than straightforward improvements on existing models. Also, the aircraft firms were reluctant to spend much Research and Development time on bold innovations – especially as the Air Ministry usually ordered only one prototype at a time from a firm – and so there was a distinct company tendency to play safe, gradually developing well-tried designs that were less likely to fail than machines with too many innovative features. Thus the British fighters of the next two decades continued the tradition of fixed undercarriage, wire-braced, biplanes with radial, air-cooled engines which, because of their lightness and compact shape, were, a least, exceptionally manœuvrable.

However, the penalty of this type of machine, with its attendant struts and wires, began to have an increasingly inhibiting effect on performance: the drag of a particular airframe increases, theoretically, as the square of its speed – consequentially requiring an eightfold increase in engine power to double an aircraft's speed. The following table reveals the increasing discrepancy between increases in engine power and speed in representative aircraft of this period, despite progressively greater attention to the streamlining of the newer aircraft:

Type	Date	Power	Top speed	Increase over previous type in	
				power	speed
Bulldog II	1929	490 hp	178 mph		
Gauntlet II	1936	640 hp	230 mph	30.6%	29.2%
Gladiator I	1937	830 hp	253 mph	29.6%	10%

One notable exception to the general pattern of fighter development was the Hawker Fury which in 1931 showed a 16% speed increase over the Bristol Bulldog for only 7% more power – the result, in particular, of using a closely cowled, in-line, water-cooled engine. It has been mentioned earlier that the Fairey Fox light bomber, using the American Schneider Trophy type of engine in a similarly streamlined cowling, could not be overtaken by British contemporary fighters. This embarrassment had led to the Fury, fulfilling the F.20/27 Specification which, significantly, called for a fighter aircraft 'capable of overtaking in the shortest possible time an enemy aircraft which is passing overhead at 20,000 feet at a top speed of 150 mph'. (Things were not much better as late as 1937 when a pilot of a Gloster Gauntlet, with a top speed not far behind that of its immediate successor, the Gladiator, had the new Bomber Command Handley Page Hampden (top speed 265 mph) come alongside and then pull away from the 'interceptor fighter', the pilot having made a familiar but rude gesture.)

The Hawker Hornet, the predecessor of the Fury, had also shown the advantage of an in-line power unit, but the new Air Ministry specification, F.7/30, gave no specific steer towards any particular engine configuration – contrary to the impression given by the employment of the in-line Rolls-Royce Goshawk engine in five of the competing designs, including Mitchell's.

The Gloster Gamecock which had entered squadron service in 1926 had a top speed of 155 mph and the Bristol Bulldog of 1929, still powered by a radial engine, was only about 20 mph faster – an average increase of about 7 mph per year. Thus, when the Air Ministry released its new fighter requirement, no more than, say, 185 mph might be expected if manufacturers responded to the new specification with the usual biplane configuration, air-cooled radial engine, and fixed undercarriage. It can thus be seen that the F.7/30 requirement for a top speed of 'not less than 195 mph', whilst carrying twice the number of machine guns, which was issued for the replacement of the Bulldog, represented a significant challenge (although tempered by realism) to an unadventurous aircraft industry.

The RAF had begun large-scale air exercises in 1927 and, as a result, it was appreciated how the biplane formula limited the fighter pilot's view of 'invading' bombers. In this respect, the F.7/30 requirement was quite specific: 'the pilot must have a clear view forward and upward' and 'particular care is needed to prevent his view of hostile aircraft being blanked out by top planes and centre sections'. One suspects a strong hint that monoplane prototypes might now be welcome but it is a reflection of the 1930 design scene that only three of the eventual eight entries were not biplanes. Mitchell's entry was, unsurprisingly, a monoplane but the only aircraft that was submitted with a retracting undercarriage also was not his. Again, there had been no specific Ministry requirement for such a feature.

Meanwhile, also in 1930, the Boeing 200 had appeared – a cantilever monoplane (with semi-retractable undercarriage) followed a year later by the Lockheed Orion – also a cantilever monoplane and with a fully retractable undercarriage. Later, in 1933, the Martin B-10 bomber appeared, a cantilever monoplane with retractable undercarriage and another American aircraft, the Douglas DC2 airliner of 1934, had similar features. The situation had not gone unnoticed outside the Air Ministry for, after the 1934 MacRobertson Trophy England–Australia Air Race had taken place, the *London Post* commented: 'It has suddenly and vividly been brought home that ... British standard aeroplane development, both commercial and military, has been standing still while America has been going ahead. It has been realised with astonishment that America now has, in hundreds, standard commercial aeroplanes with a higher top speed than the fastest aeroplane in regular service in any squadron in the whole of the Royal Air Force'.

With the performance requirements of F.7/30 given earlier, it is evident that the Air Ministry was feeling its way towards responding to developing American practice – and fortunately so as, by 1935, other leading aviation nations were beginning to produce or design monoplane fighters: the Russian Polikarpov II, the Japanese Mitsubishi A5M, the French Morane-Saulnier MS405, the American Curtiss P-36 and, of course, the Messerchmitt Bf 109. Some of these had retractable undercarriages and, in this respect at least, it was a pity that the F.7/30 did not contain a specific retractable-undercarriage requirement.

In respect of future monoplane design, Mitchell's Schneider Trophy experience should have stood him in good stead: one recalls that Gloster had only gone over to a monoplane design for the 1927 competition whereas Mitchell had produced four such models, beginning as early as 1925. However, the first Supermarine fighter design turned out to be more significant for prompting the company to enter the bidding rather than for actually providing a successful formula for winning a contract.

As we shall see, much of its under-performance was not entirely of Mitchell's making but it, at least, gives the lie to the still popular assumption that the legendary Spitfire emerged directly from the Schneider Trophy machines or rose as some single conceptual leap after its designer returned to work at the end of 1933. Any culturally biased view of Mitchell as a doe-eyed consumptive, gaining inspiration by looking at seabirds (*see* the film *The First of the Few*), could not be further from the truth.

In the first place he had to turn his mind to a military aspect of aviation that he had only briefly been engaged upon with the Sea King II fighter of 1921 – and that aircraft was a flying boat, albeit a fast and manœuvrable one. Additionally, armament on his slower reconnaissance flying boats was provided via gunners in cockpits not via guns which would now probably need to be buried in the wings. There exist some general arrangement drawings of a 'Proposed Single Seater Fleet Fighter' in response to Spec. 21/26 but nothing came of this and Alan Clifton, who joined Mitchell's design team in 1922, presumably had no knowledge of this project as he later recalled that Mitchell was uneasy about 'his first venture into military aircraft', recognising that he was 'no expert in the field'. After all, his racing machines had virtually no forward view, were designed purely for short spells of very high-speed and required an extremely long landing and take-off run, only feasible on large stretches of water. Thus, with typical pragmatism, Mitchell asked 'Mutt' Summers, Vickers chief test pilot, to arrange a visit to the Martlesham test centre to find out what the RAF pilots considered to be most important in a fighting machine.

TYPE 224 – THE FIRST SPITFIRE

The outcome of Mitchell's thoughts was the Type 224 design and it was submitted in response to the F.7/30 specification. It was, unsurprisingly, to be an all-metal monoplane and it was to be powered by a Rolls-Royce Goshawk engine. The Air Ministry's only requirement of the type of power plant for the new fighter projects was that it should be an 'approved British engine'; Supermarine, along with Bristol (Type 123), Hawker, Westland and Blackburn, chose this Rolls-Royce engine which was predicted to produce an impressive 660 hp (compared with the 450 hp of the Bristol Jupiter VII radial engine of the current fighter, the Bristol Bulldog). As the Goshawk was an in-line engine, its low drag profile might also be expected to contribute to an impressive performance, comparable to that which had been achieved by the Fairey Fox with the Curtiss D-12 engine.

Additionally, the new engine was designed to work with an elegant new cooling system, the so-called evaporative method, which was expected to bring with it significant reductions in drag – by not requiring conventional radiators to keep the engine coolant below normal boiling point. (Water boils at 100 °c at standard atmospheric pressure, but this temperature decreases with altitude – thus its ability to carry heat away from the engine drops, to the point where at high altitudes an extremely large conventional radiator would be needed. In the new system, the water in the engine is kept under pressure by pumps, allowing it to heat to 150 celsius and then the superheated water is released to turn to steam in a suitable container, with sides exposed to the airflow, where it would condense on cooling and be returned to the engine. The greater efficiency of this cooling method would therefore require an aircraft to carry less water and could operate via, and under, the skin of the aircraft, resulting in a zero-drag cooling system.)

Mitchell might have been permitted a wry smile at a return to something akin to his early experience as an apprentice at the Fenton locomotive works and Supermarine would have felt particularly confident about this aspect of the F.7/30 project, as their Type 179 large passenger-carrying aircraft was to have employed steam-cooled engines, using the leading edge of the wing for the condensers – in its turn, a development of the low-drag surface radiators used by the S.5 and S.6s. Their submission to the Air Ministry made specific reference to their experiments in connection with the earlier Type 179 wing structure and cooling system in order to strengthen their case. In the Type 224 proposal,

Supermarine TYPE 224 (1934)

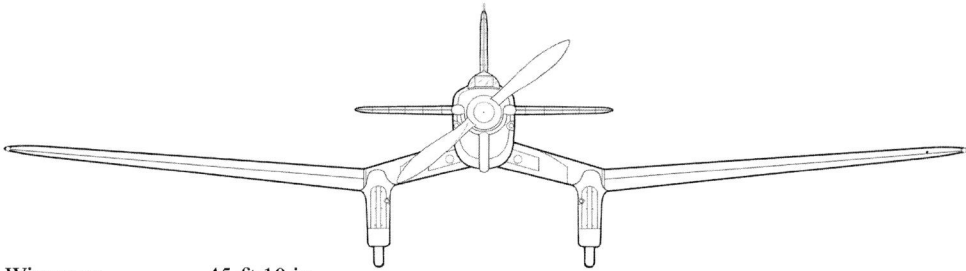

Wingspan	45 ft 10 in.
Wing area	295 sq ft
Loaded weight	4,743 lb
Maximum speed	228 mph

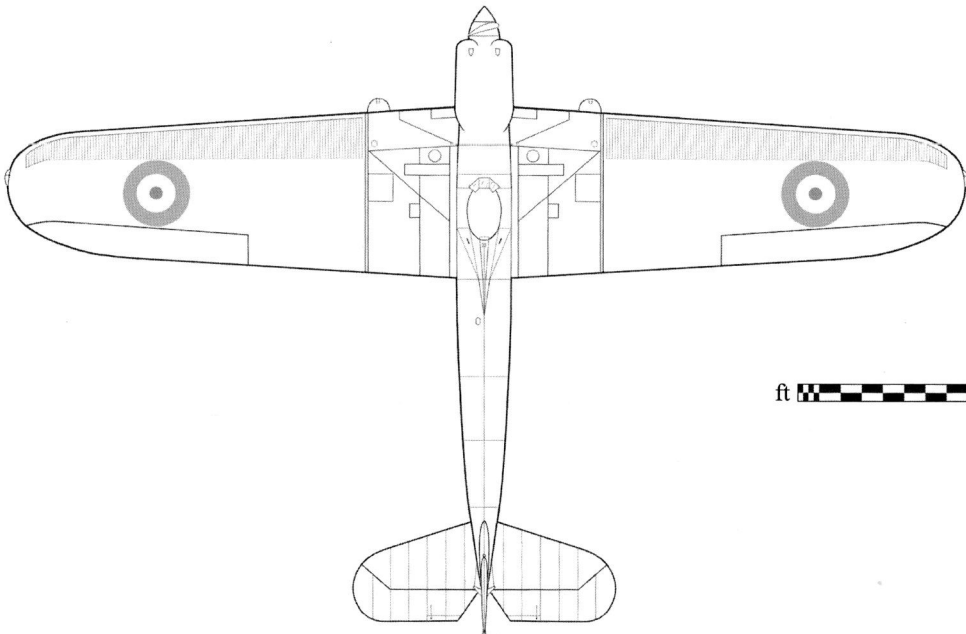

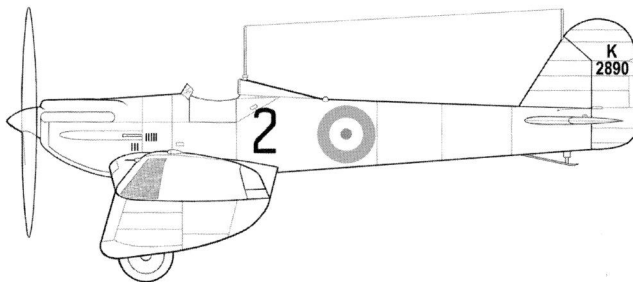

ft

K 2890

2

however, the leading edge condensers were to be made more thermally efficient for their size by having fore-and-aft corrugations exposed to the airflow, although this resulted in some drag penalty – an obvious design compromise, given Mitchell's move away from the radiator corrugations of the American Curtis CR-3 racers in 1927.

The Supermarine submission also made reference to information gained from the Vickers wind tunnel even though, by now, Mitchell was becoming sceptical of this relatively modest facility and its small scale test models as predictors of how full scale aircraft would perform. Although straight wings and a retractable undercarriage had been considered, the eventual inverted gull wing, fixed under-carriage configuration had nevertheless been arrived at in conjunction with com-parative wind tunnel tests. As Supermarine had been devising undercarriages which could be raised ever since the Commercial Amphibian of 1920, it might have been expected that a fully retractable, drag-reducing, arrangement would have been incorporated in their Type 224 entry in the early 1930s. However, the cranked wing meant that a fixed undercarriage would be reasonably short and light and would be advantageous for certain other new requirements.

One of these was that a gun could be housed in each undercarriage fairing without attempting to bury the armaments in the wings – whose leading edges were to be used for evaporative engine cooling – or without incurring any extra drag by more external wing mountings (the method to be adopted by the Gloster Gladiator). In addition, tanks for the condensed water coolant could also be situ-ated in the fairings and, in conjunction with the cranked wing configuration, produced two very convenient low points where the coolant could collect. Super-marine's submission also drew attention to how the cranked wing would allow a short, low drag undercarriage and, additionally, pointed out how it would pro-vide a wide track for easy taxiing, would give a low centre of gravity, and pres-ent 'exceptional' visibility for the pilot 'for fighting, formation work, or landing' (F.7/30).

Supermarine Type 224 under construction.

The wind tunnel was additionally used on the wing configuration to find the best combination of anhedral for the inner wing panels and dihedral for the outer ones, in order to achieve lateral stability; and, rather unusually, a full scale mock-up of the open cockpit section was used to test the effect of various shaped wind-screens upon cockpit draught – evidence that Mitchell was also paying due atten-tion to the F.7/30 requirement that 'the cockpit is to be adequately screened from the wind' (although, by this time, the use of covered canopies was becoming more widespread as speeds increased – despite many pilots' preference for an open cockpit).

With the fuselage generally, the company was on more familiar ground, using Schneider type constructional techniques, including flush riveted panels, in order to provide 'a rigid mounting for the tail unit and for the prevention of tail flutter' (*pace* the S.6B). Their submission to the Air Ministry also drew attention to the cantilever tail unit (which had been a feature of the S.5 and S.6 racers) and to the use of mass balances in the control surfaces (a neater solution than the ad hoc application of bob weights to the S.6 machines). By this time, flutter had become an increasing preoccupation as speeds had increased and Supermarine were obviously at pains to assure their potential customer that they knew what they were doing in this respect.

In view of the Ministry's slow landing speed requirement, Mitchell's Schneider experience was not much help as the wing loading of the S.6B had been 42 lbs per sq ft. A large air brake was now employed and this could be lowered from the underside of the fuselage but the Air Ministry was, nevertheless, concerned that the projected wing-loading of only 15 lb/sq ft was too high: their requirement was not just for a day fighter but also for a night fighter, which would have to land with very limited assistance from the ground. The Air Ministry requirement that aircraft submitted must 'be suitable for operation from small, rough-surfaced and enclosed aerodromes' had an inhibiting effect upon designing a top speed machine, especially where increased streamlining was leading to flatter and longer gliding approaches. As a result, the wing span of Type 224 was eventually fixed at a generous 45ft 10 in. and, in combination with a fuselage about the same length as the 30 ft span S.6, it looked somewhat out of proportion.

Before Mitchell's operation in 1933, Supermarine received the official 'Instructions to Proceed' with the Type 224 as detailed above and the aircraft was ready for testing early in 1934, just before the first flight of the Stranraer. On the 19th of February, 'Mutt' Summers took it up for its first flight and it duly appeared at the RAF Display at Hendon on June the 30th.

Type 224 with 'New Type' number 2, Hendon, 1934.

But its performance proved a disappointment and well below the company's estimated top speed of over 245 mph and a climb to 15,000 feet in 6.6 minutes. Also, in flight, the leading edge condensers expanded and the distortion of the wing caused the ailerons nearly to jam, an eventuality which would have spoilt the test pilot's day. Unfortunately, this problem was not the main one.

The real difficulty was with the cooling system as a whole. It had been arranged for the water which had condensed in the leading edge and had collected in the tanks in the undercarriage fairings to be pumped up to a header tank behind the engine; however, at the low pressure side of the pump, the water would often turn into steam again, the pump would cease to operate, and plumes of steam would be seen escaping from wing tip vents (Mitchell's early apprenticeship to a locomotive manufacturer did not include high altitude problems!). Jeffrey Quill, who had by now joined Vickers as a test pilot, has recorded how he did not exactly please the Chief Designer when he commented on these problems:

> The evaporative cooling system was a real pain in the backside, with the red [warning] lights flashing on all the time. I once made a jocular remark to Mitchell about the system. I said that with the red lights flashing on all over the place, one had to be a plumber to understand what was going on. He didn't say anything, he just looked very sour. He was rather sensitive about the aeroplane and obviously I had trodden on his toes.

As the cooling difficulties occurred particularly during rapid climbs, this meant having to level off until the system was working normally again; one can appreciate why Mitchell was not happy with his prototype, designed for the fastest possible climb to 15,000 feet in order to intercept enemy bombers.

Rival F.7/30 Aircraft

Competing fighter prototypes (*see* drawings overleaf) had not faired too well either, particularly those fitted with the evaporatively cooled Goshawk engine, which had produced similar cooling problems as with Type 224.

Thus the steam condensers of the Westland F.7/30 could not cope with varying flight and atmospheric conditions and the Blackburn F.7/30 exceeded maximum temperatures within a few minutes. Additionally, the latter had ground handling problems, caused by its high centre of gravity. Repeated taxiing tests caused cracks and dents in the metal fuselage skin to the extent that the Air Ministry withdrew its support for the project; it never flew, thus justifying the opinion of its test pilot, 'Dasher' Blake, that 'the little beast has no future'. The performance of the Westland entry had been diagnosed as 'woeful' as its top speed was a disappointing 146 mph and it took nearly 19 minutes to reach 20,000 ft – indeed, it had some similarities in appearance with the company's army cooperation Lysander but this was no recommendation for a fighter proposal. Two other prototypes powered by the Goshawk engine, the Hawker P.V.3 and the Bristol Type 123, fared little better. Type 123 was laterally unstable, finally put down to wing overhang flexing, and was withdrawn; and the Hawker entry paid only a brief visit to Martlesham as its design was essentially merely a re-engined Fury – but, at least, it led Camm later to the Hurricane.

Two of the above four aircraft, the Blackburn and the Westland appeared to have been unduly influenced by the Ministry concern for a good pilot position and, like the other two, were, basically, conventional biplanes with fixed undercarriages. Of the two monoplane rivals to Mitchell's Type 224, the Vickers Type 151 Jockey had begun as a rival to the Hawker Fury for Spec. 20/27 and was using the Wibault-Vickers corrugated metal skinned method that had too high a drag penalty at the speeds now being demanded. The intention to power the Jockey with a supercharged Mercury engine was never realised after the sole prototype was lost to a flat spin in June 1932.

But at least the other monoplane might very well have attracted favourable Ministry support, instead of the Spitfire: Bristol decided to improve upon its Type 123 entry by substituting the Goshawk for a Bristol Mercury engine. This new Type 133 was a considerable improvement on its predecessor in other ways: despite now having an air-cooled engine, it was the first proposed RAF fighter with retractable wheels and its stressed-skin construction, employing the recently

Blackburn.

Westland.

Hawker.

Bristol (Type 123).

Vickers.

Supermarine.

Bristol (Type 133).

F.7/30 Entrants.

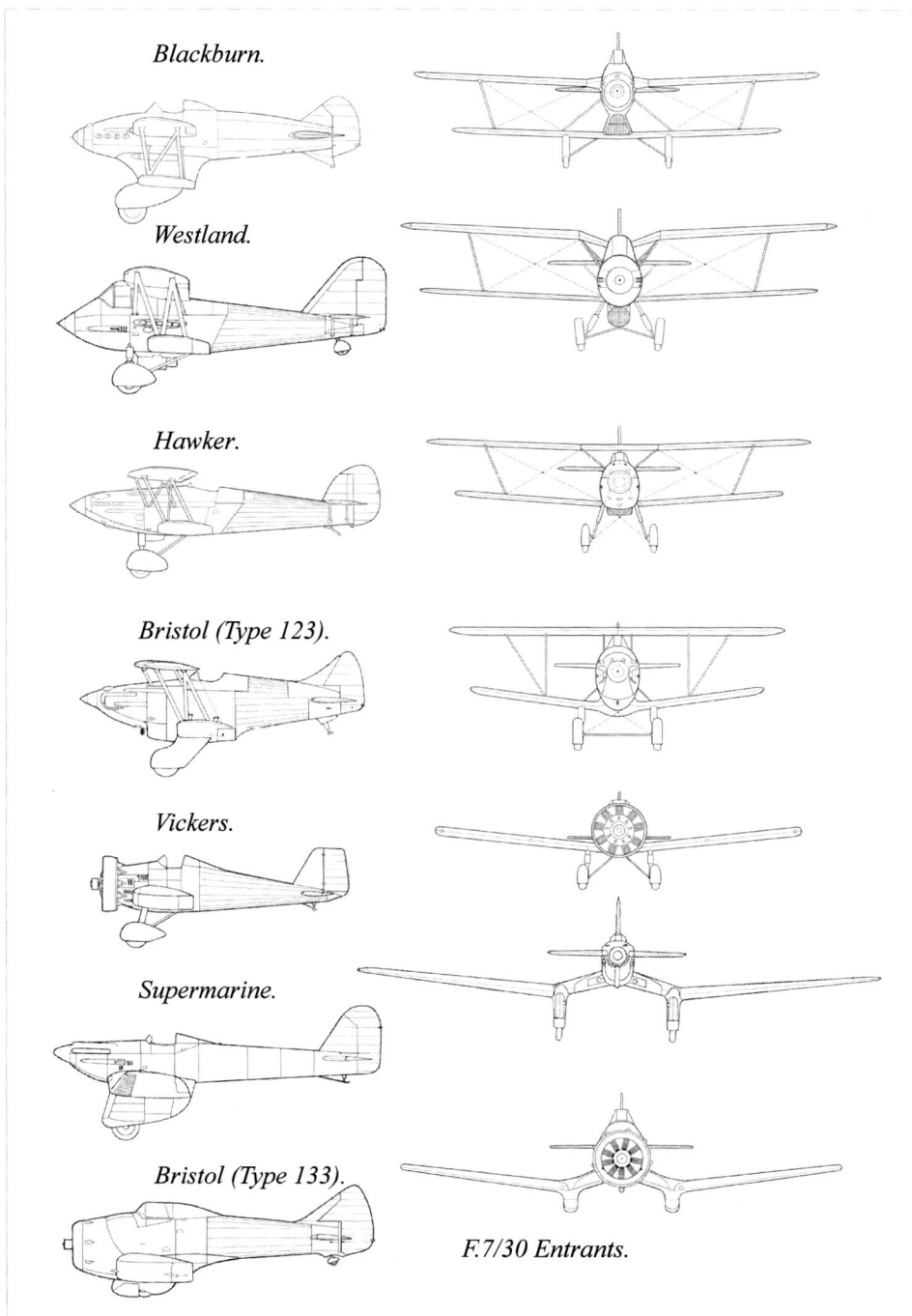

invented Alclad sheeting, allowed for the design of a forward-looking cantilever monoplane. Its test pilot Cyril Uwins was very impressed by its performance and the top speed of 260 mph but, when the aircraft was almost ready to go for competitive tests at Martlesham Heath, it also entered into a flat spin which was irrecoverable and the test pilot had to abandon another one-and-only prototype.

Meanwhile, the F.7/30 competition had one late entry, the Gloster SS.37, of a proven, fabric covered, biplane configuration. Its obsolescent airframe included a fixed undercarriage and was powered by an air-cooled radial engine, yet it still attained the requirements of F.7/30. One remembers Gloster's adherence to the biplane configuration in the Schneider Trophy contests, when others were turning to monoplane layouts, and so their racing experience had also proved beneficial and gained them an Air Ministry contract – for the aircraft which was to become known as the Gladiator.

The much larger wingspan of Type 224 made for more lethargic handling than that of the Gladiator. This fact might not have been too critical in view of the main Ministry requirements of a fast pursuit aircraft – speed and climb – but, in these respects also, Mitchell's aircraft was not competitive. The best performance figures that the Supermarine prototype had finally been able to achieve were a maximum speed of 228 mph and a climb to 15,000 feet in nine and a half minutes; the Gladiator could manage 242 mph and a climb to the same height in six and a half minutes – indeed, its predecessor, the Gloster Gauntlet II, which was about to enter squadron service, had a maximum speed of 230 mph and could climb to 20,000 feet in 9 minutes.

Thus it was that Mitchell's Type 224 design was to remain part of the history of unsuccessful experimental aircraft although it was to have been called 'Spitfire' – as Supermarine indicated via a brief announcement in 1934:

> The 'Spitfire' is a single-seat day and night fighter monoplane built to the Air Ministry specification. It is a low-wing cantilever monoplane with the inner sections sloping down to the undercarriage enclosures. It has a Rolls-Royce 'Goshawk' steam-cooled engine with condensers built into the wing surfaces. Armament consists of four machine guns. No further details are available for publication.

* * * * *

Mitchell's Illness.

Despite the potentially debilitating effects of the colostomy bag he had to wear for the rest of his life, Mitchell was determined to continue as before at the end of his convalescence. Sir Henry Royce, who had had to undergo the same procedure, also continued working but did so from his home. Mitchell, keeping his condition private, returned to his office at the end of 1933 and embarked on a work rate even greater than before – assessed by Alan Clifton, who later became Chief Designer himself, as a pouring out of new design proposals at an average rate of one per month: these designs also continued the wide range that was typical of the output that he oversaw – six proposals related to amphibians, eighteen to flying boats, three to fighters and one to a bomber. Typically, he also designed an improved colostomy bag.

Incidentally, his work rate was *not* driven by visiting Germany and seeing their aviation industry gearing up for war – as depicted in the wartime film, *The First of the Few*. It is fairly certain that he never visited that country and it is not known if Mitchell had heard that Lady Houston (as early as 1931) had sent her secretary-companion to the home of Neville Chamberlain, at that time Chancellor of the Exchequer, with a £200,000 cheque to finance a squadron of fighters for the defence of London against Germany (*see* Appendix Six). On the other hand, he would most probably have been aware of Churchill's warnings about Germany's developing air power. Armed with secret information, his message could hardly have been clearer when, on 28 November, 1934, in the House of Commons, he asserted that, by the end of 1935, Germany's air forces would be at least as strong as Britain's, that by the end of 1936 they would be nearly 50% stronger, and that 'Germany has between 200 and 300 machines with a speed of 220 to 230 mph, now carrying mails and passengers but convertible into long distance bombers in a few hours'. And Mitchell was obviously close enough to the aviation fraternity, who were well aware of the situation – in particular Bulman who had toured German aviation establishments twice and had reported back his concerns about what he had learned.

Mitchell had to hope that his bowel cancer (not mentioned in the film biography) would not return or spread if he were to play a full part in responding to the growing threat. As mentioned in Chapter One, boating had long been a part of his creative make-up and now, despite his condition, he added a third dimension to his sailing by taking up private flying. He gained his pilot's licence on 1 July, 1934, in a de Havilland D.H.60 Moth and continued flying until very close to his final illness in 1937. Climbing into the cockpit of a 30s biplane is none too straightforward at the best of times and it throws further light on Mitchell's determination not to be defeated by his condition. Indeed, he won the 1936 Hampshire Aero Club spot-landing competition – the trophy, inscribed with his name, is now in the possession of the Solent-Sky Museum.

* * * * *

Attempts to improve Type 224.

After Mitchell returned to work in 1933, new ideas for Type 224 were submitted in July, 1934, and these included a retractable undercarriage, the removal of the corrugations in the leading edge of the wing and elimination of its cranked configuration (as described in Supermarine Specification Number 425a – see opposite). The proposed modifications were expected to improve the maximum speed of Type 224 by 30 mph but the official response was lukewarm: ideas at the Air Ministry were changing, as will be discussed below.

Mitchell had already made some alterations to improve the performance of Type 224; in particular (and in respect of the future Spitfire): wider span, slimmer chord ailerons were fitted and various sorts of wing root fillets tried out. None of the Schneider racing machines had had any fairings even though such an improvement had been evident as early as 1930 with the Northrop Alpha. This aircraft was further developed two years later into a fast transport, the Northrop Gamma, which also featured a fully enclosed cockpit. Both these features, which were lacking in Mitchell's original 1931 design of Type 224, were also evident in two designs for entry in the MacRobertson Trophy England–Australia race of 1934. The sleek winning de Havilland D.H.88, was one and another was the Bellanca 'Irish Swoop' which Mitchell inspected closely when it was flown to Southampton for the fitting of its cowling. (It was reported that Supermarine had wanted to enter the Type 224 in this race but the Air Ministry would not allow its new, one-and-only, prototype to take part.)

Type 224 with tufts fixed to observe effect of airflow over experimental wing root fillet.

The following proposals for considerable modifications to Type 224 show that progress towards the eventual Spitfire was by no means direct or immediate:

SUPERMARINE SPECIFICATION No.425a
26th July 1934.

Supermarine Day and Night Fighter to Air Ministry
Specification F.7/30
Proposed Modifications

Arrangement
    It is proposed to modify the existing aeroplane by building a new pair of wings incorporating a retracting chassis. The new wings are of reduced area, the present inner sections of negative dihedral being dispensed with. Split trailing-edge flaps are provided to increase the lift and thus retain the same landing speed. The existing aircraft is shown on Drawing No.22400, Sheet 1 and the proposed arrangement on Drawing No.300000, Sheet 2.*

Construction
    Construction is greatly simplified by making each wing in one piece. Other features which simplify construction are the substitution of lattice for a plate web, enabling the riveting of the nose to be more easily carried out, and the provision of smooth in place of corrugated covering.

Pilot's View
    As a result of eliminating the downward-sloping wing roots, the view for a pilot is not quite so good close in to the fuselage, but is improved further out by the reduction of span.

Main Particulars and Dimensions

	Existing Machine	Modified Machine
Span	45 ft 10 ins	39 ft 4 ins
Length Overall	29 ft 10 ins	29 ft 4 ins
Wing Area (gross)	295 sq ft	255 sq ft
Wing Loading	16.8 lb/sq ft	18.4 lb/sq ft
Power Loading	8.25 lb/bhp	7.85 lb/bhp

Weight and Performance
    It is estimated that the proposed modifications will result in a saving of 250 pounds weight, and an improvement in top speed of 30 mph, the climb remaining practically unaltered. The following features giving the comparisons between the existing and the modified aeroplanes, are estimated. Performance tests so far carried out are incomplete, but present indications are that the estimates are reasonably correct.

	Existing Machine	Modified Machine
Weight	4,950 pounds	4,700 pounds
Max. Speed	235 mph	265 mph
Climb to 15,000 ft	8 mins	8¼ mins

[*Drawing 300000 referred to forms part of the diagram of the development of the Spitfire wing given later (*see* p.242).]

    The proposed reduction in wingspan and the retractable undercarriage had led to an abandonment of the Type 224 inverted gull wing. Meanwhile, Junkers were producing the similarly configured Stuka; it notably featured an inverted cranked wing and fixed undercarriage; in the case of the early models, the wheels were similarly encased:

Junkers Ju 87 Stuka.

It first flew on 17 September 1935 and, in the same year, Blohm und Voss produced their Ha 137, a ground attack aircraft which also had a cranked wing and a 'trousered' undercarriage. This similarity can also be seen in the Kawasaki Ki-5 – not surprisingly as it came from the same designer.

The Japanese fighter had a span about 11 ft less than the Supermarine prototype and, with over 200 hp more power, was expected to reach 240 mph. Despite Mitchell's disappointment with his fighter, it can be seen that its top speed of 228 mph speed was actually quite creditable, given its lower engine power and the imposed penalty of the much larger wingspan. It might also be pointed out that Mitchell's design appeared contemporaneously with the Kawasaki Ki-5 and before the two German aircraft, thus showing that there was nothing eccentric or derivative about Mitchell's basic approach to the F.7/30 specification.

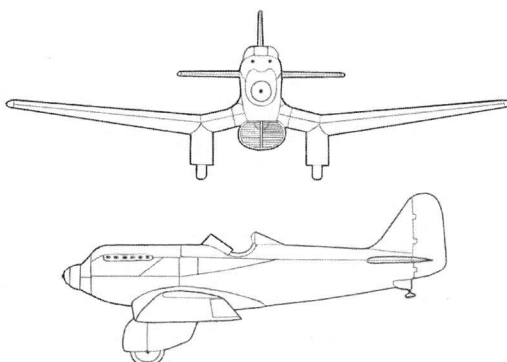

Ki-5.

Nevertheless, had Mitchell only lived to 1935, he would have been known for designing the first RAF standard coastal flying boat after the end of the first World War, the standard World War Two fleet spotter amphibian, and the Schneider Trophy winning and World Airspeed breaking floatplanes between these two wars. The name Spitfire might very well have been allocated only to Type 224 and to a footnote in aviation history.

THE REAL SPITFIRE

By the time of Mitchell's return to work, there had been external developments to energise the aircraft industry and, eventually, to give more focus to fighter development. Previously, the only remote threat to Britain was France and her Empire and, by the early 1920s, the French Air Force was becoming quite sizeable. Then, the Japanese invasion of Manchuria in 1931 and the failure of the League of Nations to respond effectively led to the end of the 'ten year rule'; and then, of course, by 1933 Germany began to emerge as a distinct threat to peace.

The reaction of the British Government, however, had not immediately given rise to vigorous fighter rearmament – influenced by the view, expressed by the Prime Minister, Stanley Baldwin, that 'I think it is as well for the man in the street to realise that there is no power on earth that can save him from being bombed. Whatever people may tell him, the bomber will always get through. The only defence is offence'. This led to a focus upon bomber development as a strong opposing force, also able to 'get through', which was considered to be a sufficient deterrent to possible aggressors. However, it was only to be expected that the nation would be happier in the knowledge that there could be a defence against the bomber and it was thus to the credit of Neville Chamberlain that, during this time, he was advocating the need for effective, that is fighter, defences.

Personnel were also changing in the Air Ministry. The Chief of the Air Staff, Sir Hugh Trenchard, who had favoured the bomber approach, had retired in 1927 and his successor, Sir John Salmond and his Deputy, Sir Charles Burnett, were more sceptical of this approach. They were very actively supported by Hugh Dowding who joined the Air Staff in 1930 as Air Member for Research and Supply; as he later became the Commanding Officer of RAF Fighter Command, it is not surprising that he held the view that 'the best defence of the country is fear of the fighter'. In addition, from 1934, an Operational Requirements Committee was formed for the purpose of bringing together the views of the Air Staff, the revised Research and Development Committee and the operational Commands of the RAF.

The following pages will describe the unusual way in which the eventual Spitfire came into being and the support that was required when doubts began to arise about its practical viability. It was thus fortunate for Mitchell that the Air Ministry and its departments were headed by RAF officers who had by now come to the conclusion that fighter development had to be significantly stepped up. Whilst F.7/30 can be seen as the milestone specification which brought our seaplane designer into the reckoning, the future support of some Air Ministry officials in certain difficult days ahead for the Spitfire-to-be had much to do with the close and mutually respectful relations between Supermarine and the RAF since the Southampton and S.5/6 days. It might be noted that two particular Ministry men were active at this time: Group Captain Cave-Brown-Cave, the Director of Technical Development from 1931 to 1935, who had led the Far East Flight in one of Mitchell's Southamptons which had never let his team down, and Major Buchanan who had been the Air Ministry representative at the 1925 Schneider Competition and vocal afterwards in his support for British participation in these events.

Type 300 – An Uncerain Start

When the Type 224 'Spitfire', was failing to satisfy the Air Ministry F.7/30 specification, there was another machine of forward-looking design which might very well have attracted favourable Ministry support instead: the Bristol Type 133, was another monoplane and was also the first British fighter design with both retractable wheels and stressed-skin construction. It additionally had a more streamlined sliding cockpit canopy and the wing roots were aerodynamically faired into the fuselage with fillets (which Mitchell had only experimented with during the later days of Type 224). However, as mentioned previously, the one-and-only prototype crashed and so time was available for Supermarine to try to improve upon their proposal – although, in view of the British manufacturers' disappointing responses to F.7/30, there was even talk in the Ministry of purchasing Poland's all-metal monoplane, the PZL P.24. This monoplane was faster than any of the British prototypes and many were equipped with the more impresssive firepower of two cannons as well as two machine guns.

Also, in view of more urgency being given to fighter development and even before trials of the F.7/30 prototypes had been concluded, Specification F.5/34 was issued, representing even more stringent demands on the aircraft designers. The first paragraph made very clear the current unsatisfactory situation as regards the contemporary two-gun fighter and, indeed, proposed more armament than did F.7/30:

> This specification is issued, therefore, to govern the production of a day fighter in which speed in overtaking an enemy at 15,000 ft, combined with rapid climb to this height, is of primary importance. The best speed possible must be aimed at for all heights between 5,000 and 15,000 feet. In conjunction with this performance the maximum hitting power must be aimed at, and 8 machine guns are considered advisable.

In later paragraphs, an enclosed cockpit was regarded only as 'admissible' and radial or evaporative cooling was not ruled out; but, on the other hand, the under-carriage was 'required' to be retractable.

It was also specified that 'the maximum speed at an altitude of 15,000 feet shall not be less than 275 mph and at 5,000 feet not less than 250 mph'; and 'the time taken to reach 20,000 feet is not to exceed 7½ minutes'. This new specification was needed to provide a replacement for the Hawker Fury, Britain's most advanced fighter, whilst no doubt also implying serious disappointment with the F.7/30 examples currently testing at Martlesham. As Mitchell's 425a submission had only promised a top speed of 265 mph and a climb to 15,000 feet in over eight minutes, it is not surprising that these Type 224 modifications had only received a lukewarm response from the Air Ministry and had sent Mitchell back to the drawing-board in earnest.

The successor to Type 224, was at first still to be designed around the evaporatively cooled Goshawk engine as, whatever the problems, it offered a considerable reduction in drag. The new machine was now to have a reduced wingspan of just over 37 ft, an enclosed cockpit, and a retractable undercarriage. The Air Ministry, responding to the prospect of an eventually successful Mitchell fighter in October 1934, requested Supermarine to submit a quotation for a modified machine, identified at that time as Type 300; they also suggested that it should be fitted with a Napier Dagger engine, expected to develop about 800 hp – 140 hp more than the Goshawk. However, Rolls-Royce had, by then, decided that their current engines were not capable of being developed into the sort of power plant needed for the next generation of fighters. Something between their 21 litre Kestrel and Goshawk engines and their 37 litre Schneider Trophy 'R' engine, was thought to be appropriate and so the company had begun design work on a further 12-cylinder Vee engine, initially expected to deliver 1,000 hp. This proposal was, of course, to become the famous Merlin but was first known as the PV12, in which the initials stood for 'private venture' and clearly indicated the engine company's own appreciation of the need for Britain to develop new aircraft.

As this new engine had passed its 100 hour test in July, 1934, the board of Vickers (Aviation) Ltd decided on 6 November to finance the design of a PV-12 powered Type 300. Sir Robert McLean, the chairman of the Vickers board, some time later, and perhaps exaggerating with hindsight, described how he had decided that Mitchell and his team should design a 'real killer fighter' in advance of any Air Ministry specification and that 'in no circumstances would any technical member of the Air Ministry be consulted or allowed to interfere with the designer' – no doubt with the unhappy history of Type 224 in mind.

It has been recently revealed that the chairman had not, in fact, been unwavering before his support for the Supermarine designer finally won out. The alternative was the parent company's Venom, a development of the promising but ill-fated F.7/30 entry, the Type 151 Jockey, which had also succumbed to a flat spin. Like the future Spitfire, the Venom had a stressed skin cantilever wing, retractable undercarriage and a metal monocoque fuselage. In fact, when it did fly, three months after the Spitfire prototype, it attained a top speed only 37 mph lower than the Supermarine prototype – and with a less powerful engine. Clearly, this machine could also have developed into a more successful fighter than the Hurricane and could have been a very serious challenge to the Supermarine project; indeed, Beverley Shenstone, Mitchell's aerodynamicist, later reported that, in his opinion the Spitfire would not have been born 'if Mitchell had not been willing to stand up to McLean, particularly in the era when McLean clearly preferred the Venom concept to the Spitfire concept because it was cheaper and lighter'.

Once Mitchell's proposal had overcome this hurdle, the combination of a Vickers/Supermarine/Rolls-Royce/Mitchell design must have warmed up the new blood within the Air Ministry, for events then moved very quickly. On 1 December, £10,000 was allocated for Supermarine to build a prototype and, when a full design conference was called at the Air Ministry on the 5th of the same month, it was headed by Air Marshall Hugh Dowding.

The modifications to Type 224 proposed by Supermarine had been deemed too late and too extensive to qualify for re-entry into the F.7/30 exercise but, fortunately, the Air Ministry, in their concern to improve the fighter breed, agreed to Supermarine proceeding independently with their new Type 300 design and a special specification F.37/34 was drawn up:

> SPECIFICATION F.37/34
> 3rd January 1935
>
> Experimental High Speed Single Seat Fighter
> (Supermarine Aviation Works)
>
> 1. General
>   This specification is intended to cover the design and construction of an experimental high-speed single seat fighter substantially as described in the Supermarine Specification No 425a and drawing No 300000 Sheet 13, except that an improvement in the pilot's view is desirable. The aircraft shall conform to all the requirements stated in Specification F.7/30 and all corrigenda thereto, except as stated hereunder.
>
> 2. Power Unit
>   (a) The engine to be installed shall be the Rolls Royce P.V.XII ...
>   (d) A cooling system is to be of the evaporative cooling type using wing condensation in association with an auxiliary radiator ...
>
> [*The full text is conveniently reproduced in Alfred Price's* The Spitfire Story; *above are the paragraphs considered most relevant to the present narrative.*]

It should be noted that this new specification was headed 'Experimental High Speed Single Seat Fighter (Supermarine Aviation Works)' and it stated that, basically, 'the aircraft shall conform to all the requirements stated in Specification F.7/30' – that is, Mitchell was to design a four-gun aircraft but without other firms being invited to tender in the usual way. The word 'experimental' might very well have reflected decreasing confidence in Supermarine amongst some Air Ministry officials after the experience of Type 224 or that the three successive Schneider Trophy wins by the Rolls-Royce/Supermarine combination had not been forgotten by the new Air Ministry officials and was used to deflect criticism that the normal method of aircraft procurement was being bypassed.

Whatever support there might be for a Supermarine project, it was not lost on the company that Hawker had also been encouraged to substantially modify their Super Fury F.7/30 entry with a rival 'experimental' offering. And the Air

Ministry additionally issued F.10/35, three months after the new Supermarine requirement, calling for at least six, and preferably eight, guns to 'produce the maximum hitting power possible in the short time available for one attack'. They also issued specification F.37/35, for a single-seat day and night fighter armed with four cannon. The Westland response, the Whirlwind, when it first flew in 1938, would have been a formidable aircraft that might have soon replaced a Supermarine fighter. It had an extremely low frontal area, an excellent pilot view, a top speed of 360 mph, and its four 20mm cannons promised to make it the most heavily armed fighter aircraft of its era. Also, as the armament was mounted in a nose cluster, there were no convergence problems as with wing-mounted guns. Its production was, unfortunately for Westland, frustrated by engine problems. Meanwhile, the whole contract situation regarding Type 300 was finally regularised when specification F.37/34 was formally signed on the third of January, 1935.

TYPE 300 and the Elliptical Wing

A whole year had now elapsed since serious rethinking about the fighter had begun but the famous elliptical shape of the future Spitfire was neither an immediate choice nor some 'visionary' piece of aerodynamic sculpturing. Price quotes a Supermarine Drawing No. 300000 Sheet 2 which shows how the Type 224 wingspan was reduced by nearly ten feet with an almost straight trailing edge and a swept back, straight leading edge with a rearwards sloping main spar; and Supermarine Drawing No. 300000 Sheet 11 also shows a continuance of the main spar positioning but some movement can be seen towards the broader, more elliptically shaped wing that was to characterise the Spitfire:

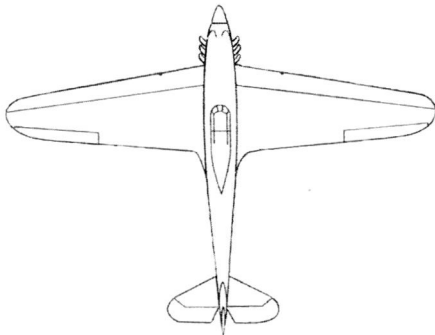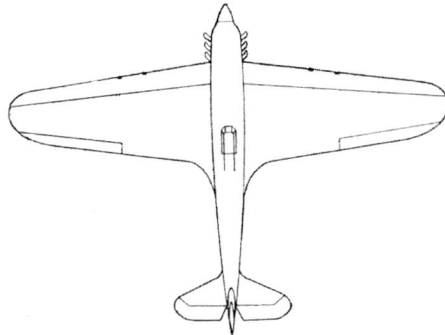

Sheet 2.   *Sheet 11.*

An elliptical configuration would, theoretically, also have required the optimum main spar position to curve or, at least, to slope backwards. However, in order to accommodate the heavier PV12 engine, the previously projected backward slope of the wing leading edge had to be substantially modified. Thus Mitchell, typically, moved to a less complex arrangement whereby the main spar was set at right angles to the fuselage centre-line and would have the effect of bringing the centre of lift forward; such an arrangement would also allow a more straightforward construction and therefore improve weight/strength considerations. Additionally, this structural consideration had the advantage of making it easier to align the wing ribs which were to be set at progressively decreasing angles of incidence as they approached the wing tips (*see* 'washout' p. 247). It also was able to follow more closely the straighter wing leading edge required by the advent of the new engine. The final result of these design compromises was the modified elliptically shaped wing that was to become, pre-eminently, the distinctive and familiar feature of the Spitfire.

Another important determinant upon the final configuration of the wing was the decision to employ a retractable undercarriage. The necessary arrangements for its housing where the wing was thickest meant that all the machine guns had to be placed further outboard. In this respect, an elliptical form of wing was particularly attractive as it tapers towards the tip very slowly at first, so allowing the siting of guns in the necessarily wider positions. However, whilst the F.7/34 agreement that Mitchell was working with referred to a four-gun fighter, the Air Ministry requirement F.10/35 had been issued and it repeated the earlier F.5/34 call for at least six, and preferably eight, guns to 'produce the maximum hitting power possible in the short time available for one attack':

REQUIREMENTS FOR SINGLE-ENGINE SINGLE-SEATER DAY AND NIGHT FIGHTER (F.10/35)
April 1935

1. General
The Air Staff require a single-engine single-seater day and night fighter which can fulfil the following conditions:
    (a) Have a speed in excess of the contemporary bomber of at least 40 mph at 15,000 ft.
    (b) Have a number of forward firing machine guns that can produce the maximum hitting power possible in the short space of time available for one attack. To attain this object it is proposed to mount as many guns as possible and it is considered that eight guns should be provided.
The requirements are given in more detail below.

2. Performance
    (a) Speed. The maximum possible and not less than 310 mph at 15,000 ft at maximum power with the highest possible speed between 5,000 and 15,000 ft.
    (b) Climb. The best possible to 20,000 ft but secondary to speed and hitting power.
    (c) Service Ceiling. Not less than 30,000 ft is desirable.
    (d) Endurance. ¼ hour at maximum power at sea level plus 1 hour at maximum power at which engine can be run continuously at 15,000 ft. This should provide ½ hour at maximum power at which engine can be run continuously (for climb etc.), plus 1 hour at most economic speed at 15,000 ft (for patrol), plus ¼ hour at maximum power at 15,000 ft (for attack). To allow for possible increase in engine power during the life of this aircraft, tankage is to be provided to cover ¼ hour at maximum power at sea level plus ¼ hours at maximum power at which engine can be run continuously at 15,000 ft.
    (e) Take off and landing. The aircraft to be capable of taking off and landing over a 50 ft barrier in a distance of 500 yards.

3. Armament
    Not less than 6 guns, but 8 guns are desirable. These should be located outside the airscrew disc …
5. View
    (a) The upper hemisphere must be, so far as possible, unobstructed to the view of the pilot to facilitate search and attack. A good view for formation flying is required, both for formation leader and flank aircraft and for night landing …

6. Handling
    (a) A high degree of manœuvrability at high speeds is not required but good control at low speeds is essential
    (b) A minimum alteration of tail trim with variations of throttle settings is required.
    (c) The aircraft must be a steady firing platform …

[*For the full text, again, see Alfred Price,* The Spitfire Story; *above are the paragraphs considered most relevant to the present narrative.*]

One suspects that Supermarine and Vickers were already looking well beyond their designated four-gun model and towards these latest requirements for greater speed and armament when their elliptical wing shape was finally decided upon – the thin wing, which would give the speed, would only accommodate any increase in weaponry via the broad chord of an elliptical wing, given the need also to allow for a retracted undercarriage.

Squadron Leader Ralph Sorley, an ex-Martlesham Heath pilot and at that time in charge of the Operational Requirements section, later recalled that, towards the end of April 1935: 'I was soon busy convincing Camm and Mitchell of the vital necessity of building the eight-gun concept into their designs, and, from that moment, I had their willing and enthusiastic cooperation.' He reported back that 'Mitchell received the Air Staff requirements for the 10/35 while I was there and is naturally desirous of bringing the aircraft now building into line with this specification. He says he can include 4 additional guns without trouble or delay'.

It could very well be that Mitchell felt it politic not to reveal that his wing choice had already anticipated such a request although the actual detail of the modifications would require some working out – as reported by Price:

> Jack Davis, working in the Supermarine drawing office, was given the task of redesigning the wing to take the extra four guns. He recalled: 'It did not take me long to work out where the guns had to go. The rib positions had all been decided so it was just a question of fitting the guns in between them. But as one went further out the wing became thinner and the ammunition boxes had to be longer to accommodate the 300 rounds required for each gun. That meant the outer guns had to be quite a long way out. In fact, to get them into the wing, I had to design very shallow blisters to fit around them. The aerodynamics people did not like the idea but they accepted it: there was no alternative if we were to get the eight guns in without redesigning the entire wing.'

The elliptical wing would even allow for their ammunition containers to be so positioned that, when emptied in action, they would not adversely alter the trim of the aircraft.

Of course, a plank-shaped wing, like that of the Air Yacht, rather than the more complex structure of an elliptical one, would have suited many of these volumetric considerations but there were the important *aerodynamic* factors to be considered. By this time, the elliptical wing was coming to be regarded as the most efficient shape for the sorts of speeds and altitudes that were now being contemplated: as Beverly Shenstone, Mitchell's aerodynamicist, said: 'Aerodynamically it was the best for our purpose because the induced drag, that caused in producing lift, was lowest when this shape was used'. (Incidentally, this decision was to prove extremely fortunate, aerodynamically, when, during the later stages of World War Two, operations at much higher altitudes than originally contemplated in the original Spitfire design were common – *see* 'The Thin Wing' below, p. 246 ff.)

Apart from the Kalinin K-7 mentioned earlier, there was another precedent for the elliptical wing approach – the two seat light aircraft, the Baumer Sausewind, notable for its all cantilever structure as early as 1925. It was designed by the Gunter brothers before they joined Heinkel and produced the He 70 mentioned later. But, again, one finds that, in the same year, Mitchell had also produced something approaching an elliptical wing with his S.4. – which also featured all cantilever flying surfaces but which was required to withstand much greater loads than those of the Baumer light aircraft.

These overseas aircraft featured symmetrical ellipses where structural considerations were the main preoccupation. Other somewhat elliptical wing shapes closer to home, where speed was the foremost consideration, were familiar to the

Supermarine design team, before Type 224 flew: we cannot know whether any more immediate precedents might have been an influence when he sent his later Type 300 sketches for drafting, but both the Short Crusader of 1927 and the Gloster VI of 1929 employed shapes which approached the elliptical apart from the narrower chord close to the fuselage:

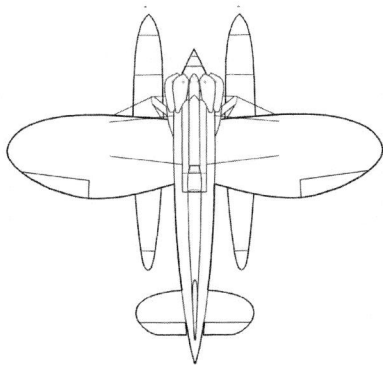

Short Crusader.   *Gloster VI.*

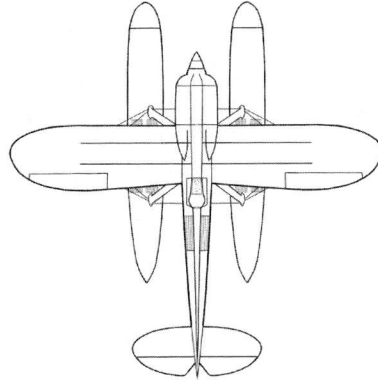

(The question arises: why did his S.5/6 of racers not follow this precedent? It may very well be that, with the short times available before the Schneider competitions, the introduction of wing-surface radiators with the S.5 was sufficiently novel for Supermarine, without the added complexity of following curving leading and trailing edges; and the move to metal wings with the S.6s was, again, perhaps enough to be going on with. In addition, the wire bracing reintroduced with these later aircraft made the strength/weight advantages of a more complex wing shape less obvious. It is, however, instructive to note that their uncomplicated and unbraced cantilever tailplanes were perfectly elliptical in shape.)

However, in the present context, the most intriguing shape was that of the Italian Piaggio P.7. Whilst it never progressed beyond its taxiing stage, the general arrangement of this rival Italian design would most probably have been known to Mitchell and the wing shape is uncannily similar to that which was developed for the Spitfire – which is notable for a similar departure from more regular elliptical configurations.

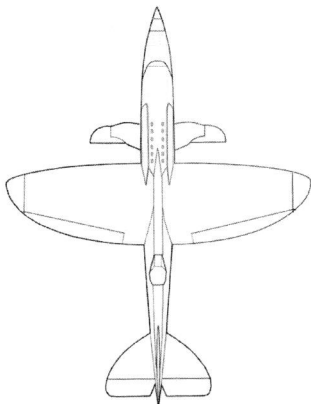

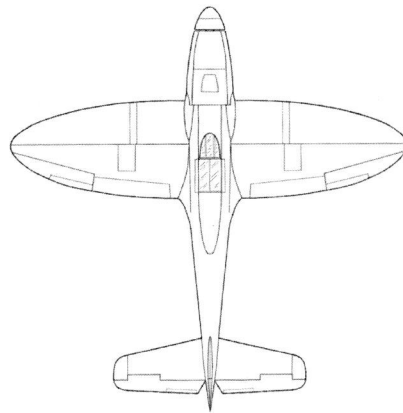

Piaggio P.7.   *Heinkel He 112.*

Whilst it is clear that Mitchell had appreciated advantages of the elliptical wing before or at the same time as other early exponents of the shape, a return to

Heinkel aircraft is especially called for – in respect of the He 112 which was a contender for the contract which produced the Messerschmitt Bf 109. This latter was chosen for the Luftwaffe in October 1936 as it was smaller, lighter, and therefore offered a better performance at the particular point in time when a decision had to be made (when it became known that the Spitfire had been ordered into production – it is not always appreciated that it was the first orders for the Spitfire which influenced the decision to produce the Messerchmit Bf 109.) Thus, when a much improved He 112B appeared in July of the following year, it did not go into service with the Luftwaffe; as its wing shape is very similar to that of the Spitfire, one wonders how formidable a fighter it might have been developed into. The present concern, however, is to note that Mitchell's design predated that of this rival aircraft, having been arrived at as a solution to various requirements – for both the weight distribution reasons and for the aerodynamic reasons mentioned above.

The Thin Wing
The first flight of the Stranraer in July 1934, with its thinner aerofoil, now clearly vindicated the view, particularly held by the aerodynamicist, Beverley Shenstone, that the way forward was not represented by the thick, relatively lightly loaded wing of previous designs. By the autumn of 1934, Type 300 Drawing No 300000 Sheet 11 also showed that Mitchell had now moved towards the use of a thinner wing but this development embodied experience gained much earlier, before Shenstone joined Supermarine – as C. G. Grey had noted:

> An interesting point about those Curtiss biplane racers [of 1923–26] was that the wings came almost to a knife-edge in front [producing an extremely low thickness/chord ratio of 6%]. One of the American technical people told me at the time that they had come to the conclusion that, at the speed which these machines reached, the air was compressed so much in the front of the leading edge that it paid to cut it. I passed the information on to R. J. Mitchell of Supermarine's who went into the idea quite deeply, and though he could not quite put a cutting edge on his Schneider Trophy monoplanes of 1927–9 and 1931, he used the thinnest possible wings, and won every time.

And Ernest Mansbridge recounted: 'Choosing the thick section wing was a mistake when we could have used a modified, thinner section as used on the S.5 floatplane …. But the Type 224 was to be an unbraced monoplane, and there were not many of these about at the time.'

In this connection, Clifton's later comments on Mitchell's doubts about information derived from model testing deserve recording:

> I think that Mitchell decided to make the wing as thin as he did, and I wouldn't like to be positive about this, but my recollection was that it was against some advice from the National Physical Laboratory in that case where wind tunnel tests, I believe, showed that there was no advantage in going below a thickness chord ratio of 15%, whereas, the [Spitfire] wing was 13% at the root and 6% at the tip. I believe that this was due to the fact that at that time the question of the transition from laminar to turbulent flow in relation to the difference between model and full scale wasn't understood and subsequently it was found that when you made proper allowance for that, there was an advantage, as the testing could be shown to prove, in going thinner.

(At about that time, Hawkers had been advised by the National Physical Laboratory that wind tunnel results had shown no drag penalty with the thicker Hurricane wing; however, the Laboratory scientists later found this advice to be incorrect – they attributed their earlier views to high wind tunnel turbulence, not appreciated at that time.)

If lack of rigidity had been the reason for the crash of the S.4, it would not be a factor in the new wing despite Mitchell's wish to keep drag to a minimum by not only returning to a cantilever but also to an unprecedentedly thin cross-section throughout. A new approach to the main spar was called for and the solution that Mitchell, and Joe Smith in particular, arrived at was a spring-leaf sort of arrangement which contributed significantly to the strength/weight ratio of the thin wing and which, at the same time, was a novel approach to progressively lightening the spar as it approached the wing tip. The two booms that made up the spar began as a concentric arrangement of square tubes, each longer than the next as they increased in cross section, and with the outer laminations changing to channel section nearer the tip, with a final termination in angle section:

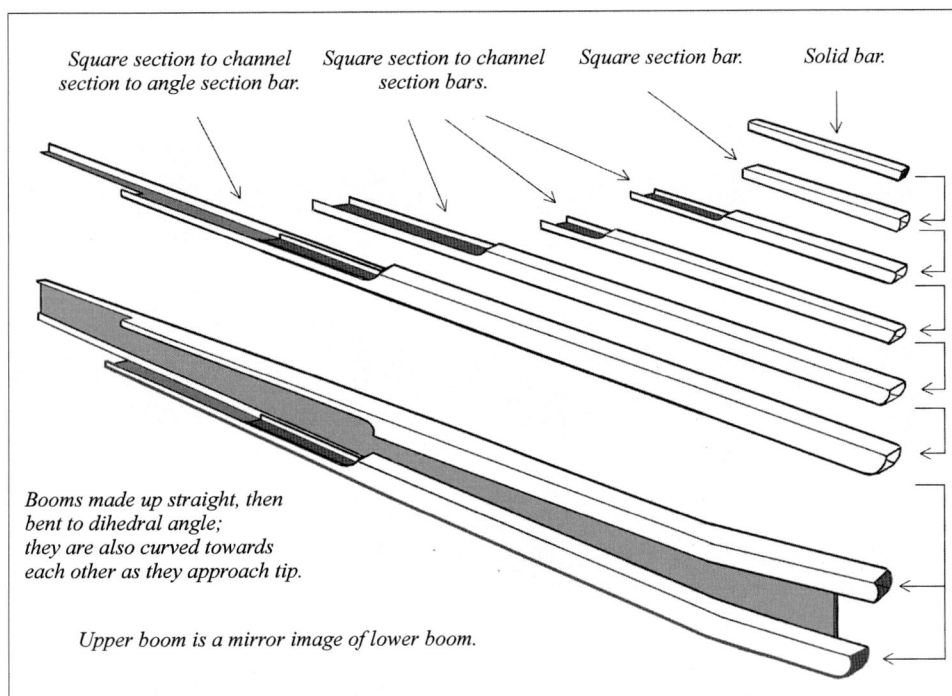

Square section to channel section to angle section bar.

Square section to channel section bars.

Square section bar.

Solid bar.

Booms made up straight, then bent to dihedral angle; they are also curved towards each other as they approach tip.

Upper boom is a mirror image of lower boom.

Schematic view of Spitfire wing spar.

Some Final Thoughts on the Spitfire Wing

If, on the other hand, a high speed stall had been the reason for the S.4 crash, then the 'twist' to be built into the new wing would be an additional safety factor as, otherwise, the entire span of the wing would stall with little warning. From plus two degrees incidence at the root, the new wing went progressively to minus half a degree at the tip. The result of this 'washout' was that the inner section of the wing would stall before the outer, thereby giving future fighter pilots (who might be otherwise engaged during tight turns!) advance warning of full stall while there was still some aileron control. Wing Commander Stanford Tuck, speaking after the war, said: 'Put into as hard a turn as possible … [the Spitfire] would flop about on its wings and then give this terrific judder which to a pilot in combat is indeed a very good indication that if he goes on pulling harder he will fall out of the sky'.

Another wartime pilot, Geoffrey Wellum, explained the matter very well: 'In a tight turn you increase the G loading to such an extent that the wings can no longer support the weight and the plane stalls, with momentary loss of control. However, in a Spitfire, just before the stall, the whole aircraft judders, it's a stall warning if you like. With practice and experience you can hold the plane in this judder … and you don't need to struggle to regain control because you never lose it'. Incidentally, this washout principle had been a feature of the Type 224 wing. In the concern to make Type 224 safe for night landings, Supermarine drew attention to the fact that washout had been introduced, 'to give adequate control and stability below the stall without the aid of slats' (the method adopted in the rival Messerschmitt Bf 109).

Two members of Mitchell's design team were particularly associated with the development of the Spitfire wing which turned out to be not only thinner but also lighter than any competing design yet with exceptional strength: Beverley Shenstone, the aerodynamicist, has already been mentioned and he, in turn, gives credit to Joe Smith for the structural features of the wing; he also described to Price in what way the Heinkel He 70 actually was a specific influence:

> The Type 224 had had a thick wing section and we wanted to improve on that. The NACA 2200 series aerofoil section was just right and we varied the thickness-to-chord ratio to fit our own requirements: we ended up with 13 per cent at the root and 6 per cent at the tip, the thinnest we thought we could get away with. Joe Smith, in charge of structural design, deserves all credit for producing a wing that was both strong enough and stiff enough within the severe volumetric constraints.
>
> It has been suggested that we at Supermarine had cribbed the elliptical wing shape from that of the German Heinkel 70 transport. That was not so. The elliptical wing shape had been used on other aircraft and its advantages were well known. Our wing was much thinner than that of the Heinkel and had a quite different section. In any case, it would have been simply asking for trouble to have copied a wing shape from an aircraft designed for an entirely different purpose. The Heinkel 70 did have an influence on the Spitfire, but in a rather different way. I had seen the German aircraft at the Paris Aero Show and had been greatly impressed by the smoothness of its skin. There was not a rivet head to be seen. I ran my hand over the surface and it was so smooth that I thought it might be constructed of wood. I was so impressed that I wrote to Ernst Heinkel, without Mitchell's knowledge, and asked how he had done it; was the aircraft skin made of metal or wood? I received a very nice letter back from the German firm, saying that the skinning was of metal with the rivets countersunk and very carefully filled before the application of several layers of paint. When we got down to the detailed design of the F.37/34, I referred to the Heinkel 70 quite a lot during our discussions. I used it as a criterion for aerodynamic smoothness and said that if the Germans could do it, so, with a little more effort, could we. Of course, the Heinkel's several layers of paint added greatly to the weight; we had to do the best we could without resorting to that.

It will be seen later that the continuous development of the Rolls-Royce engines contributed significantly to the Spitfire's long development history but Mitchell's decision to pursue the thin wing approach must be given its full credit. Harry Griffiths, writing in 1992, put the matter thus: 'He had one strong fetish, namely that for maximum performance the frontal area of an aircraft had to be as small as possible, hence … his insistence on the thin wing on the Spitfire against the advice of the experts at Farnborough.' Its thickness/chord ratio of 13% at the root and only 6% near the tip, compared very favourably with the contemporary Messerschmitt Bf 109 and the later Typhoon, which were 14% and 9% and 14.5% and 10%, respectively, at the same positions. Its overall shape also underwent relatively few changes whilst engine power, fuel weights, armour plating, and armament all increased up to threefold and maximum speeds rose by nearly 30%.

Some indication of the Spitfire's design longevity can be gained from the fact that, as late as 1944, this type was chosen instead of later fighters for high speed diving trials during which it reached a maximum speed of 606 mph (Mach 0.89). The aircraft concerned was a specially prepared PR XI (*see* Appendix

p. 313) but it should be noted that the pilot's notes for the *standard* machine gave a 'never exceed' figure which was the equivalent of Mach 0.85.

Mention has been made of the input of Beverley Shenstone, who joined Supermarine at the end of 1931 and who soon became Mitchell's Chief Aerodynamicist. Very substantial (and, it would appear, largely unsupported) claims have been made in a recent book for his influence upon the shape of the Spitfire wing – particularly the straightened leading edge component, the shape of the trailing edge and the aerofoil selection. One must certainly expect that the advice of this brilliant young man would not have been ignored: in Chapter One it had been reported how Mitchell used to call in the leaders of relevant sections and get them arguing among themselves, making sure that everyone had said what he wanted to, and then either make a decision or go home and sleep on it. Joe Smith has also indicated this quality that he considered contributed to Mitchell's leadership: 'in spite of being the unquestioned leader, he was always ready to listen to and to consider another point of view, or to modify his ideas to meet any technical criticism which he thought justified ...'

One might thus speculate that Mitchell felt confident to pursue the very thin wing, against the technical advice mentioned above, having been supported by detailed and persuasive theoretical submissions from Shenstone. And the very *final* shape of the Spitfire wing might also owe a great deal to the younger man's views, who had had direct experience of German aerodynamic theory that was in advance of contemporary British practice. It would fit with Mitchell's habitual management style that he soon recognised that the new man might well help Supermarine to progress beyond their already acknowledged lead in high speed design; we should give the Chief Designer credit for not being so flushed by his earlier Schneider Trophy successes that he did not appreciate what the younger man Shenstone might contribute.

On the other hand, it has been shown that Mitchell had been considering elliptical or thinner wings since the middle 1920s, and that gun and undercarriage housing (as well as aerodynamic arguments) would have been important considerations for the whole design team. Thus, recently reported diary entries by Shenstone about this period in his career are interesting:

> the elliptical wing was decided upon quite early on ... The ellipse was simply the shape that allowed us the thinnest wing with sufficient room inside to carry the necessary structure and the things we wanted to cram in ... Joe Smith, in charge of structural design, deserves credit for producing a wing that was both strong enough and stiff enough within the severe volumetric constraints.

It is noteworthy that the Chief Aerodynamicist is here generous in his praise of Smith's structural input and perhaps he is too self-effacing about the importance of his own contributions, speaking impersonally about 'our' purposes and what 'we' wanted to achieve.

Forty years later, at the Mitchell Memorial Symposium in Southampton, Shenstone once again makes no special claims for his own input:

> I don't think R.J. cared at all what the Germans were doing but he did care about the shape of wings, but he didn't copy anything. I think all of us at the time realised that the thinnest wing can often be the best, whereas earlier, people were afraid of very thin wings in case they broke off. I think the essential thing is that Mitchell took advantage of everything he could which would improve his aircraft. Certainly Mitchell always did the thing which should be done.

At the same symposium, Clifton said: 'Meanwhile Mitchell was moving on to ... a very thin wing against expert advice ... Mitchell was trying to put the thing together to get the maximum possible result.' Shenstone was in the audience and one might have expected some gracious reference to the importance which has recently been claimed for him, but none is recorded. Also, in the published account of the meeting, C. F. Andrews submitted a letter he had received from Shenstone in which the latter emphasized their volumetric considerations:

> I do not think that the He 70 had much direct influence on [the] Spitfire's elliptical wing. Various wing plan forms were sketched for [the] Spitfire, and the real down to earth reason for the elliptical wing was the fact that the elliptical taper is gradual near the fuselage and can be less than that for a straight taper wing, thus giving more space for retracted undercart, and in this case also for guns.
>
> I remember that I pointed out to Mitchell that the elliptical wing was optimum for induced drag, but he said he didn't care whether it was elliptical or not as long as it had room for guns and undercart ... the real advantage of the elliptical wing turned out to be its low induced drag at very high altitudes, *such altitudes not having been considered during the design, but realised during the war* ... [my italics, *see* Appendix on the PR19, p. 313–14].

It remains a matter of conjecture as to what interpretation one should put on Shenstone's statement that he 'pointed out' the advantage of the elliptical wing and the thin wing or how far Mitchell had already made up his mind on these matters. We have seen that Mitchell had appreciated the value of the elliptical wing and of the thin wing before Shenstone joined Supermarine but, at least, we can surely accept that Shenstone supplied detailed aerodynamic calculations which Mitchell took careful note of and it may be that credit should be given to Shenstone for not trying to deservedly share the limelight with his famous Chief Designer.

Beyond that, as we have seen, the Mitchell Symposium discussion does raise questions about the precise influence of Shenstone on the complex of considerations which led to the eventual shape of the Spitfire (its importance as a document in the Mitchell story seems hardly to have been noticed elsewhere). Perhaps one can do no better than quote from a Southampton RAeS member's summing up at this Symposium:

> During the discussion, Mr Clifton was asked the origin of the elliptical wing form. No authoritative reason was put forward. I am inclined to think that it was the logical result of integrating aerodynamic and structural requirements. Comparing the F.7/30 development with the Spitfire, changes are evident which must have been consciously made during the project stage. The main spar, previously swept back, was set normal to the fuselage axis. The span, wing area and thickness to chord ratio were reduced. The straight tapered wing gave place to the elliptical form of lower aspect ratio. Thus the greater and more constant chord in the inner regions of the wing gave more space for the landing flap, undercarriage, radiator and gun installation, and provided sufficient thickness for a good structure. For optimum bending strength the spar should have been placed at 30% chord but, as this would have encroached on installation space, the 25% chord position was a better choice. This must have been intentional as it was also the aerodynamic datum for the varying incidence which was progressively reduced from root to tip. From the unswept spar at 25% chord the familiar asymmetric ellipse naturally followed. The choice of a common aerodynamic and structural datum simplified work in the drawing office and hence manufacture. The unswept spar with the ribs at right angles was aerodynamically and stucturally good, and simplified manufacture. The simple basic structure was the first step to low structural weight, for otherwise all the refined detail design would have been less effective ...
>
> Considering these points as a whole and remembering the lack of precise aerodynamic date in those days, so many imponderables could

> only have been resolved by R. J. Mitchell's intuitive judgement; as wind
> tunnel work was limited to tests on spinning and the ducted radiator. As I
> see it, the elliptical wing is to the Spitfire as the ogee wing is to Concorde.
> They look right – and are right.

In this last respect at least, it might not be too fanciful to describe Mitchell's concept as 'visionary', particularly as the now customary preliminary wind tunnel tests were not employed; by this time, Mitchell had become even more distrustful of the data derived from the small-size models that were employed in the Vickers facility. Instead, there was the depth of hard practical Schneider experience of high speed flight. The following extract from an internal memorandum, again quoted by Price, reveals this experience and also the clear, incisive mind of the Chief Designer:

> It is agreed that the manœuvrability of a monoplane is likely to be worse
> than its equivalent biplane, but we hope in this design to get it very
> nearly as good, as the span of the aircraft is very little more than that of
> a biplane conforming to the same specification. I entirely disagree with
> the suggestion that the ailerons are too small. It is stated that the ailerons
> are certain to be fairly heavy at top speed, and yet it is suggested that they
> should be increased in size, which is obviously the easiest way to make
> them heavier still. We have been through the experience of large chord
> ailerons too many times to be caught again. Even on our present fighter
> [Type 224] we found that by practically halving the chord of the ailerons
> we got a very much lighter control which is as, if not more, effective.
> With very high speed aircraft it is obviously essential that the aileron
> controls must be particularly light. The only way of attaining this is to
> have them with a very narrow chord and well balanced. To attain very
> light operating loads on large chord ailerons is very dangerous, as a very
> high degree of balance has to be attained, and this leads to the possibility
> of overbalance being obtained in the dive due to small deflections. I
> believe this is the cause of several accidents involving ailerons. [The S.4
> in particular?] Furthermore, the general manœuvrability of the machine is
> not affected by the size of the ailerons, but rather by the ease with which
> aileron is applied. This is obviously the case since full aileron movement
> is never used in ordinary manœuvring.

Apart from different skinning, the first basic design modification to the ailerons was not made until seven years later when engine power had increased by about 70%. (*See* Appendix p. 308.)

Mitchell's memorandum, as well as addressing some in-house doubts about the manœuvrability of the new design, also dealt with the question of pilot view and the right amount of dihedral. In the latter case, he reveals his well-known willingness to listen to the advice of others:

> The CG position is always over the chord in low wing monoplanes, and
> even though this is agreed to be an undesirable feature, it is very difficult
> to see how it can be avoided in this type of aircraft without seriously
> impairing other features. An increase of 1.5 degrees in the dihedral angle
> of the machine has already been carried out as a result of discussion and
> comparison of statistics from Weybridge [Vickers] machines. There is no
> evidence to shew that the machine as at present laid down with increased
> dihedral will not be perfectly satisfactory on lateral control and stability.

The increased angle of 6 degrees became the standard dihedral for all the successive marks of Spitfire.

Given the Air Ministry's concern with pilot view, Mitchell was taking something of a gamble by giving the cockpit a low profile (compare the more humpbacked appearance of the Hurricane) and he met in-house concern about possible poor forward vision in a typically simple way:

It is agreed that the view can be improved, but only with the sacrifice of performance by increasing the size of the body. In the design of this aircraft the performance has been considered of paramount importance, and various sacrifices of other requirements have been made to obtain this object. It is considered desirable not to depart from this policy. If at a later date it is thought necessary to improve the view at the nose, this is best done by merely raising the pilot, and can easily be done at a later date if considered essential.

This apparently obvious way of avoiding what would otherwise have been a complex series of design modifications bears out the view of a later Supermarine test pilot, Jeffrey Quill, about Mitchell:

He was a man of enormous common sense, a great technician who could take all the mystery out of a subject when he explained it. At that time I was only 23 and had not trained as an engineer, so I had something of an inferiority complex when it came to dealing with the aerodynamics and structures experts at Woolston with slide rules sticking out of their pockets. But Mitchell put me right about this one day, in his usual very direct way. He said, 'Look, I'll give you a bit of advice, Jeffrey. If anybody ever tells you anything about aeroplanes which is so bloody complicated you can't understand it, take it from me it's all balls!'

One is reminded of Mitchell's comment to Shenstone that his concern about the Spitfire wing was essentially that it had 'room for guns and undercart'. The designer was clearly acknowledging that the practical considerations of the professional engineer and the necessary design compromises must shape the final outcome of a project – but one suspects that Mitchell's disclaimer was only half serious. After all, his Southampton had been described as 'probably the most beautiful biplane flying boat that had ever been built' and Joe Smith was surely referring mainly to aesthetic considerations when he recorded the following description of Mitchell at the drawing-board:

He was an inveterate drawer on drawings, particularly general arrange-ments. He would modify the lines of an aircraft with the softest pencil he could find, and then remodify over the top with progressively thicker lines, until one would be faced with a new outline of lines about three sixteenths of an inch thick. But the results were always worthwhile, and the centre of the line was usually accepted when the thing was redrawn.

And, of course, it is the particular and classic shape of the resultant Spitfire wing which has contributed to its lasting popular appeal. The question arises as to whether its elliptical shape justified the complexities of its production. The later Hawker Typhoon and Republic Thunderbolt fighters adopted modified elliptical forms which were easier to manufacture and the Focke-Wulf Fw 190 followed the Messerschmitt Bf 109 with a straight taper, as did the very success-ful North American Mustang. Thus, despite Mitchell's claiming that he was only concerned to house the guns and undercarriage, one suspects that æsthetics and intuition had quite a lot to do with the final choice of the Spitfire wing shape. Perhaps the last words on the overall design of the fighter might be a quotation from the normally factual account in Supermarine Aircraft since 1914 which aptly, even poetically, sums up the design history of the Spitfire: 'It is a matter of history that there emerged from the ungainly angular Supermarine Type 224, like a butterfly from its chrysalis, one of the most beautiful aeroplanes ever de-signed'.

Uncertainties Finally Overcome.
The prototype Spitfire, K5054, finally emerged from the works (if not from the chrysalis) at the beginning of February, 1936. In fact, it had been built in a corner of the Woolston workshops behind a tarpaulin screen for security: in

Supermarine SPITFIRE prototype (as at March, 1936)

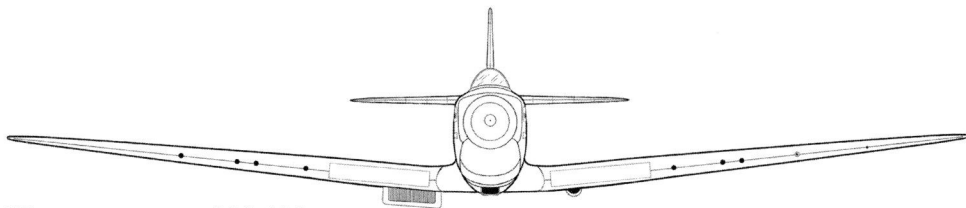

Wingspan	36 ft 10 in.
Wing area	242 sq ft
Loaded weight	5,359 lb
Maximum speed	367 mph (Mk.1)

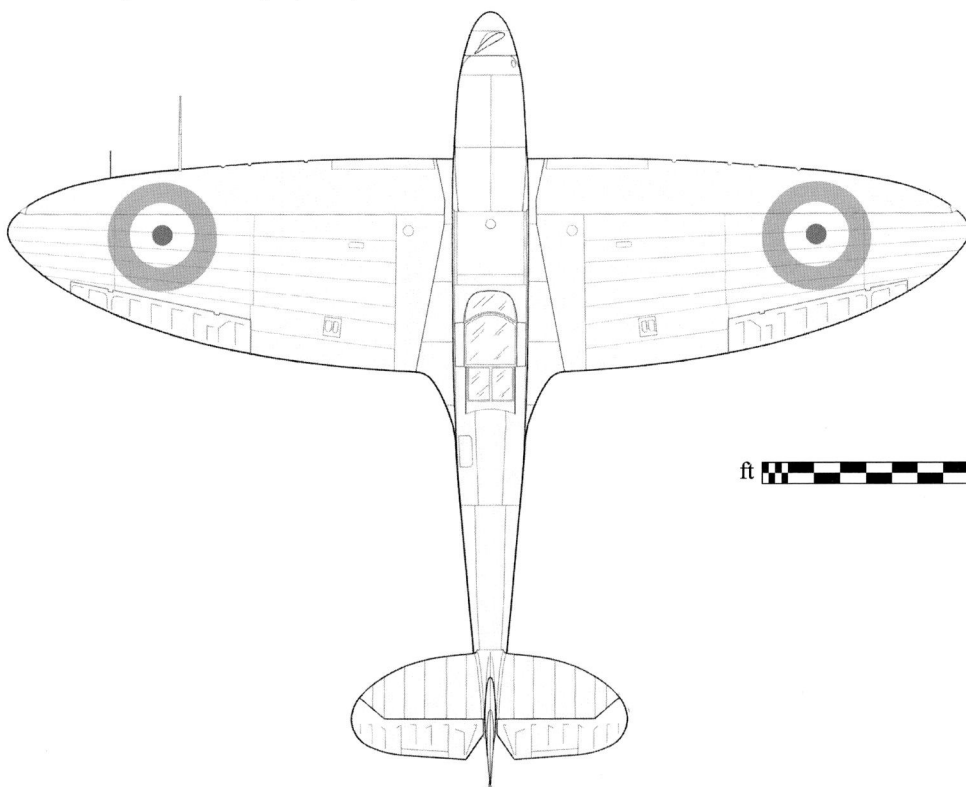

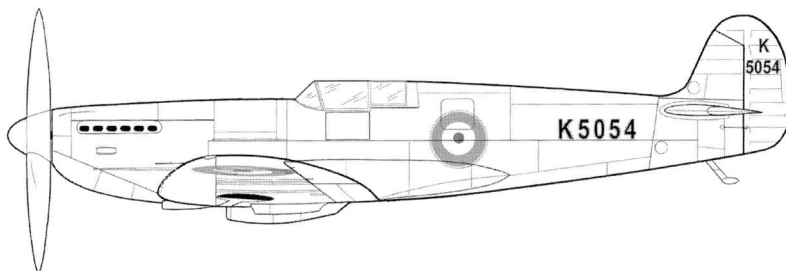

ft

K 5054

K5054

particular, because Lufthansa seaplane pilots used the Supermarine slipway for refuelling and for having their mail checked by British customs. Webb also remembered that 'the first engine run was carried out at night for security reasons, with the tail skid lashed to a holding-down ring on the quay normally used for tethering down flying boats. [The foreman, Gerry] Scrubby, recalling the occasion, said that the flames from the exhaust were spectacular, as was the noise.'

The new machine now had a ducted under-wing radiator of the type recently devised by F. Meredith of the Royal Aircraft Establishment, Farnborough. This new radiator made no difference to the basic lines of the machine and actually used the heat exchange of the radiator to produce some thrust at high speed. The new system was operated in conjunction with a new coolant, ethylene glycol, which had a much higher boiling point than water – which Bulman had first seen used in America. Thus the proposed evaporative system had been made redundant. Apart from the problems mentioned earlier, water loss during inverted flight in Type 224 had never been entirely solved and, with hindsight, one wonders about the wisdom of the Air Ministry's accepting the design of fighter prototypes having wide areas of condenser exposed to enemy fire. One also wonders what would have been the consequences if the new type of radiator had not come along at this particular moment. The reduction of the endurance requirement from 2½ hours at full throttle (F.7/30) to 1½ hours (F.10/35) was an additional help to the designer.

K5054, before its first flight, unpainted and without wheel covers.

The tail configuration of the original design had also been altered. These changes to the tail were a consequence of wind tunnel testing for spinning characteristics. Whilst Mitchell, by now, had sufficient faith in his experience and intuitions to dispense with most of such time-consuming testing, spin had been the cause of too many accidents in the history of aviation – recently, with the Bristol and Vickers low wing fighter prototypes – for any possibly useful information to be ignored. As a result of these tests, the fuselage length had been increased by nine inches and the tailplane had been raised: an increase in height of 12 inches had been recommended but a compromise of 7 inches was reached with Mitchell. Perhaps æsthetics was a factor in the final outcome; certainly this relatively small change and the lengthening of the fuselage had little effect upon the overall visual appeal of Mitchell's general arrangement.

After initial engine runs had been completed, the airscrew and wings were removed and all the components taken by lorry to Eastleigh aerodrome for re-erection in the Supermarine hangar (where one of the S.6 aircraft, less engine, was stored). As Supermarine reports show, the first flight of K5054 was on March 5th, 1936. This maiden flight was made with the undercarriage locked down; according to Webb, there was some trouble with the 'up' locks. Jeffrey Quill reported that:

> the aeroplane was airborne after a very short run and climbed away comfortably. Mutt [Summers] did not retract the undercarriage on this first flight – deliberately, of course – but cruised fairly gently around for some minutes, checked the lowering of the flaps and the slow flying and stalling characteristics, and then brought K5054 in to land. Although he had less room than he would probably have liked [because the wind was across the airfield], he put the aeroplane down on three points without much 'float', in which he was certainly aided by the fine-pitch setting of the propellor ... When Mutt shut down the engine and everybody crowded round the cockpit, with R.J. foremost, Mutt pulled off his helmet and said firmly, 'I don't want anything touched'.

As it was by no means unknown for prototypes to have to undergo costly modifications immediately after their flight characteristics were first discovered, this comment must have been a matter of great satisfaction for all the design team; it meant that there were no major control or stability problems calling for urgent attention and that the aircraft could be next tested without any alterations. The aircraft was flown again on March 10, with the undercarriage operating normally, and, after further test flights, Summers reported that 'the handling qualities of this machine are remarkably good' – the first of a procession of such comments from a multitude of Spitfire pilots. In view of the designer's memorandum quoted earlier (*see* p. 251), Mitchell must have been pleased to read that the ailerons were powerful but 'quite light'. (Incidentally, in the test pilot's log, the aircraft was described as the Spitfire II.)

By the end of March, Jeffrey Quill made his first two flights after a briefing from Summers:

> He stressed the need to make a careful approach during the landing. The flaps could be lowered only to 57 degrees on the prototype. With the wooden prop ticking over there was very little drag during the landing approach and she came in very flat. If one approached too fast, one could use up all of the airfield in no time at all.
>
> Then it was my turn, and off I went. Of course, at that time I had no idea of the eventual significance of the aeroplane. To me it was just the firm's latest product, running in competition with Hawkers and a highly important venture. And if I bent it I would probably be out on my neck!
>
> I made my first flight, getting the feel of the aircraft, and landed normally. Then I decided to taxi back to the take-off point and do another take-off. That second time did not feel quite right, and only when I was airborne did I realise I had left the flaps down. I retracted the flaps, flew around a bit, then went back and landed. Of course, everyone had noticed my faux pas. But Mitchell was very kind about it. He just grinned and said, 'Well, now we know she will take off with the flaps down'.

Quill also recollected the significantly higher cruising speed and the outstanding stall performance, which was another feature of the aircraft's characteristics so often commented upon later:

> ... the aircraft began to skip along as if on skates with the speed mounting up steadily and an immediate impression of effortless performance was accentuated by the low revs of the propeller at that low altitude. The aeroplane just seemed to chunter along at an outstandingly

higher cruising speed than I had ever experienced before ... it wasn't until after the touchdown that the mild buffeting associated with the stalling of the wing became apparent. 'Here,' I thought to myself, 'is a real lady'.

Nevertheless, it was discovered quite early that the rudder horn balance was too large and resulted in directional instability at high speed. More seriously, Mitchell confessed to being very disappointed that the top speed was 'a lot slower than I had hoped for' – being well below the 350 mph predicted. Quill again: 'unless the Spitfire offered some very substantial speed advantage over the Hurricane, it was unlikely to be put into production. Thus the disappointing speed performance of our prototype at that early stage was something of a crisis and Mitchell was a very worried man'.

The prototype was taken into the works for modification and a new paint job and, by 9 May, 1936, re-emerged with a revised rudder balance and a very smooth light blue-grey finish, thanks to automobile paint supplied by Rolls-Royce. In the afternoon, it was flown by Summers and Quill took Mitchell up in the company's Miles Falcon, along with the photographer of *Flight* magazine, to observe his creation from the air (*see* photo p. 261).

The Spitfire prototype, painted and fitted with wheel covers and revised rudder.

Despite the smooth new finish, the speed of K5054 was still less than hoped for, as its top speed, now 335 mph, was still too close to that of the Hurricane, rumoured to be achieving 330 mph. Notwithstanding the need to respond urgently to German rearmament, Shenstone also reported how little information was still shared between aircraft constructors: 'there was absolutely zero intercommunication between designers in different firms, not even very much between Mitchell and the Pierson-Wallis combination at Vickers Weybridge. We never knew our opposite numbers in other firms. Gouge [Shorts] never discussed with Mitchell, nor Mitchell with Camm [Hawkers].'

Just after the prototype Hurricane had flown, Mitchell saw it for the first time. Quill reported that 'He did not see it close up but only at a distance. He came back to Itchen very worried, and walked into the erection shed and looked at the first incomplete Spitfire. He said, "Camm's got a tiny little machine. Ours looks far too big".' In fact, the Hurricane had three feet more span than the Spitfire but one can understand Mitchell's sensitivity to size after the Type 224 fighter project; also the supposed narrow margin between the top speeds of the two new aircraft might very well have resulted in a contract going exclusively to the company which had already supplied the RAF with the Hart, Demon and

Fury fighters and which had recently absorbed the company which had just supplied the successful Gladiator. Luckily, the fitting of a new propeller (Quill recalled the previous flight testing of 'some 15 to 20 different designs') on 15 May produced a dramatic increase to 349 mph.

This improvement is even more dramatic when viewed against the fighter development figures given earlier in the chapter: the Gladiator's increase in power had produced a 12% advance in speed whereas a similar increase in power eventually gave the Supermarine aircraft the very impressive leap of more than 100 mph:

Type	Power	Power increase	Top speed	Speed increase over previous type
Gladiator I	830 hp	30%	253 mph	12%
Spitfire I	1079 hp	30%	**362 mph**	**43%**

It is worth noting that as early as 1933, the German Reichluftministerium had called for a new breed of fighter, capable of at least 250 mph at 19,690 feet, whereas the British F.7/30 and F.37/34 which resulted in the eventual Spitfire, had only required a top speed of at least 195 mph at 15,000 feet. Thus the 362 mph figure achieved by the Spitfire prototype, reveals how far in advance of initial requirements was Mitchell's fighter, thanks mainly to its thin, elliptical wing.

Penrose's memory of his first test flight in the Spitfire admirably illustrates the startling advance of its performance:

> So swiftly were we climbing that I was surprised to find the altimeter already showed 5,000 ft. I leveled off and felt the controls grow firmer with increasing speed; tried a gentle turn, and then steeper. Nothing in it! ... Suddenly a Gladiator appeared 1,000 ft above me, offering opportunity of a mock dogfight with this latest contemporary biplane fighter of the RAF. I drew the stick back in manner long accustomed, unprepared for the lightness of response A vice clamped my temples, face and muscles sagged, and all was blackness. My pull on the stick relaxed instantly yet returning vision found the Spitfire poised almost vertically and the Gladiator 2,000 ft. below. Ah! If the fighter boys could cope with a machine like this it was going to be an ace ...

Naming the Spitfire and RAF Acceptance
When did R. J. Mitchell say that 'Spitfire' was 'a bloody silly name' for his fighter? Dr Alfred Price noted, via the logbook of test pilot George Pickering, that the earlier Type 224 was known as the Spitfire some time before July 1935 and so Mitchell's reported remark might have been made about this time or earlier, when this fighter was so designated, in the brief announcement by Supermarine in 1934. (*See* p. 235.)

Thereafter, the revised project, following the disappointment of Type 224, was usually referred to in the works as 'the fighter' – after all, his most beautiful racer was only ever known as the 'S.4'. Gordon Mitchell's book copies a Supermarine document of 29 February, 1936, in which the soon-to-fly aircraft was referred to merely as the 'Modified Single-seater Fighter K5054' and he also noted that his father, on occasions, erroneously referred in his diary to his machine as F.37/35.

When the new design was officially named 'Spitfire', at the end of April, 1936, it is just as possible that this was the time when Mitchell made the well known comment – perhaps he did not want reminding of the disappointment of the first Spitfire or he was now voicing his general dislike of PR names for his machinery. However, despite the views of its designer, Supermarine publicity for 1936 read:

THE SUPERMARINE 'SPITFIRE I.'

The 'Spitfire' is a single-seat day and night fighter monoplane in which much of the pioneer work done by the Supermarine Company in the design and construction of high-speed seaplanes for the Schneider Trophy Contests has been incorporated. [The company is silent about Type 224.] The latest technique developed by the Company in flush-rivetted stressed-skin construction has been used, giving exceptional cleanliness and stiffness to wings and fuselage for a structure weight never before attained in this class of aircraft. The 'Spitfire' is fitted with a Rolls-Royce 'Merlin' engine, retractable undercarriage and split trailing-edge flaps. It is claimed to be the fastest military aeroplane in the world.

No further details of the machine are available for publication.

It is interesting to note how, for the very first time when announcing an entirely new Supermarine design, the aircraft had been designated a Mark One. It might be that the company was merely wanting to avoid any further references to Type 224, and thus to draw a line under that less than successful machine but, in view of the many variants to be produced in the next nine years, one likes to think that the designation was prophetic.

Vickers' suggested name for the new fighter, and accepted by the Air Ministry, was in all probability inspired by Ann McLean, the chairman's daughter, who had habitually been referred to as 'a right Spitfire'. As well as the comment 'It's the sort of bloody silly name they would give it', Mitchell was also reported as saying that it could be called 'Spit-Blood' for all he cared. Mansbridge's daughter has recorded that 'Shrew', 'Shrike', and even 'Scarab' had been considered, and it is unlikely that our Chief Designer would have preferred any of these. But, by this time, the Aircraft Nomenclature Committee was no more and names were now selected, in discussion with the manufacturer, by the Air Member for Supply: for fighters, especially, words like Gamecock, Woodcock, or Siskin were now replaced by ones indicating speed and aggression – for example, 'Fury', 'Gladiator', 'Gauntlet, 'Whirlwind', Hurricane' – and 'Spitfire' more or less fell into this general category.

However, the Supermarine name also had a 'British' pedigree as well: it had been applied in previous times to cannons emitting fire, to angry cats, and to anyone displaying irascibility or a hot temper, especially women – as evidenced in 1762 when Lord Amhurst is quoted as saying to his mistress: 'Not so fast, I beg of you, my dear little spitfire'; and Shakespeare echoed the general sentiment when King Lear defies the elements: 'Rumble thy bellyful! Spit fire! spout rain!'. In 1778, a Royal Navy vessel was named 'Spitfire' – a euphemistic version of 'Cacafuego', a Spanish treasure galleon captured by Sir Francis Drake; thereafter the Navy used the name nine other times up to 1912. It was also to be seen in the titles of several 1930s films and thus at that time was not just a relatively obscure part of the English vocabulary. By now, the word would probably have become obsolete – had it not been for the Battle of Britain, and many wartime actions thereafter, involving Mitchell's fighter.

* * * * *

Now that the newly christened Spitfire had achieved its better performance figures, the company considered that it was safe to send their machine to Martlesham Heath for evaluation and service testing (and, as it turned out, an inspection by King Edward VIII in Mitchell's presence). Whilst the use of flaps was, by this time, quite common, the relatively novel retractable undercarriage was now to produce some disconcerting moments. It has been mentioned earlier how the prototype Walrus was landed in the sea with the undercarriage still down and turned turtle. Thus the Spitfire prototype had also been fitted with an audible warning to prevent similar sorts of accidents. This precautionary measure nearly failed to prevent costly delays to the testing programme when the one-and-only prototype was flown by an AAEE pilot for the first time.

Fl. Lt (later Air Marshall Sir) Humphrey Edwardes-Jones' account to Price of this first flight also indicates something of the impact of Mitchell's new design upon the test centre:

> Usually the first flight of a new aircraft did not mean a thing at Martlesham, they were happening all the time. But on this occasion the buzz got around that the Spitfire was something special and everybody turned out to watch – I can remember seeing the cooks in their white hats lining the road. I took off, retracted the undercarriage and flew around for about 20 minutes. I found that she handled very well. Then I went back to the airfield.
>
> There was no air traffic control in those days and I had no radio. As I made my approach I could make out a Super Fury some way in front of me doing 'S' turns to lose height before it landed. I thought it was going to get in my way but then I saw it swing out to one side and land, so I knew I was all right. But it had distracted my attention at a very important time. As I was coming in to land I had a funny feeling that something was wrong. Then it suddenly occurred to me: I had forgotten to lower the undercarriage! The klaxon horn, which had come on when I throttled back with the wheels still up, was barely audible with the hood open and the engine running. I lowered the undercarriage and it came down and locked with a reassuring 'clunk'. Then I continued down and landed. Afterwards people said to me, 'You've got a nerve, leaving it so late before you put the wheels down'. But I just grinned and shrugged my shoulders. In the months that followed I would go quite cold just thinking about it: supposing I had landed the first Spitfire wheels-up! I kept the story to myself for many years afterwards.

With hindsight, one wonders how a crash-landing of the sole, unknown and untried prototype would have affected its future. The concern of Edwardes-Jones and of Quill before him that they should not damage a new prototype was foremost a matter of the professional test pilot's self-esteem but the usual Air Ministry practice of only ordering one prototype from a firm was hardly wise, given the urgent need to equip Britain with a 300 mph plus fighter – for the German contract exercise, equivalent to the British F.7/30 requirement, four firms had each been authorised to build three prototypes.

Events in Europe were certainly beginning to create an even more urgent need to find an adequate replacement for the standard RAF fighters of the day and thus it was that this first flight at the AAEE took place as soon as the aircraft had been delivered by Summers and he had briefed the service pilot. The usual preliminaries were dispensed with and Edwardes-Jones was instructed to telephone the Air Ministry as soon as he got down:

> Normally, a firm's test pilot would bring in a prototype aircraft for service testing, and it would be first handed over to the boffins who would weigh it very carefully and check that the structure was as it should be. It was usually about 10 days before it came out for its first flight with us. With the Spitfire prototype, it was quite different. Mutt Summers brought her over, and orders came from the Air Ministry that I was to fly the aircraft that same day and report my impressions ...
>
> Once down I rang the number at the Air Ministry I had been given, as ordered. The officer at the other end [Wilfred Freeman, Air Member for Research and Development] said ... 'All I want to know is whether you think the young pilot officers and others we are getting in the Air Force will be able to cope with the aircraft'. I took a deep breath – I was

supposed to be the expert, having jolly nearly landed with the undercarriage up! Then I realised that it was just a silly mistake on my part and I told him that if there were proper indications of the undercarriage position in the cockpit, there should be no difficulty. On the strength of that brief conversation the Air Ministry signed a contract for the first 310 Spitfires on 3 June, eight days later.

This extract provides an interesting subtext to the conclusion of the Martlesham report on the handling trials of K5054:

The aeroplane is simple and easy to fly and has no vices. All controls are entirely satisfactory for this type and no modification to them is required, except that the elevator control might be improved by reducing the gear ratio between the control column and elevator. The controls are well harmonised and appear to give an excellent compromise between manoeuvrability and steadiness for shooting. Take-off and landing are straightforward and easy.

The aeroplane has rather a flat glide, even when the undercarriage and flaps are down and has a considerable float if the approach is made a little too fast. This defect could be remedied by fitting higher drag flaps.

In general the handling of this aeroplane is such that it can be flown by the average fully trained service fighter pilot, but there can be no doubt that it would be improved by having flaps giving a higher drag.

The prototype before the undercarriage bottom fairing doors were discarded.

The prototype, with flaps lowered (at original, shallower, angle).

The extract also reflects the report in McKinstry that Sorley had recommended the aircraft's being ordered 'without the delay of prototype testing' and the view of the Director of Technical Development, Commodore R. H. Verney, that as soon as the aircraft had flown that 'would be the time for a production gamble'. Thus it was that, after K5054 had been sent to Martlesham in mid-May, an order for 310 Spitfires was placed in June – before the formal A&AEE report, dated 'September 1936', had been produced.

At last, the Air Ministry could now envisage the possibility of supplying their fighter squadrons with an aircraft equipped with eight guns, unproblematical to fly and capable of a maximum speed some 120 mph faster than its namesake two years earlier. However, the acquisition of a trendsetting new fighter was not to be without its production problems as the very large order required different sections of such a machine, embodying advanced structural methods, to be subcontracted out to more inexperienced firms. The problems were, not surprisingly, most acute in respect of the wings although an incident recalled by Quill revealed the general novelty of the advanced technology of K5054: after an unscheduled stop at an RAF airfield in December, 1936, Quill was disconcerted to hear tapping noises coming from the rear of the aircraft: 'I checked I had shut everything down but the tapping sound continued. Then as I climbed out I saw the reason. Several mechanics were standing around the rear fuselage, tapping it with their knuckles disbelievingly. "My God," one of them exclaimed, "it's made of tin!".'

Believed to have been taken during Mitchell's last airborne viewing of his fighter.

* * * * *

And so, after the lean years of the early 1930s, Supermarine was suddenly engulfed with work. The Australian order for 24 Seagull Vs in 1934 had been followed by one for 17 Stranraers and an initial order for 48 Walruses in 1935. The unprecedentedly huge requirement of 310 Spitfires in June of the following year was followed a month later with an order for 168 more Walruses. A visibly impressive demonstration of the previous five years' work was the Vickers Press Day on 18 June, also in 1936, when the parent company's prototype Wellington and the Wellesley bombers were joined by Mitchell's Scapa, his Stranraer, his Walrus, and the Spitfire prototype. As Alan Clifton later commented, not entirely jokingly: 'One might have thought the Air Power of Britain was intended to consist almost entirely of Supermarine aircraft. At that time, remember, there were about a dozen aircraft firms'.

It is a nice illustration of Mitchell's especial pride in his fighter creation that, when an oil leak was discovered, he took the view of the test pilot that there was enough oil in the machine to risk a five minute flight for the large gathering of pressmen. Against the (predictable) advice of the expert from Rolls-Royce, he had agonised for a few minutes and then said to Quill, 'Get in and fly it'. Almost as soon as the pilot was airborne, an oil pipe finally fractured and only some very skilful airmanship saved the one-and-only prototype once again.

Of all the aircraft at the Press Day, Mitchell followed the progress of the Spitfire the most closely, as Quill recalled: 'Whenever the new fighter was flying Mitchell would get into his car and drive from his office to Eastleigh. As I was coming in to land I would see his yellow Rolls-Royce parked and know he was there. He kept a close eye on things.' On 7 April, 1936, he also visited Martlesham with his son, Gordon, to see how the service testing of the Spitfire was going. On the 27th of the same month, it was put on public display in the Royal Air Force Pageant at Hendon and two days later was demonstrated at the SBAC display at Hatfield with slow and fast runs, finished off by a display of fast aerobatics. The *Flight* report on this last display aptly summed up the sort of machine that Mitchell had been working towards: 'It is claimed – and the claim seems indisputable – that the Spitfire is the fastest military aeroplane in the world. It is surprisingly small and light for a machine of its calibre (the structural weight is said to have been brought down to a level never before attained in the single seat fighter class), and its speed and manœuvrability are something to marvel at'. For security reasons, Supermarine's publicity, had again to be content with claims that the company were not allowed to substantiate with facts.

By the time that the Martlesham verdict on the Spitfire prototype was received in September, 1936, the Company was fully engaged in preparations for

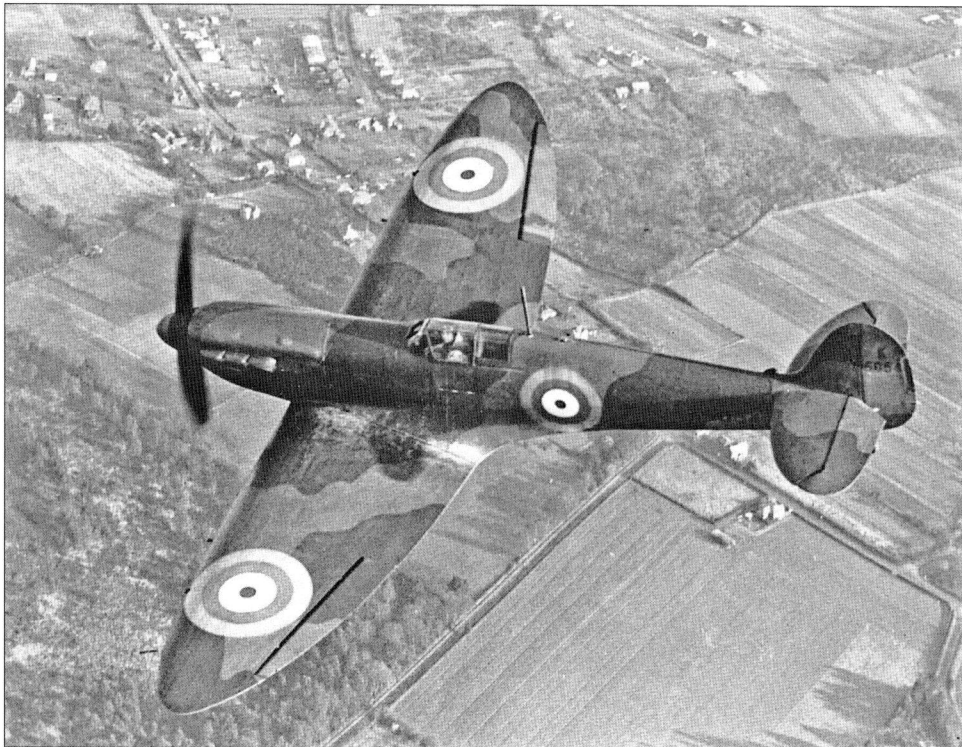

Jeffrey Quill flying the Spitfire prototype with the camouflage finish it carried for its later development trials.

the production of the actual Mark I version for squadron service. Meanwhile, the prototype Spitfire was extensively modified to bring it in line with the forth-coming production standard, including the fitting of the 1030 hp Merlin II, and its high gloss finish was replaced by the Air Ministry dark green and earth day fighter camouflage scheme. The prototype, having fulfilled its main purpose, was then subject to a series of accidents: a wheels-up landing on 22 March 1938 due to oil pressure failure; brake failure on 15th May, resulting in its nose in gorse beyond the airfield; and a ground loop nine days later, due to the undercarriage collapsing. It was finally written off after a fatal landing accident on 4th September, just one day after England had declared war with Germany.

The first production Mark I Spitfire made its maiden flight on the 14th May, 1938, and the first service machine, K9789, was delivered to No. 19 Squadron on August 4th of that year – just over a year after Mitchell had died on the 11th of June, 1937).

Supermarine Spitfire F.IX floatplane conversion, 1944.

— *Chapter Nine* —

After
Mitchell

Whilst engaged upon the Spitfire, Mitchell had been involved with other projects which, if they had been completed, would have appeared after his death. One particular project was for a two-seater version of the Spitfire to meet the Air Ministry specification F.9/35. This was drawn up in response to a particular theory about combating bombers: instead of a diving approach from behind, there was an almost naval concept of aircraft flying alongside the enemy and firing across from turrets situated immediately behind their pilots. (A compromise between these two methods was incorporated in the F.10/35 document where it was contemplated that some or all of the guns might be 'mounted to permit a degree of elevation and traverse with some form of control from the pilot's seat' but, unsurprisingly, nothing came of this somewhat vague idea – Mitchell and others had their hands full trying to produce the basic multi-gun fighter to satisfy the main Air Ministry requirements.)

This F.9/35 Specification required a powered turret as the main armament in order to improve on the Hawker Demon where the slipstream over the exposed rear cockpit produced helmet and goggle 'flutter' and made it difficult for the rear gunner to operate, even after the fitting of a protective shield. The Supermarine response was a typically low frontal area proposal to place the gunner under a low blister canopy and to accommodate the guns further back in a barbette which could be kept to quite small, low drag, proportions because of the separation of gunner from his remotely controlled guns (*see* similar considerations in respect of Mitchell's Bomber, below).

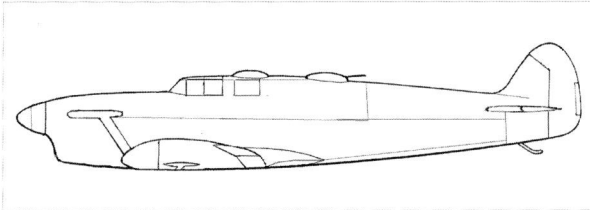

Two more conventional aircraft were produced to this Specification, the Hawker Hotspur and the Boulton-Paul Defiant, both designed to take a more conventional Boulton-Paul semi-powered, enclosed, four-gun turret. The latter went into production but its top speed of just over 300 mph compared unfavourably with the estimated 315 mph of the Supermarine proposal. Whether it would have fared much better than the Defiant, when the fall of France allowed higher performance enemy fighters to protect the bombers attacking the British Isles, remains a matter of speculation as the Supermarine bid was not taken up.

The B.12/36 Bomber and the Bombing

By the end of 1935, Mitchell confided to his diary that he feared his cancer had not been eradicated but when a new specification for a bomber was issued in July of the next year, he involved himself fully in a response – perhaps in defiance against his mortality as much as against Nazism. Thus, in the last full year of his life and even when he was supervising the finishing touches to the Spitfire design, another major innovative project was occupying his mind. This was in response to Air Ministry specification B.12/36 for a high-speed, four-engined bomber with a maximum range of 2,000 miles, carrying a 14,000 lb bomb load or 24 soldiers. It also had to be able to be broken down into component parts for transport by the existing railway system and to lift off from a 500 ft runway, clearing a height of 50 ft at the end; for this last purpose there was added a requirement to provide a catapult take-off capability – because of the small airfields currently in use and, particularly, in order to extend the bomber's potential range and load capacity. At the same time, the wingspan was to be limited to no more than 100 ft – in respect of transport considerations and perhaps to discourage the design of very large, over-expensive aircraft rather than with regard to the size of existing hangars. The bomber also had to have a retractable ventral turret as well as nose and tail guns; and also to be capable of staying afloat for several hours in the event of being forced down in the North Sea or Channel, as might be expected if the international situation did not improve.

It is interesting that eventually Shorts and Supermarine were each awarded contracts for two prototypes rather than, for example, Handley-Page or Vickers which had extensive experience of the larger sort of land-based machines. But, in view of some of these Ministry requirements, it might be noted that Supermarine and Shorts had a great deal of experience of water-resistant hulls and Supermarine had just provided the RAF and Navy with the very efficient Stranraer and the catapult-stressed Walrus. Indeed, the proposal received from Supermarine must have confirmed the officials' regard for Mitchell's importance to the aviation industry: his tender for the bomber had a proposed wingspan of 97 feet but it, nevertheless, was to make use of a single-spar wing supported by torsion-resistant leading edge boxes on a similar principle to that developed for the much smaller Spitfire. Unusually for its time, fuel was to be carried in these leading edges, thereby saving weight, and with the tanks adding to the rigidity of the wing.

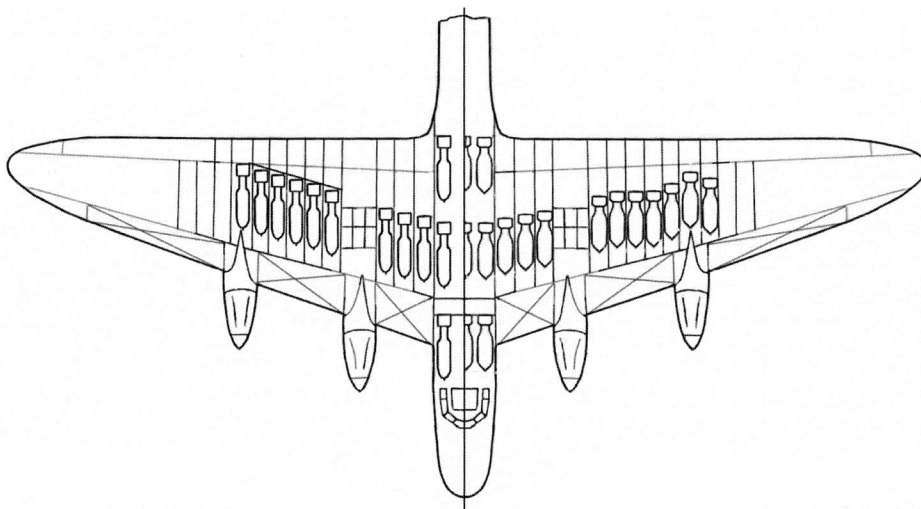

Supermarine Type 318 showing alternative stowage of 29 × 250 lb bombs and of 27 × 500 lb bombs; also leading-edge fuel tanks.

Behind this spar component, the structure allowed sufficient room for the main stowage of bombs, thus avoiding the need for conventional tiered bomb stowage in the fuselage, which would have substantially increased its cross-section and its drag component – further evidence of Supermarine's Schneider Trophy background, in giving, instinctively as it were, particular attention to the reduction of frontal area. The Polish PZL P.37 medium bomber had made use of bomb bays in the wing between the engines and the fuselage but Mitchell's more extensive proposal was not adopted in any of the other front-line bombers of World War Two. A further refinement was the proposal to place the required armament well below the eyeline of the gunners, not only giving them an improved view but also enabling a reduction in the cross-section of the turret and a more rapid traverse of the guns.

Three versions of the Bomber were proposed:

Type 316 with Bristol Hercules engines, single fin and deltoid wing shape;

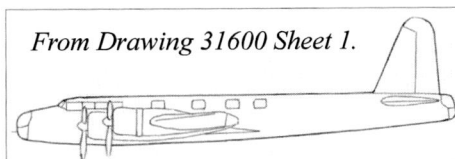
From Drawing 31600 Sheet 1.

Type 318 with Rolls-Royce Merlin engines but otherwise similar to 316;

Type 317 with Bristol Hercules engines, twin fins, and wing leading and trailing edges tapering almost equally.

Supermarine model of Type 317.

The above painting by the author is based on Type 317 G.A. and company model.

Photographs of the fuselage, almost completed before destruction by enemy bombing, do not allow one to know what fin and wing type was to be fitted although the twin fins were probably expected to be lighter and more aero-dynamically efficient than a single fin; and the Type 317 wing would have been less complex to design and build – an important consideration at a time when a new and more formidable bomber was urgently needed – hence a recent claim that the Air Ministry had instructed the company to concentrate upon this version. It might also be noted that Mitchell's aerodynamicist, Beverley Shenstone, who was familiar with advanced wing theory and who was, no doubt, a strong influence on the Type 316/8 wing shape, had left the company by 1938 and so the more conventional outline, with a slight elliptical planform outboard of the engines, had most likely prevailed.

At the beginning of 1937, Mitchell could no longer work at the office and soon was unable to contribute to the development of his bomber. His Deputy Designer and Personal Assistant, Maj. Harold Payn, assumed command and in November, 1938, Mitchell's team submitted a lighter and smaller design than their rivals for the same specification and produced a set of estimated performance figures; these make an intriguing comparison with published figures for the earliest Marks of the most well-known British four-engined bombers of comparable size:

Aircraft	Power Rating	Range	Bomb Load	Max. Speed
Supermarine (estimate)	1,330 hp	3,680 miles	8,000 lb	330 mph
Stirling I	1,590 hp	1,930 miles	5,000 lb	260 mph
Halifax IB	1,145 hp	1,985 miles	7,000 lb	282 mph
Lancaster I	1,390 hp	2,530 miles	7,000 lb	287 mph

One of the bomber fuselages before their destruction by enemy action.

More often than not, Supermarine estimates were actually achieved when their designs flew (one remembers Orlebar's comments about the correct estimate on the S.6B or the Spitfire performance when the right propeller was found) and so it will always be a matter of rueful conjecture as to whether the extraordinarily competitive figures for Supermarine's proposed bomber would have been achievable when war did begin. It might, however, be maintained that, with the need for volume production – using standard gun turrets – and probably being forced to add a dorsal one too, the Supermarine figures might have turned out to be closer to those actually achieved by the other firms' products.

Unfortunately it will always be a matter of speculation as the project only reaching the stage of two unfinished prototype fuselages when they were destroyed by enemy bombing. The Southampton factory was easily within range of German aircraft, now that northern France had been overrun, leaving precious little response time for defending RAF fighters.

The first attack upon Supermarine took place on 15 September 1940, mainly causing civilian casualties in surrounding houses and with only very minor damage to Supermarine; this was followed by another attempt eight days later when even more civilians were killed and the aircraft factory was again left mostly intact – owing to the Luftwaffe intelligence maps showing the Thornycroft shipbuilding yard downstream from Supermarine as their target (even though the

Supermarine slipway was well known to pre-war German mail plane pilots). The three light anti-aircraft batteries positioned near the factory had now been supplemented by barrage balloons but, two days later, on 26 September, two successive waves of German bombers delivered an estimated 36 tons of bombs more accurately. This time many Supermarine workers were killed or seriously injured and damage to the factory was extensive. Thirteen Spitfires were destroyed or badly damaged, although most of the machinery and jigs were still useable; however, as the two half-completed bomber fuselages were badly damaged, it was decided that Supermarine should concentrate upon its other current products, especially the Spitfire.

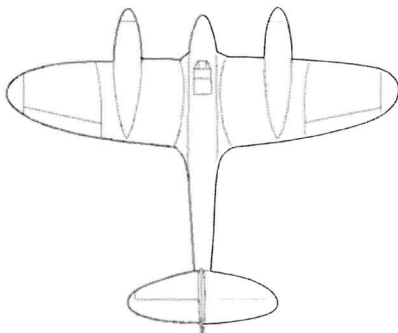

There was another project destroyed by the bombing, a mock-up of a twin-engined six cannon fighter with the then novel provision of a tricycle undercarriage. The Specification that had given rise to this proposed machine was designated F.18/37 and, in fact, led to the Hawker Typhoon. As the Air Ministry decision after the bombing resulted in the termination of Supermarine's interest, the question of Mitchell's possible involvement is academic but its proposed elliptical and rounded flying surfaces are, at least, a suggestion of the continuing influence of Mitchell on his design staff.

* * * * *

Mitchell's death

Our Chief Designer never lived to learn of the fate of his last projects, just as he never actually saw his Spitfire go into squadron service before the start of World War Two. Towards the end of February, 1937, he went into a London hospital. The prognosis was not good and a stay at the Cancer Clinic in Vienna was arranged in the April. Letters testify to his dismay at not being able to continue his input into the design of the bomber but it had become clear that this was not possible. Mitchell returned to Southampton on 25 May 1937, the very day that his first Spitfire, Type 224, was finally retired, to eventually become a ground target at the gunnery range at Orfordness. He died on 11 June, aged 42.

As he invariably gave full credit to his design staff in his speeches, it was fitting that he requested they be given first place at his funeral and it is thus significant that Harry Griffiths' recollection of Mitchell's death prompts a memory of his relationship with this team:

> On the day he died Arthur [Black, chief metallurgist] and I were standing at the bench discussing a problem when Vera Cross, R.J.'s faithful secretary over many years, came in and just said, 'It's all over.' Arthur looked at me and shook his head then he turned away and was silent for a long time.
>
> At the annual dinner that year we stood in silence in his memory and then drank a toast: 'To a very gallant gentleman.'
>
> The Christmas before his untimely death he arrived late for the annual design staff dinner, and in spite of a place having been kept for him at the head of the table he insisted on sitting at the other end with us lads and sharing a joke and some wine.

Before his death, Mitchell had had the satisfaction of attending the Vickers Press Day where his Scapa, Walrus, Spitfire, and Stranraer were all on display together. With the last three, he had produced, respectively, the slowest and the fastest aircraft for the RAF and the fastest biplane flying boat. After the 15 production Scapas and 17 Stranraers, he then saw his company receive initial orders for a total of 217 Walruses and 310 Spitfires. Whilst there had been a lean period, with no production runs, between the Southampton and the Scapa, this was no mean an epitaph for a designer, but he could not have imagined the extent of the future wartime requirements and varied duties of these last two aircraft, particularly the Spitfire. Nevertheless, he had created a team which could carry on from where he had had to leave off and so a description of just how they did so with his initial concepts will be a fitting part of this epilogue to his design career.

* * * * *

The Walrus in Wartime

In the case of the Walrus, the initially perceived need for fleet spotting and maritime reconnaissance generally was supplemented by the anticipated need to rescue 'ditched' aircrews and also by the need for a MkII training version. As a result, the number of aircraft ordered eventually rose to a grand total of 746 and by January, 1940, the Mark I aircraft formed the major part of the equipment of No. 700 Squadron, which embodied all the catapult units aboard British warships; the Walrus also served on most capital ships of the Australian and New Zealand Royal Navies. At the same time, new Spitfire orders were coming in so that, by the end of 1939, the Walrus production had to be transferred to the company's old rivals, Saunders-Roe, including 191 Mark IIs, which saw a reversion to wooden hulls. This apparently backward step was a means of saving on priority light alloys in this training version; it could also be produced and repaired by workers who were in less demand than those who worked in metal.

A demonstration of air-sea rescue by Walrus.

In addition to fleet spotting and naval reconnaissance duties, the Walrus was used for photoreconnaissance, for artillery spotting, for anti-submarine convoy patrol, and for communication work. Mitchell had produced an aircraft which was to operate in most theatres of war from the African deserts to the Pacific Ocean; it also carried secret agents and landed on jungle airstrips and on ice floes and was variously fitted with bombs, depth-charges, air-sea rescue containers, and ASV (air-to-surface vessel) radar for reconnaissance. (After the war, one

Walrus made an unlikely appearance as the winner of the 1946 Folkestone Trophy Race at Lympne and another was used for whale-spotting aboard SS *Balaena* in the Antarctic during 1947.) The varied wartime duties of the Walrus said as much about the exigencies of war as they did about the suitability of the machine but its low speed was a bonus in many circumstances. Its varied employment was also a testimony to the basic integrity of the structure created by Mitchell and his team and to its culmination of a succession of very sturdy smaller flying-boat designs from his company.

Sea Otter and Seagull ASR.I

Whilst Mitchell could not have guessed at the very wide use of his Walrus in later years, he had been involved in a requirement for an improved version and had met the Air Ministry Director of Technical Development as early as February, 1936, to discuss this project. On 17 April, the company received 'Instructions to Proceed' with a heavier machine which was to have a longer range and a dive-bombing capability. A greater load capability was also a requirement which proved to be justified, as its predecessor, the Walrus, would have, on occasion, to taxi rather than fly back to base because of the number of survivors rescued.

The hull of the new design, in particular, showed a marked similarity to the Walrus, although it was slightly more rounded at the front; and this less angular feature was carried through more thoroughly in respect of the flying surfaces. However, the most noticeable change was a reversion to a tractor-engined configuration with the nacelle raised to occupy most of the centre-section of the top wing. This neater engine mounting also permitted a tidier trailing-edge for the folding wing arrangement and the offsetting of the engine was avoided by giving the fin an aerofoil section. Because of the other pressing demands of the time, the design and the construction of the prototype aircraft was not completed until September, 1938, and thus Mitchell never saw this eventual development of his Walrus. Named Sea Otter, two hundred and ninety-two were eventually built, of which all except the two prototypes were again subcontracted to Saunders-Roe. Entering service at the end of 1944, it represented the last biplane flying boat from the long line of Supermarine aircraft to have come from Mitchell's drawing-board.

The prototype Sea Otter.

There was one further seaplane, a monoplane this time, also required for air-sea-rescue, which deserves a mention. As it appeared as late as 1948, it might more than most of those mentioned above be called a posthumous Mitchell machine. It had a cantilever, variable incidence wing with full-span flaps and leading-edge slats, mounted on a central pylon which also supported a Rolls-Royce Griffon engine with contra-rotating propellers.

The first Seagull ASR.I.

These features, and the butterfly tail with end-plate fins, revealed how far thinking had outstripped the old amphibians that Mitchell knew but its single-step planing surfaces and aggressive nose still embodied the basic Walrus sea-going formula. Only three prototypes were built which, in the context of the story of Mitchell's designs at least, was a pity in that a production order for the ASR.I would have prolonged a venerable Supermarine name – it was designated 'Seagull'. Also in the Supermarine tradition, it set up a 100 km closed-circuit record, in 1950, for amphibians with a speed of 241.9 mph.

The Stranraer in Wartime
The other Mitchell flying-boat category, the larger long-range patrol seaplane, did not have any such successors, although his last design did contribute to the war effort. The Stranraer served with No. 228 Squadron when it was needed to patrol the North Sea and some of the Stranraers of this unit were transferred to No. 209 Squadron and, in particular, assisted in patrols between Scotland and Norway. Fitted with extra fuel tankage under one wing and bombs under the other, they conducted patrols against enemy shipping until replaced in April 1941 by the Lockheed Hudson. No. 240 Squadron was also equipped with the Stranraer and made the last operational patrol of the type on 17 March, 1941, after which it was replaced by the Catalina.

A Stranraer of No. 240 Squadron touching down.

In addition to the 17 production Stranraers for the RAF, the RCAF also adopted them and 40 were built by Canadian Vickers. These saw a great deal more service than their British counterparts, in the battle against the U-boats in the Atlantic, and they operated with bomber reconnaissance squadrons, Nos. 4, 5, 6, 7, 9, 117 and 120. They were finally replaced on active service by the Canadian Catalina, the Canso, in March 1941. In view of their original low maintenance specification and their associated anti-corrosion features, it is worth recording that the last RCAF Stranraer was retired as late as 20th January, 1946 and that fourteen of the aircraft were sold to the civil sector, especially to private airline companies in Canada where the lakes of the Northern Territory provided ready made runways – just as in the pioneering flying-boat days. The last one of these Stranraers served in these spartan regions until 1958.

First of the Canadian built Stranraers, with bombs under wing.

Nevertheless, the longevity of the Stranraer does not conceal the fact that, in this larger flying-boat category, the influence of Mitchell was relatively short-lived – in many ways a result of lack of official encouragement of his Air Yacht (monoplane) approach to the reconnaissance flying boat type and of the cancellation of the enormous Type 179 Giant cantilever monoplane. Meanwhile, the early and single-minded approach of Shorts to all-metal aircraft had paid dividends and their response to the R.24/31 requirement was a stressed-skin, cantilever monoplane with a wingspan of 90 feet. Known unofficially as the 'Knuckleduster' because of its sharply angled gull wings, it had an excellent maximum speed of 150 mph (thanks to its clean lines) and was faster than its Supermarine counterpart, the Stranraer. However, because of engine and fuel supply problems, it remained in prototype form.

Nevertheless, the Knuckleduster led to the military development of another cantilever monoplane type, the Sunderland, which dominated the wartime long-range sea patrol effort with a total of 741 being built. In addition to its military counterpart, Short's clean, streamlined civilian version had monopolised the flying-boat provision on the Imperial Airways routes just prior to the outbreak of the war and represented a major step forward in flying-boat design without a rival aircraft from Supermarine.

One of the other seaplane firms, Saunders-Roe, was also contracted to produce an all-metal cantilever monoplane flying boat, in 1938. With an 81 foot wingspan, it was the sort that might have been developed by Mitchell, had the military origins of the Air Yacht been given more official support. Ironically, this Saunders-Roe Lerwick did not attract many orders as it was laterally unstable, had poor stall characteristics and a poor take-off performance (its lack of success

provided the capacity for Saunders-Roe to take over the surplus orders for Supermarine Walruses and Sea Otters). The American Catalina was the successful
alternative, instead of an equivalent which Mitchell might have developed from
the Air Yacht. (After the war, Saunders-Roe did make something of a stir in the
aviation world with a jet flying-boat fighter and with the enormous Princess civil
flying boat. The first of these represented a fascinating return to the N.1B Baby
formula which Mitchell developed when he first joined the about-to-be-formed
Supermarine company but, like its Supermarine predecessor, the Saunders-Roe
machine did not go into service production. Similarly, the Princess did not go
beyond the prototype stage and it also echoed another Supermarine venture
which came to nothing – the Type 179 Giant.)

* * * * *

Joe Smith Takes Over

Thus it came about that the most enduring memorial to Mitchell's career as
Chief Engineer and Chief Designer was to be not a development of his many
large flying boats but the landplane Spitfire: it continued in service in front-line
fighting situations long after any of Mitchell's other aircraft and was still being
improved upon long after the Stranraer, the Walrus, or the Sea Otter, had ceased
to be regarded as having any further potential.

After the death of Mitchell, Maj. H. J. Payn took over but, as his wife had
German connections, it was considered that he might be a potential security risk.
Joe Smith was moved from the post of Chief Draughtsman to that of Chief Designer in 1941 and the credit for fully realising the potential of Mitchell's fighter
design must go to him and his team. Joe Smith had faith in the development
potential of Mitchell's design, saying that it would 'see us through the war'. His
Schneider Trophy experience with Mitchell of developing the 900 hp S.5 into
the similar, but larger, 1,900 hp S.6 and then into the strengthened 2,350 hp S.6B
must surely have pointed the way; Quill points to one other particular event that
might very well have been of considerable influence upon Smith's faith in the
long-term potential of Mitchell's design. This event was also associated with a
speed competition – in this case, Supermarine's planned attempt upon the World
Land Speed record then held by the Howard Hughes H-I racer.

A specially prepared 'Speed Spitfire' airframe with a Merlin engine developing 2,000 hp promised a top speed of 425 mph at sea level although, in the
end, the venture was abandoned after Germany put up the record to 469 mph
– with an Me 209. Unable to compete with this 'special' version of the Bf 109,
with a wingspan reduced by 7 feet, Smith must nevertheless have appreciated
that this development of the basic Spitfire airframe clearly revealed the future
potential of the Mitchell design, given the current Mark I aircraft's 367 mph
with a Merlin II rated at only 990 hp. J. D. Scott wrote in his history of Vickers,
'By 1940 Joe Smith … had reached the conclusion that the Spitfire design was
capable of the most extensive, and indeed of almost infinite, development'. The
various Marks of Spitfire which were produced demonstrated this was the case,
to the extent that Rolls-Royce, when designing the Griffon engine to succeed the
Merlin, tailored the new engine to fit the existing fighter's airframe.

Nearly thirty main variants of the the Spitfire, as well as numerous minor
modifications, followed from Mitchell's prototype and it is a measure of the
contribution of Smith and the design team to the war effort that an average of
over four distinct Marks of Spitfire per year were developed. Whether Mitchell
would have pursued a similar course, would have come up with some ingenious
stroke of lateral thinking, or would have been permitted to go for a completely

Joe Smith.

new design must always remain a matter of speculation; certainly the pressures of war would not encourage the tooling-up necessary for a new type as long as modifications of existing aircraft could conceivably meet the changing wartime requirements: as Smith said, justifying his continuous modifications to the Spitfire type, 'the hard school of war leaves no room for sentimental attachments and the efficiency of the machine as a fighter weapon is the only criterion'. However, as we shall see, producing the realisation of Mitchell's prototype was by no means straightforward.

Spitfire Production Problems

It was mentioned earlier how RAF ground staff were surprised at the metal fuselage of the prototype Spitfire and, thereafter, full-scale production of the sophisticated all-metal machine presented many challenges and, unfortunately, many delays. As a result, Air Ministry frustrations with the first stuttering supply of the Spitfire might very well have resulted in an early termination of the type. Lack of frankness on the part of the Chairman of Vickers, Sir Robert McLean, and mismanagement at Lord Nuffield's huge Castle Bromwich aircraft factory, caused growing concern at the Air Ministry and, thereafter, in the Government. Fortunately for Mitchell's reputation, at least, the difficulties with the Rolls-Royce Peregrine engine in the very promising four-cannon Westland Whirlwind and, later, with the Napier Sabre engine in the proposed Hawker Hurricane replacement, left the Air Ministry with no alternative but to persist with getting the Spitfire produced in large numbers.

At Supermarine, the order for 310 Spitfires just eight days after the Martlesham test of the prototype presented immediate problems, not only because Supermarine lacked sufficient capacity and skilled local labour for the initial order but also because virtually a whole new set of drawings was needed – in-house arrangements sufficient for the production of a 'one-off' prototype would not do for the firms to whom work would have to be subcontracted. Some of these companies had no previous experience of aircraft manufacture and so the parent firm was further burdened with a high level of supervision and with replying to a multitude of queries.

Nor were the problems confined to the satellite firms, as Webb has well documented: in particular, there was no proper recording system at Supermarine and parts were often stored in the nearest bin available, thus often involving subsequent time-consuming searches. It ought also to be noted that a labour force, used to working around large flying boats, would now have to change its procedures for a relatively tiny fighter into which its equipment had to be far more tightly packed.

The dismay at the Air Ministry over the non-appearance of the fighter, contrary to the Chairman's assurances, was such that the Secretary of State for Air, Lord Swinton, was considering requiring the Air Council to see the whole Vickers Board in order to 'put an end to what can only be described as an intoler-able situation'. Meanwhile, the effect of Hawkers having produced a fighter of more traditional construction was that the first Hurricanes were delivered in December, 1937 and they had equipped five squadrons before the first Spitfires entered service in August the following year.

Trevor Westbrook, Supermarine's fierce works manager, had begun to significantly improve matters at Woolston but he was recalled to the parent firm at Weybridge early in 1937 to be replaced by a far less forceful character. Mitch-ell, having turned his last energies to the B.12/36 bomber design, was about to go to the Vienna Cancer Clinic and was, at least, spared the distraction of the mounting concern about Supermarine's lack of capacity, chaotic subcontracting arrangements and poor management. The replacement of Sir Robert McLean by Sir Charles Craven as Managing Director of Vickers Aviation was eventually to achieve the required results but a turn-round in production figures could not be achieved overnight. It thus emerged that the first Spitfire would not be delivered by December 1937, as promised, and was not likely to appear before February of the following year.

By the end of 1937, just six fuselages had been completed and were awaiting the delivery of wings from subcontractors. And so Craven had to revoke McLean's unrealistic production estimates but he could only promise fourteen aircraft in the first quarter of 1938. Yet, even this modest response to the original order for 310 Spitfires was not met – the first production aircraft first flew on 14 May, 1938, and it was two months before the second was ready. Meanwhile the concern over German militarisation produced a *Times* editorial on 18 July which stated that 'the delay in the production of the Spitfire has been one of the most disappointing features in the progress of the air programme' – in despite of which, Chamberlain's Cabinet on 7 November decided that bomber production should only be such that jobs would not be lost whereas it set an ambitious target of 3,700 fighters.

In anticipation of the enlarged requirement for fighters, Lord Nuffield had already been approached with a view to employing his skills as a mass producer of cars to oversee the building of a massive new factory with an initial order for one thousand Spitfires. It is clear that, at this time, Mitchell's Spitfire was re-garded as the vital element in Britain's defences but serious problems with the Castle Bromwich factory soon began to erode this view. Whilst it was only to be expected that the extensive jigging required for the new production lines would take time to be put in place, it was discovered that, by the end of 1939, the aircraft plant still looked like a construction site and that Lord Nuffield appeared to be no longer up to the job; he seemed often vague and unable to grasp detail yet his autocratic manner made him unwilling to accept Ministry 'interference' – that is, their extreme concern that none of the promised sixty Spitfires per week from April 1940 had appeared. It was thus fortunate that, on 14 May during a telephone exchange with Lord Beaverbrook, the new Minister of Aircraft Production, he is reported to have said, 'Maybe you would like me to give up control of the Spitfire factory', to which Beaverbrook replied, 'Nuffield, that's very generous of you. I accept'.

Meanwhile, by 4 August, 1938, No. 19 Squadron received its first operational Spitfire from Supermarine, and by September, 1939, the company had produced almost all the original order of 310 machines. But 36 had been lost in accidents and by March, 1940, the number produced had risen to over 500 but 98 had been written off in accidents or were being repaired. As the Spitfire was still regarded as the best machine to oppose the Messerschmitt Bf 109, the need to conserve stocks was such that Air Vice-Marshall Dowding persuaded Churchill not to send any abroad to assist in the defence of France. And Vickers was tasked with reorganising the West Bromwich facility: a recent account by Leo McKinstry of the company's findings showed a lack of organisation of the building work and fitting out, extensive underutilisation of machine tools, poor record keeping of supplies and, more seriously, a management that had not effectively tackled poor timekeeping and frequent petty labour disputes by the work force.

It was not good for general morale or propaganda purposes for this situation to become widely known, especially the labour disputes, but many dismissals followed; production gradually improved thereafter with nearly two hundred aircraft being produced by the end of October, 1940, rising to over six hundred by the following February. Meanwhile, during the actions to protect the troops being evacuated at Dunkirk in the May and June of 1940, 386 Hurricanes were lost whilst the much smaller number of Spitfires available was depleted by 76. This engagement left a mere 331 available Spitfires and Hurricanes but, at least, Supermarine was working well, and delivered its 1000th fighter in August, 1940. Thus, whilst more Spitfires would have been very welcome when Britain's war began in earnest in this year, a sufficient number was available to begin the development of the image of Mitchell's fighter in the public imagination.

The Rise of the Spitfire Legend
A large number of these early fighters was paid for by the emergence of a scheme which played a significant part in placing the Spitfire in the forefront of the popular imagination: after the mining millionaire, Sir Harry Oakes, gave £5,000 to fund one of these aircraft, other individuals took up the idea of buying a Spitfire with enthusiasm, including the National Federation of Hosiery Manufacturers and a lady called Dorothy who collected funds from others with the same christian name. Towns and cities also organised the funding of aircraft and thus, by the spring of 1941 when the fund closed, £13,000,000 had been raised, financing over 2,000 aircraft. The owner of the Vienna Cancer Clinic to which Mitchell went during his last days was also a contributor.

Had production gone smoothly both at Southampton and Castle Bromwich, a significantly larger force of Spitfires would have been available during the Battle of Britain in the summer and autumn of 1940 and there would most likely have been fewer deaths of British civilians and RAF personnel. However, in the context of the Mitchell legacy, the fact that 'the few' caused Hitler to abandon his invasion plans by the October contributed significantly to the developing popular reputation of the Spitfire – even though over 60% of the fighter force consisted of Hurricanes. Whilst any immediate assessment of the relative merits of the Spitfire over the Hurricane could only be made by the Air Ministry and by the few pilots who had experience of both types, the final shape of Mitchell's fighter which emerged from the thick pencil lines on his drawing board embodied an aesthetic which contributed in no small way to the Mitchell legend.

As Rendall has pointed out, the Spitfire was the aircraft which, par excellence, was a by-product of the new 'Streamline Moderne'. In architecture, as a reflection of the austere economic times of the thirties, the 'streamline moderne' represented a rejection of unnecessary ornament; it also favoured relatively uncluttered curving forms and long horizontal lines. The Bentley and Bugatti racing cars now looked distinctly old fashioned beside the highly streamlined Mercedes and Auto Unions and beside Sir Nigel Gresley's streamlined Class A4 locomotives, which culminated in the record breaking Mallard, with its wind-tunnel-tested, aerodynamic body. Manufacturers of clocks, radios, telephones, furniture and numerous other household appliances began to embrace the new design concept (even though the speed-squared formula for drag mentioned earlier was hardly relevant to such items!).

The classic, yet 'moderne', shape of the Supermarine fighter.

Twenty days after the first flight of K5054, a report in The *Southampton Evening Echo* showed an immediate reaction to the 'moderne' shape of the Spitfire: 'Even the uninitiated have realised when watching the streamlined monoplane flash across the sky at five miles a minute (300 mph) and more, that here is a plane out of the ordinary'. Lieutenant Colonel William R. Dunn, ex-No. 71 (Eagle) Squadron, had a similar response to the outward appearance of Mitchell's design: 'The Spitfire was a thing of beauty to behold, in the air or on the ground, with the graceful lines of its slim fuselage, its elliptical wing and tailplane. It looked like a fighter, and it certainly proved to be just that in the fullest meaning of the term'.

Whilst Mitchell had professed not to be interested in the shape of his fighter's wing, as long as he could get the guns in, there is little doubt that its elliptical shape, echoed by the curved tail surfaces, was in keeping with the new shapes of speed; the thinness of his wing further enhanced its appeal. Whilst the

Air Ministry F.7/30 specification had called for a good 'fighting view' and, as a result, other prototypes had cockpit enclosures which were placed quite high on the fuselage, Mitchell, true to his Schneider experience, integrated his cockpit more completely into the streamlines of his fuselage – which was, like those of his Schneider Trophy planes, especially slender.

In previous chapters, it has been shown how good fortune had, at times, favoured the career of our designer and this has also been apparent in the emergence of the Spitfire as a national icon, founded not in the theoretical calculations of contemporary aircraft design but in an aesthetic movement appealing to a much wider, popular, audience. Following the onset of the daylight bombing of London, the British people were thankful for any response by the RAF and began to focus their gratitude on Mitchell's design which seemed so logically to combine power with modern, curvaceous lines. Illustrators of the time, responding to the aesthetically pleasing shape of his design, tended to single out his machine in their drawings of combat and, whilst pilots over Dunkirk and in the Battle of Britain had come to appreciate the fighting qualities of the Spitfire, the general public, 'knowing nothing of performance figures … have instinctively chosen one particular type as the paragon of protective types and they have guessed correctly' (*Flight*).

It is noticeable how contemporary observers of aerial engagements tended to report only Spitfires in action. The more numerous Hurricane, accounted for more successes than did the Spitfire and yet the Hawker machine, unfairly, has never been the popular image of the Battle of Britain – even though it was, if possible, employed against the slower, less manoevrable bombers, which obviously posed the real physical threat to the British populace. Nevertheless, the Battle of Britain became the defining moment in the developing legend of the Spitfire; indeed, it is, perhaps, the only encounter in British history that is primarily remembered by a weapon rather than by its commander-in-chief or the place where the event took place. On 15 September, 1945, on the fifth anniversary of the battle, the RAF staged a flypast over London of 300 aircraft; it was led by Douglas Bader in a Spitfire.

The reputation of Mitchell's fighter was further enhanced by the eventual lifting of the siege of Malta. In early 1941, the Luftwaffe had had to take over the action from the Italians and so reinforcements of Spitfires had to be flown in from HMS *Eagle* (they did not have the range to go directly from Gibraltar). Eventually, sufficient numbers of Spitfires, better organised ground support, and the deployment of many experienced pilots, led by the October of 1942 to the lifting of the siege and even to a developing Allied offensive strategy from the island. The Spitfire was once again seen as the significant factor in another British 'backs-to-the-wall' campaign and featured prominently in the newsreels, *Malta Shows She Can Take It*, and *Malta Convoy*.

Then, in 1953, *The Malta Story* appeared, followed by *Reach for the Sky* in 1956 and by *The Battle of Britain* in 1969. These films continued to keep the Spitfire legend alive, in no small part due to the availability of the Supermarine aircraft, thanks to its long production runs: it was only on the 20th February 1948 (almost twelve years from the prototype's first flight) that the last production Spitfire, an F. 24, left the production line and, of the two main Battle of Britain fighters, just over 14,500 Hurricanes and Sea Hurricanes had been built, compared with nearly 23,000 Spitfires and Seafires. Thus *Reach for the Sky* concentrated on Bader's Spitfire flying, rather than on his Hurricane days as, at that time, there was only one air-worthy representative left of the Hawker fighter; and when the *Battle of Britain* film was made sixteen years later, there was still only one flyable Hurricane available against twenty Supermarine fighters.

The Spitfire/Hurricane Controversy

Because of its slightly earlier appearance and the greater ease of its production, the Hurricane was available in greater numbers than the Spitfire at the beginning of the war and two recent investigations of squadron successes in the Battle of Britain by J. Alcorn in *The Aeroplane* reveal that 62% of the squadrons fully engaged in daylight combat were at that time equipped with Hurricanes, compared with 34% flying Spitfires. As the Hawker machine was responsible for a greater proportion of the enemy aircraft destroyed in the Battle, there is considerable justification for the view that the Spitfire legend has, unfairly, been at the expense of the Hurricane – despite the statistics given below which favour the Spitfire, it has to be recognised that the actions of Hurricanes and their brave pilots were essential to the defence of Britain at that critical moment in time.

Thus, in view of the present concern with Mitchell's fighter, Alcorn's statistics are worth reporting in some detail. Of the 1180 air-to-air victories credited, 521.5 were achieved by Spitfires when the percentage of squadrons equipped with Mitchell's design might, proportionally, have been expected to have only accounted for about 400 enemy aircraft. Also, this author derives an average of 27 victories per Spitfire squadron, compared with 22 for the Hurricane units – showing the Spitfire to be 1.25 times more effective in combat. It was also observed that six of the ten top scoring squadrons in the Battle, including the first three, were equipped with Spitfires. Whilst the squadron returns are unlikely ever to be precisely accurate and whilst the above statistics cannot take account of which type of aircraft was flown by the most gifted or aggressive pilots, which squadrons engaged battle on the most advantageous terms or what was the effect on the statistics of Spitfires being employed primarily to engage the enemy fighters and of Hurricanes tackling the bombers, a further statistic is worth considering: the victory-to-loss rate of all the fully engaged Spitfire squadrons during the Battle was 1.8 to 1 compared with 1.34 to 1 for all the Hurricane units (which were more often engaging the slower German bombers); and a snapshot view of a crucial engagement – 15 September – showed four percent Spitfire losses compared with six percent Hurricane losses.

The Hurricane has always been regarded as providing a particularly stable gun platform and, of course, it had a very long pedigree of Hawker fighters and fighter-bombers before it. But two distinguished Spitfire pilots, Alex Henshaw and Rod Smith, were unimpressed by the arguments in favour of the Hurricane and the latter made the following pithy comments: 'If you could hit with any fighter you could hit with a Spitfire. If any pilot in the three Spitfire squadrons I was in had said that he had got on the tail of an enemy aircraft but had failed to shoot it down because his aircraft was an imperfect gun platform, he would have been sent to tow drogues in Lysanders and laughed off the aerodrome'.

In one of the many accounts which exist in defence of the Hurricane and its fine manoeuvrability, test pilot Roland Beamont recalled how, in November 1940, a pilot of a new Spitfire II could not shake off his Hurricane I in a series of rolls, roll-reversals and 'tightest turning with wing tip vortices'. Nevertheless, this incident was at very low level where differences were, indeed, marginal. The early Spitfires also only had a slightly higher rate of climb to 15,000 feet but they became faster and more manoeuvrable up to their higher operational ceiling; and, although it has been claimed that Hurricane pilots felt more confident of the strength of their wings in high speed vertical dives, it was the case that the never-exceed diving speed of the Hawker machine was at least 70 mph below that of the Spitfire.

There also exist many pilots' subjective comparisons: William Dunn, mentioned earlier, wrote that 'it was the finest and, in its days of glory, provided the answer to the fighter pilot's dream – a perfect combination of all the good qualities required in a truly outstanding fighter aircraft. Once you've flown a Spitfire, it spoils you for all other fighters.' Another comparative view came from Pilt Off. H. G. Niven of Nos 601 and 602 Squadrons:

> Flying the Spitfire was like driving a sports car. It was faster than the old Hurricane, much more delicate. You couldn't roll it very fast, but you could make it go up and down much easier. A perfect lady. It wouldn't do anything wrong. The Hurricane would drop a wing if you stalled it coming in, but a Spitfire would come wafting down. You couldn't snap it into a spin. Beautiful to fly, although very stiff on the ailerons – you had to jam your elbow against the side to get the leverage to move them. And so fast!!! If you shut the throttle in a Hurricane you'd come to a grinding halt; in a Spitfire you just go whistling on. [*See* earlier Quill's findings in this respect after his first test flight of K5054 – p. 255.]

Another pilot with experience of both machines, Wing Commander Tom Neil, was quite blunt: on being posted to Burma in 1945, he wrote, 'after flying Hurricanes in Malta during the last months of 1941, I had earnestly prayed to the "person up there" that I would never see another Hurricane for the rest of my life'.

And there was the report by Al Deere after his actions covering the Dunkirk evacuation: 'the Hurricane, though vastly more manoeuvrable than either the Spitfire or the Me 109, was … badly lacking in speed and rate of climb … The Spitfire possessed these two attributes to such a degree that, coupled with a better turn than the Me 109, it had the overall edge in combat.' He later wrote that 'There can be no doubt that victory in the Battle of Britain was made possible by the Spitfire. Although there were more Hurricanes than Spitfires in the Battle, the Spitfire was the RAF's primary weapon because of its all-round capability. The Hurricane alone could not have won this great air battle, but the Spitfire could have done so'.

In the May of 1940, a captured Messerschmitt Bf 109 was tested at Farnborough against a Spitfire and a Hurricane where it was shown that the German fighter was superior to the Hurricane in all respects except in its turning circle and in its manœuvrability at low altitude, whereas Mitchell's fighter was at least a match for the 109 at the higher altitudes – where tactical advantages were the main deciding factors rather than relatively minor differences between the performances of the two aircraft. Rod Smith again: 'the essence of the achievement of Mitchell and his design team was coming up with a fighter which, though weighing only a few hundred pounds more than the Bf 109E, had 40 per cent more wing area and yet unbelievably had no more drag, if as much.' The Hurricane was thus best employed at altitudes around 18,000 feet and against bombers and their close escorts whereas the Spitfire was able to engage higher flying German fighters on equal terms and, in fact, also protect the Hurricanes in action below them.

Whatever the subjective views concerning the two machines, it is certainly undeniable that the Hawker machine was, in the final analysis, a Super Fury biplane with the top wing removed. The advantages of a traditionally manufactured aircraft were that only about 10,000 man-hours were required to produce a Hurricane compared with about 15,000 for the technologically more advanced Spitfire and it was more easily within the capabilities of the current workforce and easier to repair than the Spitfire. It is to the credit of its designer that the less complex Hurricanes were available in numbers during the Battle of Britain; however, it was also the case that future developments by Hawker involved producing the completely new Typhoon and Tempest designs. Rod Smith also had something to say in this respect: 'Anyone who thinks the Hurricane could have looked forward to any real future after the Battle of Britain should read Sholto Douglas's correspondence, beginning in January of 1941 [two months after he had taken over from Dowding] concerning the fighters to be chosen for that year. His remarks on the new and more powerful Mark II Hurricane were devastating: "... not good enough ... far below the 109 in speed ... inferior in climb".' Air Vice-Marshall Keith Park also wrote of the Mk.II: '... no amount of increase in engine power will have the effect of turning it into a high altitude fighter'.

The Continuation of the Spitfire Legend
When Mitchell first joined Supermarine and saw the Channel conversions that he had worked on going to Scandinavia, South America, Japan and New Zealand, even with the unbounded optimism of youth, he could not have imagined the future worldwide employment of his Spitfire, progressively, in so many theatres of war – in the Mediterranean and the Middle East, Italy, Yugoslavia, France, the Low Countries, Burma and the Pacific. Nor could he have envisaged his fighter in the service of so many countries, particularly Russia in the early years of the war when its own aircraft were outclassed by German machines: Mark Vs and, later, Mark IXs to a total of over 1,300 were sent by 1945. Another request came from Australia which received Mark Vs, which were faster than the Japanese Zero; unfortunately, they were not new aircraft and were unreliable. But in November, 1943, it was agreed to send 40 new Mark VIIIs followed by a regular supply of the newest tropicalised versions which could also out-climb and out-dive the Zero. By the end of the year, the Japanese had abandoned their raids on the Australian mainland and the Spitfires then went into action over Borneo and elsewhere in the South Pacific. Mark V Spitfires had also arrived in Burma, followed by Mark VIIIs and took part in the decisive battle of Kohima from April to June, 1944; by the end of the year the Japanese Army Air Force threat to India and Burma had largely been overcome.

Seafires, the naval version of the Spitfire, also covered the landings at Rangoon and Penang and attacked oil fields in Sumatra; they were active against kamikaze aircraft and in their last action of the Second World War, 15 August 1945, destroyed eight Zero fighters without loss. Canada flew Seafire XVIIs until April 1954 and India retained a fleet of VIIIs, PRXIs and F.19 Seafires until 1958. America also had three separate 'Eagle' squadrons (Nos 71, 121 and 133) formed from the large number of American volunteers; and these were eventually subsumed into the USAF.

And whilst the distinctive shape of the Spitfire/Seafire moved ever eastwards, Marks IX, VIII, and XII continued the Spitfire presence nearer to home and contributed significantly to allied air supremacy. In 1944, massed ranks of Mark V to Mark XIVs supported the Normandy landings on D-day and then continued operations in Europe until the end of the war, including ground attack duties alongside the Hawker Tempest against V2 rocket installations in Holland.

Despite the advent of the jet fighter at the very end of the war, the Spitfire continued post-war operational service in Malta, Cyprus, the Suez Canal Zone and the Persian Gulf. It was also deployed against communist activities in Malaya and Korea and PR19s, from Hong Kong, maintained discreet reconnaissance flights along the Chinese coast and even over mainland China. Sweden bought 50 of this Mark to watch Soviet activities in the Baltic region whilst the Soviet Union Spitfires reappeared in the People's Republic of China. As jet aircraft began to equip air forces of richer or more imperialist countries, other nations were keen to buy Spitfires as stopgaps: Belgium, Holland, Norway, Denmark, Yugoslavia, Greece, Italy, the Philippines, Thailand, Burma, Israel, Egypt, Syria,

Turkey, Portugal, India, Czechoslovakia, South Africa, Southern Rhodesia and Ireland.

A Spitfire F.22 awaiting delivery to the Royal Egyptian Air Force.

Thus the Spitfire saw in the era of the all metal, propeller driven fighter and saw it out but, as these aircraft around the world were gradually replaced, the Spitfire icon still continued to have a physical presence, thanks to various preservation enterprises, led by the Battle of Britain Memorial Flight organisation. When the Spitfire was phased out of RAF service in 1957, three machines were being flown by the Temperature and Humidity Monitoring Flight at RAF Wood-vale. These were PR19s and they were transferred to Biggin Hill on 11 July, 1957, to form the Historic Aircraft Flight. Two were made serviceable and, in April 1964, returned to what had now officially been renamed the Battle of Britain Flight. In September 1965, another Spitfire joined the ranks when Vickers-Armstrong parted with a Mark Vb and, after the filming for the 1969 *Battle of Britain*, the Flight was presented with a Spitfire IIa. This was the world's oldest airworthy example of its type and a genuine combat veteran of the Battle of Britain. In November 1997, there was yet another addition, a newly restored Spitfire LF IXe – not only was the Flight getting a variant it had not previously operated but this machine had flown during the D-day period in June 1944, supporting the Allied invasion of occupied France.

Mitchell's fighter appears therefore destined to remain in the public eye for many years to come, not just because of the Flight's activities but because of the industry that has grown up around the restoration of the type. The June 2010 edition of *Flypast* listed the five Memorial Flight Spitfires and also another nineteen airworthy machines at various bases in the United Kingdom. Additionally, twenty other Spitfires and three Seafires were currently listed as under or awaiting restoration and one Seafire and fifteen Spitfires were on static display.

Whilst this preservation and restoration scene is constantly changing, it does not seem to be diminishing and, equally, the number of flights at air shows and special events is not lessening – as evidenced by the Battle of Britain Memorial Flight statistics: after some years of relatively low-key operation, typically with 50–60 appearances per season, the 1992 participation had risen to 150 appearances, growing to 200 in 1995 and exceeding 500 in 1996. Since 2003 the Flight has put on over 700 individual aircraft appearances during each year's display season and by 2008 the BBMF were tasked with 529 separate events. That year, the Flight's aircraft appeared live in front of an estimated total audience of 7 million people, not to mention those who recognised the sound of a Merlin or Griffon engine overhead whilst a Spitfire was in transit. The Spitfire is also seen on occasions of national importance in Australia, New Zealand, India and Israel.

The Continued Development of the Spitfire
In the preceding account of the widespread activities of the Supermarine fighter, various Marks have been mentioned but, whilst Joe Smith's elaboration of the type is a tribute to Mitchell's basic design, the detail of this development is perhaps more appropriate to the present book by being placed in an Appendix. It might, nevertheless, be noted that the continuous and relatively economical modifications of the Spitfire further contributed to its reputation as a fighting machine as an impressive number of progressive changes enabled the aircraft to be kept well abreast of the developing needs of aerial warfare, competitive with improved enemy aircraft and abreast of changing technologies – in all probability better than was likely with an expensive, entirely new design being developed at one particular point in time.

Some mention ought also to be made of other factors which contributed to the longevity of the design after Mitchell's death. Whilst the Air Ministry was looking forward to the next generation of fighter, particularly in respect of the Hawker Tornado/Typhoon development, their hopes were continually frustrated by the shortcomings of the proposed Sabre engine. The aircraft was planned to appear in service early in 1941 but it did not become operational until late in 1942, and then only in small numbers. It was clearly a fast, sturdy aircraft, which proved especially suited to being armed with rockets, and it turned out to be most successful in low-level attacks. But its delayed appearance and difficulties with it afterwards meant that a gap in the development of a new fighter had to be filled – mostly by further developments of the Spitfire.

Thus when deliveries of the formidable Tempest began in October 1943, Smith's Mark VC had been adapted to take four cannons like the two Hawker successors to the Hurricane and top speed of the Tempest was not better than that of the 447 mph Spitfire Mark XIV – thanks to the newly employed Griffon engine. And, of course, there were economies of type development – a new mark of an existing aircraft, however radical, required much less time and effort to design than a brand-new type. Woolston quotes Vickers figures which revealed that no later mark required more man-hours spent on design than on the Spitfire Mark I. The highest number was 165,000 man-hours devoted to the Spitfire 21, compared with 330,000 man-hours on the Mark I – with the average time per mark being 41,000.

Naturally, the advantages of the development of Mitchell's machine had to be matched by advances which successfully countered improving enemy aircraft. Details of these improvements are to be found in Appendix One; meanwhile, the extent of these developments are summarised below:

	Spitfire Mark I	*Seafire 47*
Maximum speed	362 mph	452 mph
Engine power	1,050 hp	2,350 hp
Fuel capacity	85 gal.	287 gal. (inc. 90 gal. drop tank)
Normal loaded weight	5,820 lb	10,300 lb
Wing loading	24 lb/sq ft	42.2 lb/sq ft
Service ceiling	31,500 ft	43,100 ft
Maximum range	575 miles	1,475 miles
Climb to 20,000 ft	9.4 min.	4.8 min.
Rate of roll at 400 mph	14 degrees/sec.	68 degrees/sec.
Maximum diving speed	450 mph	500 mph

In the course of this development, the weight of ammunition carried had almost doubled and that of the protective armour now amounted to more than the weight of an average pilot. Indeed, Jeffrey Quill has pointed out that, at its maximum gross take-off weight, the Seafire 47 was equivalent to a Mark I Spitfire with the additional load of 32 'airline standard' passengers each with 40 lb of baggage.

The first production Mark I, Eastleigh, May 1938.

A well laden Seafire 47, circa 1949.

Whilst the extraordinary development of Mitchell's 'basic' fighter is well illustrated above, it would not be realistic to conclude with too triumphal an account of the Spitfire's upgrading. Many modifications were rather desperate attempts, in wartime conditions, to avoid the production of a completely new type. Jeffrey Quill, who had by far the most extensive experience of testing the various Spitfire types, has made the following comment:

> Almost every design change introduced in the course of the extremely rapid development of the Spitfire was in some way basically detrimental to the flight handling, usually in terms of longitudinal or directional stability. My main preoccupation as chief test pilot was thus to ensure that the flight handling characteristics of the Spitfire remained within manageable limits, and it wasn't really possible in the prevailing circumstances to do more than this.
>
> To keep the flight handling situation under control, many expedients were forced upon us by the immense pressures of wartime production effort. The 'elegant' solutions were usually not available to us, simply because of the pressures of time; solutions had to be found at once and therefore we improvised – generally with good success. But our expedients were not always entirely successful and the flight handling of the Spitfire sometimes left a certain amount to be desired.

Some of the factors which most affected controllability were the progressive increase in speed and weight, in propeller blade area ('propeller solidity'), and in moments of inertia due to the redistribution of the increasing weight. Quill again:

... the Spitfire was almost continually at the margins of longitudinal stability. The problem was to find ways of providing for greater ranges of centre-of-gravity movement. An aeroplane that gets on the negative side of longitudinal stability can sometimes be fun to fly but can become very, very dangerous at the high diving speeds. As time went on I became very conscious of the need to provide positive stability margins but everything militated against this. I knew, from George Pickering's accident, that Spitfires could suffer catastrophic structural failures. Unfortunately, his was not an isolated case; there were in fact something in the order of 25 catastrophic total structural failures with Spitfires in the air. There is no better way of risking overstressing an aeroplane than having negative stability margins longitudinally and then diving it up to somewhere near or beyond its maximum permissible diving speed. If you lose control of it then, that is when it breaks. Now I was only too acutely conscious of this and the problems involved in preserving positive margins as the aeroplane developed. It would not be any good ringing up and saying 'we want another 25% on the area of the tailplane as from next week please'. This would have been impossible. We had to make do with a whole series of expedients; mostly getting some returns on the stick-free stability by modifications of the elevator in order to keep this situation under control. Once or twice we had to go into all sorts of terrible things like putting inertia weights in the elevator circuit and so on, which pilots hated, but it had to be done.

There was also the major problem of converting the Spitfire into the Seafire for aircraft carrier operation – as Quill whimsically put it:

One very important aspect of the operation of the Spitfire was when it became a naval aircraft and this presented a large number of problems which completely took the Spitfire by surprise I may say. It was never intended for this sort of thing at all and anybody who has ever watched any naval operation will know that the Fleet Air Arm do the most dreadful things to aeroplanes when they bring them back onto the decks of ships.

An airman who flew Seafires throughout the war, Capt. George C. Baldwin, gave the following perspective of problems associated with aircraft-carrier landings with the early conversions (*see* also pp. 315–316):

The Spitfire was designed to have excellent air to air combat qualities, it was a very light fragile aircraft and was really very unsuitable for the rugged flying ... when not too experienced pilots were to throw it onto a flight deck and it was to be arrested in full flight by a hydraulic arrester gear ...
We had a curious little expression about the Spitfire which summed up our problems on the deck and we used to say it suffered from 'pecking, pintling and puckering'. Actually that was no joke, because 'pecking' was a phenomenon caused by the tail being thrown up as the aircraft caught the arrester-wire and the propeller touching the flight deck and, if it was a wooden propeller, pieces flew off in every direction. Believe it or not, that was cured by just taking a sharp knife and cutting three inches off the end of each blade with no noticeable loss of performance whatever. 'Pintling' was a phenomenon caused by the rather weak undercarriage. You could do a fairly reasonable, but slightly rough landing and thereafter you could either not get the undercarriage up when you took off or once you had retracted it you could not lower it again. It was because the pintle in the undercarriage had become misplaced. 'Puckering' was if you had made a successful landing, but had a little bump which might have caused pecking – you would also get puckering which meant that the tail would drop hard onto the flight deck and the fuselage would bend just in front of the empennage. Well, these were all difficult problems and they reduced serviceability somewhat considerably.

One particular photographic reconnaissance conversion also had its problems: the PRD/IV version was found to be incapable of flying straight and level for the first half hour after take-off because of the disposition of its very heavy fuel load. But, to end on a more positive note, most photoreconnaissance versions of the Spitfire were not problematical and many versions of the fighter were regarded as exceptional aircraft: in particular, perhaps the best fighter, from a handling point of view, was the Mark VIII, the most successful high altitude fighter, the Mark XIV, the most outstanding PR variant, the PR19, and the most complete naval version, the Seafire 47.

A detailed account of the main versions which went into quantity production is provided in Appendix One and reveals just how far Mitchell's basic design was capable of development. Meanwhile, drawings of their basic modifications, will give an indication of these efforts:

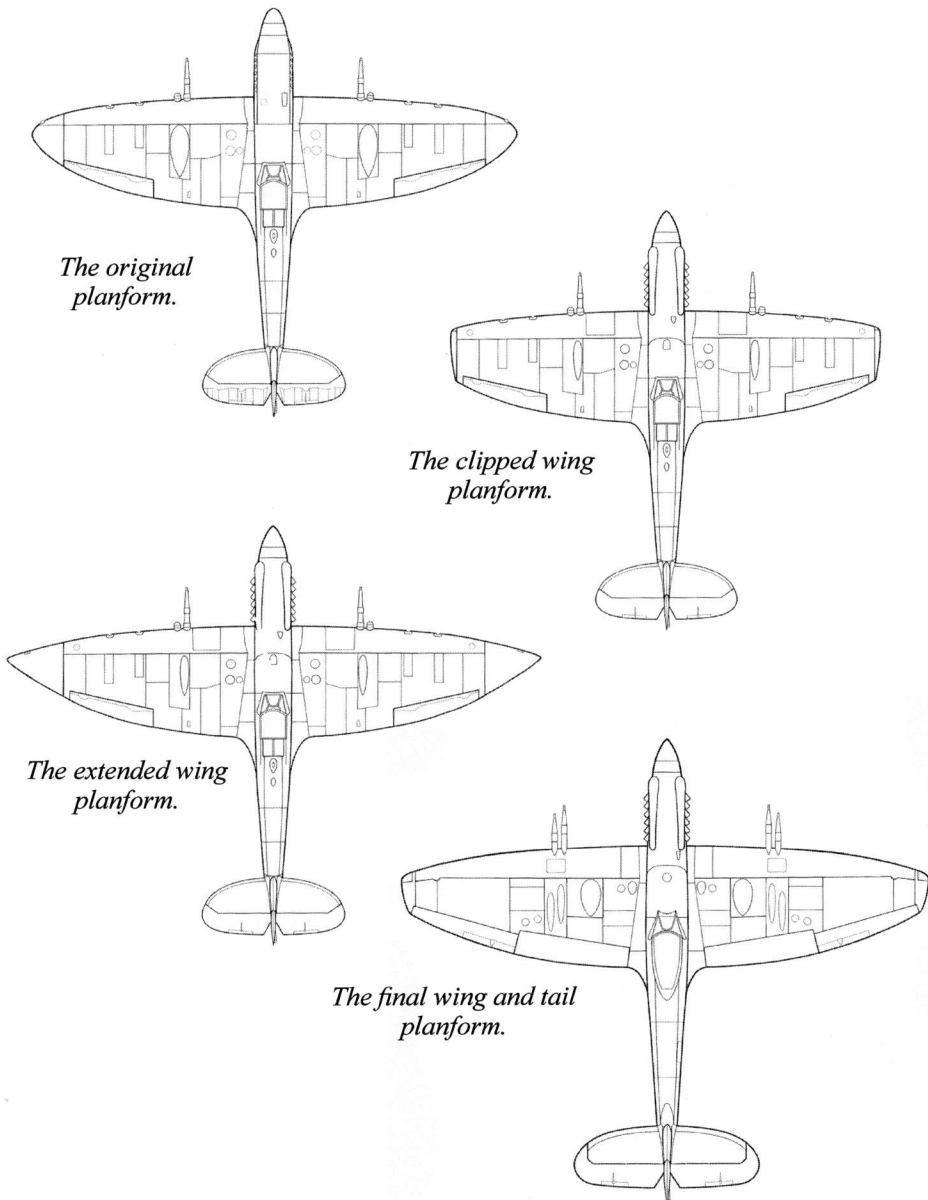

The original planform.

The clipped wing planform.

The extended wing planform.

The final wing and tail planform.

The drawings show, first, Mitchell's original semi-elliptical planform, followed by the clipped wing version for higher manoeuvrability at low level. The third

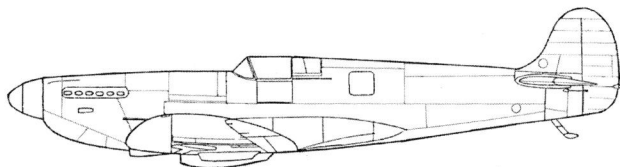

The prototype and early Mk.I with flat cockpit cover.

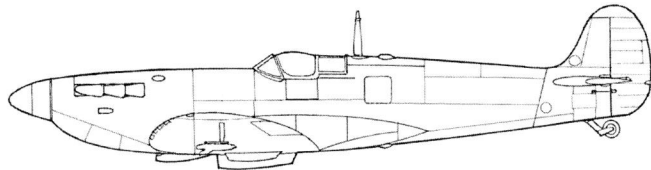

An early Mark with the more familiar domed cockpit cover.

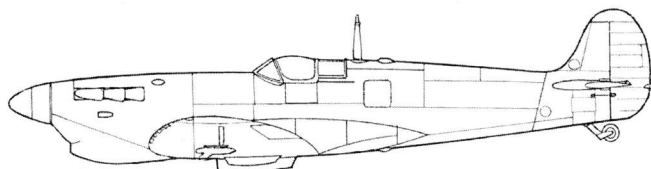

*An early Mark with the 'beard' lower engine cowling for the tropical
air filter.*

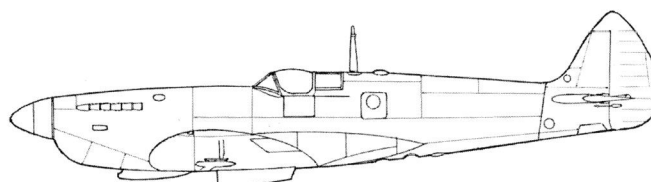

*A PR version with deeper lower engine cowling, symmetrical
radiators, retractable tailwheel and second version fin and rudder.*

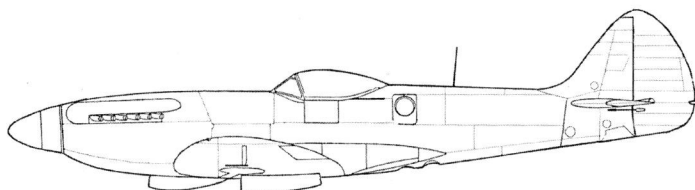

*The early Griffon-engined Mark, with balloon cockpit hood, lower
rear fuselage decking, modified engine cowling, repositioned air
intake and third version fin and rudder.*

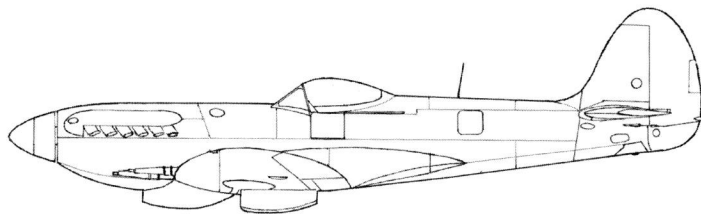

*The final fighter version with deeper radiators and fourth version fin
and rudder.*

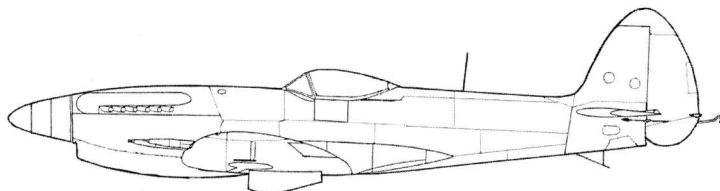

*The final Seafire version with contra-rotating propellers, further
revised air intake position and modified fourth rudder.*

illustration gives Smith's most obvious departure from Mitchell's original concept, the pointed wing version of the high altitude Marks VI, VII, VIII, and 21; this was a particularly stopgap expedient and certainly lacked a great deal in elegance. Finally, the fourth drawing shows the revised wing and tail surfaces, as seen in Marks 22 and 24 – whether they would have had a different outline if they had come from Mitchell's own drawing board can only be guessed at.

Perhaps a more graphic illustration of the continuous development of the wartime Spitfire can be seen in its changing side-views (*see* opposite). Mitchell's original design concept remained most clearly evident in the Merlin-engined fighter Mark I–II–V–IX line of development, in the photoreconnaissance Mark I–VII sequence and in the Seafire Mark I–II–III–XV series. The introduction of improved equipment brought some obvious changes: the Griffon engine and the extra oil requirement in the longer range PR variants produced somewhat changed nose shapes and increases in power also necessitated redesigns of the tail surfaces. The more powerful later Merlins had given rise to the simple but makeshift change to a pointed and broader chord rudder, whilst the even more powerful Griffons necessitated a complete redesign of the unit (for example, in the Mark 18 and PR 19).

Another factor affecting a more fundamental redesign by other hands was the later availability of the teardrop canopy which resulted in a cut-down rear fuselage decking and a great improvement in rear view.

Retractable tailwheels, deeper engine cowlings or air filter coverings, and contra-rotating propellers also contributed to the gradual alteration of the Spitfire's appearance; however the most significant change was the completely redesigned wing, as fitted to the Mark 21, 22 and 24 fighter and the Mark 45, 46, and 47 Seafire. Whilst many of the changing features were variously embodied during later stages of the production runs of other Marks, it is with these last six types that the necessary movement away from Mitchell's original concept is at its most marked – the Seafire 47, the last version of the Spitfire, combined revised tail surfaces, rear-view canopy with cut-down rear top decking, and contra-rotating propellers and the air intake filter was now more comprehensively faired into the lower engine cowling, with the duct opening positioned just behind the propellers.

The p. 284 comparison between the Mark I and Seafire 47 gave further proof of how far Mitchell's original concept had been capable of development but, setting aside the huge difference in load carrying between the two aircraft, it can be calculated that the latter's 124% increase in engine power had only produced a 25% increase in speed. Thus Mitchell's propeller-powered design could be seen to be reaching the limits of its potential; nevertheless, one can surely agree with Rendall's assessment of the Spitfire that

> few aircraft straddled the transition between the two ages of aviation so comprehensively, let alone so elegantly: the biplane and the monoplane ages, the piston age and the jet age, the subsonic and the supersonic.

* * * * *

A Selection of Photographs Showing Spitfire Development

One of the first Mark IIAs from Castle Bromwich.

The Mark V, two cannon version, with clipped wings.

An HF VII with extended wing, retractable tailwheel, and second version fin and rudder.

Mark XIIs of 41 Squadron. Redesigned cowling blisters for the Griffon engine clearly evident on EB-B.

Mark XIV, showing the five-blade r.h. propeller and slightly deeper radiators.

Mark 18, showing camera aperture, the re-designed fin and rudder, lower fuselage decking and 'tear-drop' cockpit canopy.

Mk.F.22, showing the final revisions to flying surfaces and the fourth type fin.

Seafire F.46 With contra-rotating propellers and modified fin.

* * * * *

Schneider Trophy Full Circle

Amongst the very numerous Spitfire developments outlined in Appendix One, there was a Spitfire floatplane variant which, although it owed very little directly to its Schneider Trophy predecessors, would seem to bring this particular Mitchell narrative to an appropriate conclusion – Jeffrey Quill:

> Yes, I flew the Spitfire floatplane. We developed this to meet a requirement for the Mediterranean operations … It was a very successful

experiment altogether – the floats were very good; in fact, it would have been rather surprising if they were not; they were designed by Supermarine who did understand about making floats! It handled beautifully in water. The drag of the floats was surprisingly low; the Mark IX with floats was still faster than the standard Mark V without floats. Now [the Mark V floatplanes] did go out to the Mediterranean. There was a good deal of traffic across the Mediterranean with JU 52s and such like on supply lines to North Africa and the idea of these float planes was to lurk around some of the smaller islands and then suddenly pop up and surprise these chaps and shoot them down. However the outcome of events at the end of the Western Desert Campaign created a situation where the requirement was no longer there. But is was a very successful development, I think it is fair to say … they were great fun to fly.

The first of the three Mark V float conversions with early type tropical air filter.

Appendices

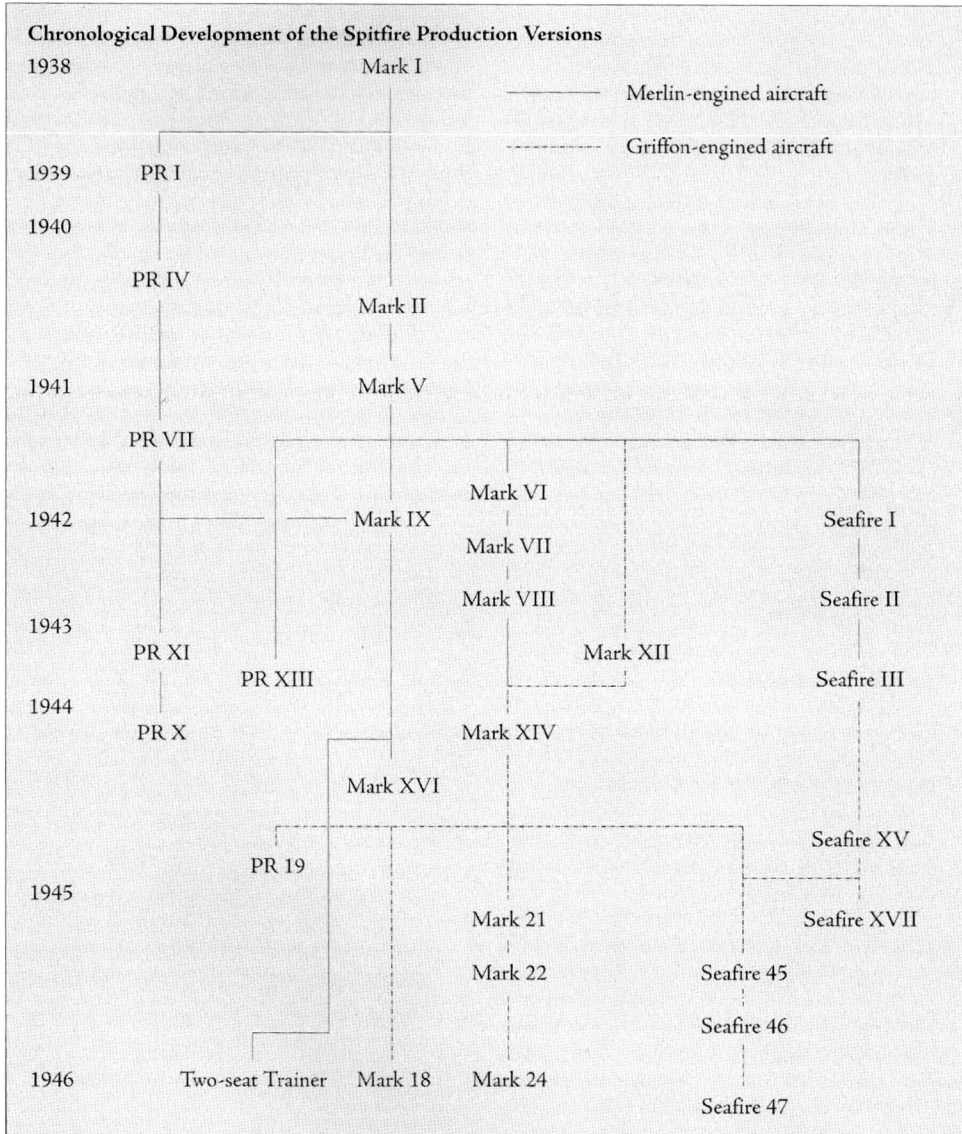

Chronological Development of the Spitfire Production Versions

1938   Mark I

————— Merlin-engined aircraft

-------------- Griffon-engined aircraft

1939   PR I

1940

     PR IV

                    Mark II

1941   Mark V

   PR VII

                         Mark VI

1942   Mark IX   Seafire I

                         Mark VII

                      Mark VIII   Seafire II

1943   Mark XII

   PR XI   PR XIII   Seafire III

1944   PR X   Mark XIV

         Mark XVI

                                         Seafire XV

         PR 19

1945   Mark 21   Seafire XVII

                     Mark 22   Seafire 45

                                    Seafire 46

1946   Two-seat Trainer   Mark 18   Mark 24

                                         Seafire 47

— *Appendix One* —
Production Versions of the Spitfire

(i) General

Full accounts of the development of the Spitfire after Mitchell's death are contained in Robertson, Morgan and Shacklady, and Price. It was felt, nevertheless, that a brief description ought to be included here for completeness sake and as, without this remarkable development, the need for a book on the career on R.J. would be far less obvious.

Also, this present narration of the various marks of Spitfire has been included in order that any existing version or any photograph of a Spitfire can be conveniently related to Mitchell's original design, which was, of course, the culmination of this book. The brief descriptions of the modifications that produced each subsequent version will enable the reader to judge how close (or otherwise) any particular aircraft is to Mitchell's original concept and will also summarise the variety of modifications and developments that were able to be made to his basic design.

Mitchell's basic fighter was eventually developed, separately, for high altitude (**HF**) and low altitude (**LF**) operation, photoreconnaissance (**PR**) work, and ship- and carrier-based duties (the **Seafires**); there were also floatplanes and a two-seat trainer. The somewhat bewildering nomenclature of these different Marks of Spitfire may also be conveniently clarified in this appendix, as well as the non-consecutive dates of their appearance and the apparent inconsistency of the mark enumerations – for example, the Seafire 45 appeared before the Seafire XVII. In the course of the account, the exact reason for some of these apparently illogical mark numbers will become clear and such detailed nomenclature as 'ASR II' or 'LF XVIE' will also be explained, as well as the modifications of these sorts of letterings as Mitchell's aircraft underwent its progressive development.

The survey also reveals which were the most numerous Marks (V IX VIII and XVI) and indicates what might perhaps be regarded as the best fighter, from a handling point of view (the Mark VIII), the most successful high altitude fighter (the Mark XIV), the most outstanding PR variant (the PR19), and the most complete naval version (the Seafire 47).

But, before describing the numerous Marks of Spitfire in some detail, it was thought useful to show a 'family tree' of the production versions of the Spitfire, particularly with reference to the Griffon-engined variants (see facing page).

(ii) The Spitfire Marks in Detail

Merlin Engined Spitfires

The Mark I differed in outline very little from the prototype; there was, however, an increase in flap angle and the substitution of a tailwheel for a skid. Mitchell had tried to resist this last modification because of the increased drag that would be created but the Air Ministry, as the possibility of war increased, was not disposed to share with him the information that all front-line airfields were to be given all-weather surfaces – for which a skid would not be suitable. And when the production model appeared, the wing plating was different and the wing leading-edge torsion box had been strengthened.

The prototype's Merlin E was replaced by the Merlin II, with ejector exhausts, but the fixed pitch, two-bladed propeller was retained. On the 75th aircraft and subsequently, a three-bladed, two-position type was fitted, powered by a Merlin III of similar power but with a standardised shaft for three-blade airscrews; and, by the time of the Battle of Britain, a three-bladed, constant speed propeller was fitted as standard and all previous aircraft retrofitted accordingly. The speed of the Spitfire had now increased by 5 mph to 367 mph and the time taken to reach 20,000 ft reduced by 3½ minutes. A slab of bulletproof glass had also been attached to the outside of the front windscreen, 73 lbs of armourplating fitted behind the pilot's seat and the straight-topped cockpit canopy given a domed cover to the relief of taller pilots. Another irritant was the damage to pilots' knuckles when operating the pump to raise the undercarriage and their often undulating flight whilst doing so; later Mark Is had an engine driven hydraulic u/c system.

By now the Spitfire in dives was exceeding the speed of the specialist Schneider Trophy racers and it had been found that the Chief Designer's ailerons virtually locked solid in high speed descents. Also, the rate of roll was affected by the ailerons becoming heavy at speed. The problem was not, in fact, due to the proportions that had been decided on, but rather to the conventional practice of keeping them light by using fabric covering. It was now discovered that this material tended to balloon out at speed and so the ailerons were returned to efficiency by being given a metal skinning. This cure involved a crash programme to modify all the Mark I aircraft that had been supplied to front-line units. This modification was undertaken by Air Service Training at Hamble and Quill wryly noted that 'the word swept round Fighter Command like wildfire and in no time the air around Hamble was thick with Spitfires of Wing Leaders and Squadron Commanders all trying to jump the queue to get their aircraft fitted with the new metal ailerons'. Quill also discovered shortly afterwards in a captured Messerschmitt Bf 109E that the problem had been just as bad on the other side.

One other concern with the skinning of the aircraft led to an investigation as to whether the extra complexity of flush riveting all the covering panels was justifed. No doubt Mitchell would have approved of the simple device of gluing split-peas over the rivet heads to simulate the standard round-headed ones and discovering the effects of progressively removing them from different parts of the skinning. As a result, production Spitfires were constructed with dome headed rivets to the fuselage while the wing surfaces continued to be flush riveted.

During the Battle of Britain it was found that German bombers could often absorb hits from machine gun fire by virtue of their armour plating and so the attraction of the more penetrative but slower firing cannon led to the development of an alternative armament of two cannons and four machine guns. The Spitfires with the original eight-gun arrangement were therefore designated **IA** and the ones with the cannons, **IB**. Joe Smith had to have the cannons mounted on their sides in the Spitfire's thin wing and this led to empty cartridge cases jamming in the breeches so frequently that the cannon version was withdrawn from front-line duties for reversion to the eight machine gun version until the problem of stoppages could be solved.

(For photoreconnaissance versions of the Mark I, see the PR section which follows. It might also be mentioned here that one Spitfire, the 48th in the Eastleigh production line, had been specially modified for an attempt on the World Landplane Speed Record. This so-called Speed Spitfire is also described in the photoreconnaissance section.)

1550 machines produced.

The **Mark II** represents the first of Joe Smith's variants and it had *largely replaced the previous version in front-line service by the end of 1940*, some in time to take part in the later stages of the Battle of Britain. It was, basically, similar to the later production models of the Mark I, with its armour plating and bullet-proof front windscreen; the **IIA** still had the eight machine guns and the **IIB** had the two cannons and four machine guns. The essential change from the Mark I was the more powerful Merlin XII with a Rotol constant speed propeller, giving an increase in speed of 15 mph and a climb of 473 more feet per minute.

When the Mark II was superseded in the fighter squadrons, some were modified to operate in an **air-sea rescue** role by dropping dinghy and survival canisters to assist airmen until they were picked up by boat or seaplane; a small rack for smoke-marking bombs was fitted under the port wing, inboard of the oil cooler, and two flare chutes in the fuselage, just aft of the cockpit, housed a small dinghy and a metal food container. This version was originally identified as Spitfire **IIC** but, with the advent of the 'C' wing (see below), was redesignated **ASR II**.

921 built

The **Mark III** first flew in March, 1940, and represented the first attempt to improve significantly on the Spitfire's performance, particularly by making retractable the tailwheel that Mitchell had objected to. Also, the wings were clipped – a modification which had been first tried out with the Speed Spitfire of 1939. In the present case, the concern was first and foremost with increasing the rate of roll, rather than speed, and looked forward to other later clipped wing versions – in particular, the low-level Mark VB which was to appear a year later. The mainwheels were raked forward by two inches to improve ground handling and were now fully covered when retracted.

The Mark III also saw the introduction of the C 'universal' wing which allowed the fitting of either the A or B gun combinations or a four cannon arrangement. It was also proposed to reduce drag further by making a neater fitting of the bulletproof glass inside the windscreen and to improve performance at altitude by installing the Merlin XX with its two stage supercharger.

The more powerful engine also required a deeper oil cooler intake, resulting in a completely circular scoop. As Rolls-Royce could initially only produce a limited number of these engines, it was decided to give priority to re-engining the Hurricane and so the Mark III Spitfire never went beyond the prototype stage. However, as we shall see, a number of the experimental features of this machine were successfully adopted in other versions of the Spitfire; in particular, the test fittings of the new Merlin 60 series also made the Mark III virtually the prototype for the Mark IX.

2 built.

(For the **Mark IV**, *see* the introduction to the Griffon-engined Spitfires section and the photoreconnaissance section, below.)

The **Mark V** was intended as a stopgap type until a more comprehensively redesigned aircraft could be produced but it ended up being the most numerous of all marks. The immediate concern was to be ready by 1941 for an anticipated second Battle of Britain, expected to be fought at a higher altitude and to counter the Messerschmitt Bf 109F (when it appeared in early 1941) which had significantly greater speed and rate of climb than the Spitfire Mark II and could even out-turn it.

Rolls-Royce had been able to move from a single stage, two-speed speed engine to a two-stage, two-speed supercharged engine without detriment to their current production line and this engine, the Merlin 45, was to be fitted to the basic Mark II airframe, with the deepened oil cooler intake introduced with the Mark III, until the specially designed, pressurised, Mark VI could be delivered. In the event, this proposed interim version was able to operate at the higher altitudes, despite the additional weight of extra armour plating around the ammunition boxes, under the pilot's seat, and in front of the coolant header tank; it proved to be largely superior to the new Messerschmitt, with a top speed of 376 mph, an operating ceiling of 36,000 ft and ability to climb to 20,000 ft in six minutes. (Later German comparison tests between a captured Mark V, fitted with a Daimler engine, and a Messerschmitt Bf 109G revealed that the Spitfire was slower at sea level but was superior in climb and altitude performance.)

Fitted with external and jettisonable fuel tanks and progressively more powerful engines, this Mark came out in three versions, according to the type of armament provided and in response to its changing role: the **VA** retained the original eight machine-gun layout; the **VB** had two 20 mm. cannon and four machine-guns (now that the jamming problems had been largely overcome) making it more powerfully armed than the Me 109E; the **VC**, which was largely intended for overseas duties, had the stronger wing of the Mark III; this allowed four cannons to be installed, a repositioned undercarriage, and provision for carrying a 250 or 500 lb bomb under the fuselage (after the Battle of Malta, Spitfires had been fitted with improvised bomb racks in order to go on the offensive.)

The two-cannon arrangement, which was the most effective and favoured by pilots, usually had the cannons in the inner positions and each outer barrel position filled with a rubber plug. This arrangement had the advantage of doubling the number of rounds carried for each cannon. As the strengthened 'C' wing also allowed for the 'A' and 'B' gun arrangement, it was known as the 'universal' wing and represented an important development, paving the way for the introduction of the ubiquitous Mark IX. The aircraft also had heating for the guns, which were now being subjected to more prolonged periods at higher altitudes, and provision for the carrying of a long-range, external, belly fuel tank.

Once the expected second Battle of Britain did not materialise, the Spitfire had been able to extend Fighter Command's largely defensive activities and take the fight to the enemy with sweeps over Northern France, particularly after the higher-flying Mark IX appeared. For this new role as a low altitude fighter and fighter bomber, the Mark V had clipped wings and the lower-altitude-rated Merlin 46, 50A and 56. Unfortunately in the first half of 1942, partly to draw some of the Luftwaffe's fighters away from the Russian front, Fighter Command lost over 300 Mark Vs, largely as a result of operating at less advantageous altitudes, being more exposing to ground fire and meeting the new Focke-Wulf 190.

The Mark V was also equipped with a bulky Vokes 'beard' air filter for tropical service and thus, with the various engines, the Mark V version was active in both temperate and tropical zones. In the latter regions, however, the filter cut down boost pressure at high levels and produced poor performance at low levels.

Earlier, the German invasion of Norway in 1940 had initiated an experiment to equip Spitfires with floats originally designed for a Blackburn Roc conversion – for use from sheltered waters, such as Norwegian fjords – but subsequent events led to an abandonment of this idea; however, the Japanese entry into the war gave rise to a later consideration of the floatplane fighter, to be based where the terrain was unsuitable for the construction of airfields or where there might be an important tactical advantage in the presence of fighters not being expected. This time, three Mark VB machines were specially equipped with floats (fittingly designed by Arthur Shirvall who was responsible for those of the Supermarine Schneider Trophy winners). In view of its Schneider Trophy predecessors, it is not surprising that the test pilot found that 'it was a most beautiful floatplane and all we had to do, predictably, was to increase the fin area to compensate for the float area ahead of the centre of gravity'. An underfin, unique to the floatplane Spitfire, was fitted, followed by the addition of a fillet added to the leading edge of the dorsal fin to achieve the extra area required and, with the floats and their well-faired vertical mountings, only reduced the speed by 30 mph. These aircraft, were sent, in the end, not to the Far East but to Alexandria. They never saw operational service, however, and no production series was ever initiated. (*See* photo p. 293.)

The Mark V was produced before America had joined the war and when Britain was heavily engaged in the Mediterranean and North African theatres of war. It was therefore *produced in larger numbers than any other Spitfire mark.; it remained the main RAF fighter until the summer of 1942 and the low level LFV remained in use into 1946.*

6,476 built.

The **Mark VI** was the first serious attempt to adapt the current Spitfire V, with a ceiling of about 30,000 ft, for high altitudes. Desperate modifications to reduce weight had been made to some Mark VCs based at Aboukir to combat high-flying Junkers Ju 86P photoreconnaissance aircraft over Egypt, which resulted in great feats of endurance and determination on the part of the pilots at heights above 40,000 ft. As the lightening modifications involved stripping out the aircraft's radio, a second Spitfire was required, in contact with ground control. Two Ju 86s were shot down and the patrols ceased. However, any thought of standard operation at such heights or for any length of time was not possible owing to the strain on pilots and had to await the successful development of a pressurised cockpit and an increased wingspan.

This increase, without all the additional production delays associated with a major redesign of the wing, was produced by the expedient of adding pointed tips. This achieved a span of 40 ft 2 in. and a wing area increase of 6.5 sq ft, although completely destroying the appearance of Mitchell's original conception.

The pressurised cockpit, also, left much to be desired as the hood had to be locked on before take-off and could not be opened in flight; whilst it could be jettisoned in an emergency, the hood had to be completely removed for normal entrance and exit. This stopgap variant, fitted with a Merlin 47 engine and a four-blade propeller, which came into production at the end of 1941, was again necessitated by the fear of German large scale production of high flying bombers, with a ceiling approaching 45,000 ft. However, the German invasion of Russia in June, 1941, caused fears of high altitude bombing to recede and so this Mark did not go into large scale production.

100 built.

The **Mark VII** represented, along with the Mark VIII, a breakthrough in Spitfire development with the fitting of the Merlin 61 with its two stage, 2 speed supercharger. It did not appear straight away after the Mark V and VI as the substantial design changes decided on took longer to incorporate – these included wings redesigned to take internal fuel tanks, 11 gal. more tankage in the fuselage, and a retractable tailwheel. The airframe was now fully stressed to take the new Merlin 61 and the pressurised cockpit (now with a sliding cabin hood but no hinged door in the side of the fuselage) was more satisfactorily sealed than that of the Mark VI.

It began appearing, rather slowly, in August, 1942, and later aircraft were fitted with an extended fin and broader chord rudder; the **HF** version, with the Merlin 71 had the expanded span wing tips. Top speed had now risen to 416 mph but, as its performance proved to be no better than that of the lightened Mark IX which was already being produced as a stopgap, no large production orders were placed for the Mark VII. *Many of these variants found a role in meteorological work.* (*See* photo p. 290.)

140 built.

The **Mark VIII,** on the other hand, which appeared a year later, was *one of the more numerous variants*, employing clipped, standard or extended wing planforms for its many roles, and eventually replacing the equally numerous Mark IX. The standard fighter version was powered by the Merlin 63, the **LF** version was powered by the Merlin 66, and the **HF** machine had the Merlin 70. This last was a non-pressurised version of the previous Mark, with the revised fin and rudder but usually without the extended span wings, as action had now tended to take place below 25,000 feet with the ending of high altitude attacks. Fuel capacity was now increased to 124 gallons thanks to extra tanks in the wing roots, the wing could take either two or four cannons, and the tailwheel was again retractable. Provision was now made for a 250 lb bomb to be carried under each wing, thus enabling the fitting of an additional fuel tank under the fuselage. The Mark VIII was also fully tropicalised without the bulky 'beard'-type carburetter air filter housing that had been fitted to the Mark V and so *it was used almost exclusively in the Middle East, India, Burma and Australia.*

As with the Mark VII, the aileron structure was stiffened by a reduction in the overhang outboard of the outer hinge and this modification, together with the Merlin 66 engine, produced an aircraft which was assessed by Jeffrey Quill as follows: '*I always thought the aeroplane which was the best from the pure text book handling point of view was the Mark VIII ... With the standard wing tip it was a really beautiful aeroplane*'.

1658 built.

The **Mark IX** was, like the Mark V, also planned as a short-term expedient until the Mark VIII could be put into production but, in the end, it became one of the most numerous of all Spitfire variants. Whilst the Mark V had become the backbone of RAF Fighter Command, a fresh response was required to the appearance of the pressurised Junkers Ju.86R bomber, to the better high altitude performance of the Bf 109G and especially, to the Focke-Wulf 190. This last had appeared over northern France in September, 1941, and had been found to out-climb by 450 ft per minute, outrun by more than 20 mph, and outmanœuvre the Mark V Spitfire, with or without clipped wings and the assistance of 'Miss Shilling's orifice': this latter was a stopgap device devised by Miss Beatrice 'Tilly' Shilling whereby a hole was punched into a metal diaphragm placed across the float chamber of the Merlin carburettors to prevent fuel starvation in negative G manœuvres (not a problem with the German fuel-injected engines). 1943 saw a permanent solution with the introduction of the Bendix-Stromburg carburettor which injected fuel at 5 psi through a nozzle directly into the supercharger.

The Mark IX, when it began to be produced in June, 1942, was, essentially, this earlier machine but fitted initially with the vastly improved Merlin 60 series

– the standard fighter had the Merlin 61 and the 63, the low altitude version was fitted with the Merlin 66, the high altitude aircraft was powered by the Merlin 70. These three versions were now officially designated **F**, **LF**, and **HF**, respectively. Some IXs were also pressed into service as photoreconnaissance machines. Lightened Mark IX versions were now able to go higher than the specially designed Mark VI and one Mark IX recorded the highest air battle of World War Two, at 43,000 feet – against one of the Junkers Ju 86 bombers. Stripped of armour plate and equipped with only two cannons, it pursued the bomber and damaged it before the guns jammed. This action was, fittingly, over Southampton.

The new Mark's top speed had now risen to 408 mph and *marked the point where the dedicated record breaking S.6B floatplane was overtaken in level flight by a fully equipped fighting machine.* The extra power of the new Merlins now required a four-bladed propeller and increased radiation which resulted in symmetrical underwing ducting for the first time.

A Mark IX showing symmetrical radiator ducting and bomb attached.

Such changes were obviously not perceptible from any distance, especially as the standard Mark V elliptical wing was still the norm, and so enemy pilots, unsure whether or not they were approaching the new higher performance aircraft were now more disposed to keep away from the older Mark V which was still very active in the skies. But, when the Mark IX was tested against a captured Fw 190 in July of the following year (Oberleutnant Arnim Faber had conveniently landed it, a little lost, at RAF Pembrey Sands on 23 June 1942), the differences in performance were too small for comfort and the German's rate of climb between 15,000 and 23,000 feet was superior. Fortunately, Rolls-Royce had been able to respond with the improved Merlin 66.

The more powerful new Merlins also gave rise to experiments with increased fuel tankage and with a six-bladed contra-rotating propeller. Other subsequent modifications included clipped wings for low altitude operations and the pointed fin and broad-chord rudder; later, as with most aircraft after 1944, it was also supplied with a 'teardrop' canopy and cut-down fuselage rear decking for better rearward visibility. Fuel capacity was also increasing so that, with the introduction of two 18 gal. fabric cells into the wings and a 72 gal. tank into the rear fuselage, the internal fuel capacity of the Spitfire had increased to nearly twice that of the prototype. With a further 45 gal. drop tank, Jeffrey Quill flew a Mark IX from Salisbury Plain to the Moray Firth and back at less than 1000 feet – the equivalent of East Anglia to Berlin and back (and thereby proving that the

Spitfire might be used if need be as a bomber escort over Europe).

The 'C' wing with two cannons and four machine guns was still standard although an 'E' wing was introduced later, which offered the two cannons, together with two larger bore 0.50 inch Browning machine guns. To signify the change from the 'C' wing, a new designation was used: **LF IXE** and **HF IXE**.

There was also, in 1944, a floatplane conversion of the Mark IX with the modified fin of the Mark V conversion but without the 'beard' type tropical air filter; the carburettor air intake was also moved forward to avoid the entry of spray. (*See* photo, p. 264.) No production orders.

Two-seater Trainer.

After the War, the Mark IX also figured in a revival of the 1941 proposal for a two-seat trainer version. It must be admitted that Mitchell's concern to give the performance of his fighter 'paramount importance' did not allow enough fusel-age volume for the creation of a very comfortable two-seater, although training accidents might have been reduced during the war if the RAF had had such a variant (a two-seat trainer version of the Hurricane had also been proposed as early as 1939). A Mark VIII Spitfire was initially converted by means of moving the original cockpit 13½ inches further forward in order to allow room for a sep-arate instructor's cockpit that was behind and raised well above the other one. In the end, 20 Mark IX aircraft were thus converted and sold to foreign air forces, giving rise to an unkind RAF speculation that 'johnny foreigner' needed two people to fly the Spitfire.

5653 Mark IXs built.

(For Marks **X** and **XI**, see the photoreconnaissance section.
By this time, the Mark numbers were being allocated consecutively, whether the aircraft was a fighter, naval, or photoreconnaissance type.)

The **LF XVIE** From September, 1944, a version of the Mark IX, with the low altitude Merlin 266 (essentially, the low-rated Merlin 66 built under licence by the Packard Motor Co. of Detroit) was built exclusively at Castle Bromwich.

Spitfire LF XVIE.

This version included many of the later modifications to the Mark IX, in particular the increased fuel tankage, the 'E' wing instead of the 'C' wing, and the modified 'rear- view' fuselage. Clipped wings were normal as the machine was intended mainly for ground attack roles. It was given the separate **LF XVIE** designation.

1053 built.

Griffon Engined Spitfires

Rolls-Royce, soon after the outbreak of the war, were developing a new engine based on the Merlin but with a capacity of 36.7 litres instead of the 27 litres of its predecessor. Despite the larger capacity, the firm was able to keep the frontal area of the new engine to no more than 6% greater and its length to no more than 3 inches longer than its immediate Merlin predecessors. Its weight was also within 600 lbs of the earlier type. Thus Rolls-Royce made it possible to adapt the Spitfire airframe to the enormously more powerful engine and Mitchell's design took on a completely new lease of life.

The first Griffon-engined prototype was seen as combining the benefits of the new engine with the improvements built into the experimental Mark III – in particular, the retractable tailwheel; on the other hand, the power of the new engine necessitated a change to a four-bladed propeller. The resultant machine was originally designated Mark **IV** but, as other stopgap variants had had to be produced before it could be fully developed, it was re-designated the Mark **XX**. Quill narrates how, in July 1942, he was asked to fly a Spitfire in a comparison test with a Typhoon and a captured Fw 190. He took the new Griffon-engined machine, DP845, which had first flown in the previous November, and caused quite a stir with its performance against two aircraft supposed to be the fastest fighters in service. In a race at 1,000 ft from Odiham to Farnborough, the Fw 190 had to drop back due to engine trouble but Quill was able to leave the Typhoon well behind.

At the same time, the Merlin 60 series was being fitted to the Mark IX and the resultant performance led to the proposed Mark XX with the Griffon engine remaining as a prototype; nevertheless the Air Ministry did order the production of a Griffon-engined Spitfire and this led to the Mark XII.

The **Mark XII** was thus *the first production version of the Griffon-engined Spitfire*, using the single-stage Griffon III, and, because it was intended specifically for low level operation, was produced with clipped wings as standard.

DP845, the Mark IV/XX prototype with six cannon mock up.

It was based on the VC airframe and so some of the earlier XIIs still had fixed tailwheels (*see* photo p. 291). It had a strengthened fuselage, the 'C' wing, and the pointed, broad-chord fin and rudder, made even more necessary by the increased power and length of the new engine and its four-bladed propeller. The top engine cowling was now modified with blisters over the cylinder banks of the new engine to assist the view of the pilot down the centre of the longer nose – which, with the new rudder, increased the overall length of the new mark to 31 ft 10 in. Its improved performance was such that the enemy in 1943, when the new silhouette was identified, was none too anxious to be drawn into low altitude battles where the Mark XII was dominant – as proven by its later success rate against the V1 flying bombs.

Nevertheless, the Mark XII was an improvised machine with a poor rate of climb because of its single-stage supercharged engine. It was only built in limited numbers and was phased out in September, 1945. *The Spitfire Mark IX was still the most useful all-round fighter*.

100 built.

(For the Mark **XIII**, see the photoreconnaissance section below.)

The Mark **XIV** was required in order to surpass the high altitude performance of the Mark IX and signified the fitting of a Griffon engine with two stage supercharging which was so flexible that *no different high- or low-altitude versions were now necessary*; however, pending the development of a 'super Spitfire', to be powered by this all-altitude Griffon engine (appearing eventually in 1946 as the Mark XVIII), *the Mark XIV was another interim type*, employing a strengthened Mark VIII airframe and the 'C' wing with two 31 gal. wing tanks. There was still some directional instability, as in the previous Griffon mark, and so the fin and rudder area needed to be further enlarged because of the increased power and the solidity of a five-bladed propeller. This was initially achieved by increasing the side area of the pointed fin version via a straightened leading edge; later the whole fin and rudder outline was redrawn to encompass the increased area and became the first thorough redesign of these components. The power of the Griffon 65 engine was now absorbed by the five-bladed propeller and the radiators were slightly deeper. Later, the Mark XIVE appeared, signifying the fitting of the 'E' wing; the new rear-view canopy was also substituted and some had clipped wings.

The Mark XIV called for vigilant flying – particularly because, as part of a move towards standardisation, the Griffon engine rotated in the opposite direction from the Merlin and the swing during take off was not only more powerful but was now to the right; and the slightest throttle movement resulted in a dramatic surge of power. Nevertheless, Quill's first impression of the new machine was expressed in his phrase 'quantum jump', for the performance was spectacular: 445 mph at 25,000 feet and a from sea-level climb of over 5,000 feet per minute.

The A & AEE later reported 447 mph at 25,600 feet and a climb to 40,000 feet in fifteen minutes; the absolute ceiling was no less than 44,600 feet. Although the next redesign of the tail surfaces, with the Mark 22, was the real solution to some control problems, the Mark XIV with the new two-stage engine was *very successful at low level against the V1s and was also the main high altitude superiority fighter until the end of the war.*

It also represented the first appearance of the Spitfire in the **FR** category, *signifying the use of standard fighters additionally equipped with cameras and operating as fighter-reconnaissance machines.* And it could be equipped with four cannons and could carry bombs. *Thus it could claim to have the best all-round performance of any current fighter apart from range.* (*See* photo p. 291.)

Later versions were fitted with lower fuselage decking and the improved rear view 'teardrop' canopy. The normal 'C' wing armament later gave way to the 'E' type arrangement of the later Mark IXs and some machines had clipped wings. A few also were engined with the Griffon 85, geared for contra-rotating propellers.

The Mark XIV formed the last *substantial* production run of any type of Spitfire and the last version to see significant action; the first Mark XIVs were produced in time for the invasion of Europe in June, 1944 and one had the distinction of being the first aircraft to shoot down the formidable Messerschmitt Me 262 jet fighter. Many hundreds were sent to India to assist with the Pacific War but this ended before many of this latest variant saw active service.

957 built.

(For the **Mark XV**, *see* the Seafire section below.)
(For the **Mark XVI**, *see* after the Mark IX above.)
(For the **Mark XVII**, *see* the Seafire section below.)

The **Mark 18** (*from 1943, it became the RAF convention to use arabic numerals*) was very similar to the Mark XIV, being also fitted with the Griffon 65 and incorporated the later features of the Mark XIV – the rear view fuselage, retractable tailwheel, and the 'E' wing armament. *Together with the PR 19, it represented the last F variant to embody the classic elliptical wing outline.* The wings, however, were in fact, not the 'C' type of the Mark XIV, but a considerably redesigned unit with a strengthened wing centre section and undercarriage – with an 11in. wider track width; fuel tankage was also increased, as was the depth of the radiators, and the FR version now carried two vertical cameras as well as the oblique type employed in the FRXIV.

The Griffon 65 was later replaced with the 67 version. It was produced two years after the Mark XIV and so too late for World War Two service; *it did, nevertheless, see action as late as 1951 with No. 60 Squadron against Communist forces in Malaya.* (*See* photo p. 291.)

(For the Mark **19**, *see* the photoreconnaissance section, below.)
(For the Mark **XX** – so designated because it was proposed before the adoption of arabic mark numerals – see the introduction to the section on Griffon engined Spitfires above.

The **Mark 21** was similar, in many ways, to the Mark 18 and also had the Mark XIV fin and rudder, and five-bladed propeller. The wings were of the extended span type but it was soon supplied with the first completely revised wing in which the familiar Mitchell outline was altered and the total area slightly increased.

Mark 21 prototype with extended span wing and third type fin and rudder.

The move to metal skinned ailerons had improved lateral control in the earlier Spitfires and metal skinning of the rudder and elevators had now been introduced. However, as speeds further increased, heaviness of the ailerons was again experienced and so a complete redesign with a 20% increase in area, was now incorporated into the new wings. Their structure had also to be further stiffened as the Griffon engined Spitfires were now beginning to experience the problem of 'aileron reversal'; that is, the torsional load applied to the wing by the ailerons at the speeds now possible, was causing the wings to twist and, in extreme cases, to reverse the expected effect of aileron movement. Apart from the internal modifications, the evident alteration to the wings was the move to half-rounded tips – a compromise between the clipped version and Mitchell's original elliptical configuration.

A typical Mark 21.

There were now two leading-edge fuel tanks in each wing and, for the first time, four 20 mm cannons were fitted and no machine guns. Small blisters in the top surface of each wing were required in order that the ammunition belt feed mechanisms were cleared; by way of compensation, however, doors were also provided to cover the lower half of the wheels when retracted. As with the Mark

XIV of the previous year, the larger diameter, five-bladed propeller was fitted to its Griffon 61 and this led, in turn, to a longer undercarriage, slightly repositioned. The take-off still needed care.

Some aircraft were fitted with the Griffon 85 (designed to take a six-bladed contra-rotating propeller unit which improved the machine's handling characteristics but the aircraft was still longitudinally unstable, with very sensitive elevators, and therefore in need of redesigned tail surfaces. RAF test pilots did not consider it an improvement on the Mark XIV as an all-round machine and that 'no further attempt should be made to perpetuate the Spitfire family'. On the other hand, Quill found the new mark to have 'a tremendous lightness of control' and 'revelled in aerobatics at speeds that would have been impossible before'. The major redesign work involved had been put in hand by 1942 but this mark was only operational at the end of 1944. It was particularly used against V2 sites in northern Holland. It was not produced in large numbers as *the war ended soon after its production began.*

121 built.

The **Mark 22** was again powered by the Griffon 85 and was also completed very late – in 1945 – and did not proceed to large scale production. The development from the Mark 21 to the Mark 22 and to the Mark 24 was so seamless that no prototypes as such were necessary and the main difference between the last two variants was the increased fuel capacity of the later mark. They both had the new wing and the four cannon arrangement of the Mark 21 but, whilst the Mark 21 still retained the balloon type of cockpit hood and its associated rear fuselage decking, these latest machines were equipped from the outset with a final rear-view fuselage and canopy configuration. A new fin was also fitted.

Very soon after the Mark 22 was produced, provision was made for carrying three 500 lb bombs, one under each wing and one under the fuselage. It equipped one RAF squadron and also two squadrons of the Southern Rhodesian Air Force, operational until 1954.

During 1946, further development work resulted in the fitting of substantially revised and enlarged tail surfaces of the Spiteful type, further distancing the Mark 22 and the Mark 24 from Mitchell's original design. These tail surface increases, the new wing and fin, and the new rear-view cockpit arrangement, taken together, produced *an even more marked departure from the original Spitfire shape than that of the Mark 21.* (*See* photo p.292.)

268 built.

[The **Mark 23** was to have a revised cross section wing but was never built.]

The **Mark 24**. Now that the stability problems associated with the increased power of the later Griffon engines had been overcome by the revised wing and tail planforms, the increased speed, rate of climb, and ceiling, combined with excellent manoeuvrability and handling, and the increased range, made the Mark 24 the '*ultimate Spitfire*' and showed how far it had been possible to develop Mitchell's prototype of 1936. In round terms, the engine power and the loaded weight had doubled, the fuel capacity had trebled and *the maximum speed had increased by 100 mph – a similar advance as that achieved between the prototype Spitfire and the previous Gladiator.* These impressive increases had all been achieved by the progressive development of Mitchell's original formulation – as indicated by the fact that, in spite of all the increases just mentioned, the final wing area was less than 2 sq ft more than that of the prototype. One does, however, tend to agree with Quill: 'the genius passed on by Mitchell had died.

The beautiful symmetry had gone; in its place stood a powerful, almost ugly fighting machine'.

The final Spitfire came off the production line in February, 1948. The mark equipped No. 80 Squadron, being handed over to the Hong Kong Auxiliary Air Force in 1952.

74 built.

Photoreconnaissance Spitfires

At the outbreak of war, the idea of photoreconnaissance by an unarmed fighter seemed like two contradictions in terms; however, the ability of such an aircraft to fly higher and faster than other standard designs was certainly worth considering for this very different role, especially after operational experience with Blenheims and Lysanders had shown the need for aircraft capable of avoiding, rather than trying to defend against, enemy machines in order to return safely with vital photographic information. Because of its high speed, the Spitfire was an obvious choice for the experiment and the earliest Mark was utilised even though there was severe competition for its services elsewhere.

As with the naval Seafire (*see* below), the photoreconnaissance version had never been envisaged by Mitchell although, interestingly enough, one of the first used for this sort of operation was the '**Speed Spitfire**', which looked back to the earlier Mitchell days of record breaking. At an international aeroplane meeting at Zurich in July, 1937, it was announced that a modified Messerschmitt, the Me 109, had become the fastest fighter in the world, having attained a top speed of 379 mph. The Air Ministry ordered a competing model and the 48th Mark I was taken from the production line for modifications which included blunter wing tips, a reduced span of 33 ft 8 in. and the fitting of a cockpit enclosure with a profile not unlike that of the de Havilland D.H. 88 Comet which won the MacPherson Robertson race from England to Australia in 1934. It had also reverted to a tailskid and increased radiator and oil cooler intakes to accommodate the greatly boosted Special Merlin engine, producing 2,100 hp. A four-bladed fixed-pitch propeller was fitted to absorb the additional power and the royal blue upper surfaces, with silver undersides and fuselage flash, were given a highly polished finish (another precedent for the future photoreconnaissance types).

The Speed Spitfire.

At one stage, to cut down drag, the radiator was completely removed and the coolant allowed to boil away but when, on 26 April, 1939, the Messerchmitt 209 put the Speed Record up to 469 mph, (smashing the absolute speed record of 440 mph held by the Italian Macchi M.72 Schneider floatplane since 1934) the whole project was dropped as the Supermarine aircraft could not exceed 410 mph. This coincided with the period of disillusion mentioned earlier when the Spitfire was beginning to be considered as no more than a stop gap until more powerful aircraft arrived. Nevertheless, valuable information was gained in respect of the PR machines described below. With the start of hostilities, N.17, as it had been designated, was fitted with an oblique camera and, as K9834 (its original production serial number), became one of the very first Photographic Reconnaissance Spitfires. Not surprisingly, the Speed Spitfire did not have a great range and so was, in reality, hardly used for photographic work; the most obvious influence of the Speed Spitfire project was that following PR aircraft were given extra streamlining by having their surface joints filled with plaster of Paris; the camouflage paint was then applied and given a high polish. In addition, these non-armed machines had their gun ports faired over. As a result, they were capable of up to 15 mph more than the corresponding standard fighter.

The next PR version was a Spitfire Mark I, **PR IA**, powered by a Merlin III engine, with the guns removed and with cameras in spaces previously occupied by the two inboard guns and their ammunition boxes. Once this minimal modification had been proved successful, it was clear that the standard 84 gallon fuel capacity of the aircraft would have to be increased. Thus the **PR IB**, 'medium range' version, was soon created by fitting a 29 gallon tank in the fuselage in place of the 40 lb of lead ballast (necessitated by the early change from the original Spitfire's two-bladed wooden propeller to the heavier three-bladed, two pitch metal one). A vertical camera was carried in a fairing beneath each wing. By 1940, the starboard blister was utilised to carry an additional 30 gallon of fuel and the port one was made to carry two cameras. This variant was regarded as the 'long-range' machine and designated the **PR IC**. Extra oxygen was carried for maximum high-altitude flying and a vertical camera was fitted in the fuselage for the first time.

It was also necessary to take low altitude photographs so at least one aircraft was fitted with oblique cameras, still mounted in and under the wings, and known as the **PR IE** (the 'D' and 'F' versions required more extensive modifications and so came later – *see* below). This was soon superseded by the **PR IG**, a Spitfire Mark I with its 'A' armament, bullet-proof windscreen, and armour plate retained because of the low level at which it operated. For this operational height, most of the PR IGs were camouflaged with a very pale pink in contrast to the high-flying variants which were painted a medium blue overall (pale green being the initial choice of colour for PR work).

The need for a more thoroughgoing development of the Spitfire for both the high and low level specialist PR roles was now being appreciated but, while this was being implemented, the **PR IF**, 'super long range' machine was produced and this carried a 30 gallon blister tank under each wing. Under optimum conditions, this variant could now reach Berlin and furnished final proof that the performance of Mitchell's fighter had also made possible the future development of an outstanding long-range reconnaissance type. The increased duration now possible meant that an enlarged oil tank was necessary; this resulted in the deeper line to the cowling beneath the engine which then became the standard shape for all the Merlin engined PR variants (*see* sideview drawing, p. 288). Alternative camera combination mountings were also fitted, additional oxygen was carried and, to improve downward vision, 'teardrop' transparencies were fitted to each side of the cockpit hood. Nearly all the PRIBs and PRICs were modified to the PRIF standard.

A PRI Special Survey Flight was deployed to Seclin, France, in November, 1939 – the first time a precious Spitfire was based or flown outside of the country; its first operational flight was on 18 December over Aachen. PRIs monitored the progress of the German invasion preparations whilst the rest of the Spitfire force was firmly deployed in England, prior to the Battle of Britain. By 1940 there were 8 PRIBs, 3 PRICs and 1 PRIE at Seclin. After the Dunkirk evacuation, reconnaissance activities were eventually organised into the Photographic Reconnaissance Unit (PRU).

The **PR Mark IV** began to appear by late 1940, following the decision to make specific production versions of certain of the PR Mark I modifications. The 'D' and the 'G' versions were chosen and Mark numbers allocated corresponding to the order of these letters in the alphabet.

The 'D' (PR Mark IV) version had had to wait for the adaptation of the wing leading edge sections to take fuel – it will be recalled from Chapter Eight that the original design of the Spitfire wing was to allow for steam to condense in these sealed D-shaped spaces (in the 1940s, this use of the wing voids was not common although its novelty would have been less if Mitchell's bomber had been put into production). There was now no need of the drag-inducing blister tanks of the previous long-range Mark IF as the gain was an extra 114 gallons in the wings which, together with the original 84 gallon tankage and the extra 29 in the rear fuselage, now increased the aircraft's endurance from the original 45 minutes to over 5 hours and the operational radius from about 130 miles to over 600. When the 'D' version, known as 'the bowser', became the Mark IV, the wing tanks had been further modified to take an extra 19 gallons, so enabling the rear tank to be removed as the aircraft was, otherwise, directionally unstable until the fuel in this tank had been used – the 'D' version was reported to be incapable of flying straight and level for the first half hour after take off. But it was able to photograph bomb damage over Germany, including a round trip to Stettin of over 5 hours duration (with unheated cockpit). Merlin 45 engine.
    The Mark IV was the mainstay of the RAF Reconnaissance Unit until 1941.
    241 built.

The **PR VI** was a special conversion of 6 Mark VI fighters sent to the Middle East in 1942 for PR work)

The **PR VII** was a more straightforward development of the PR IG as it had normal Spitfire wings, in order to retain its armament – because of the low level at which it operated – and it was likewise equipped with bulletproof windscreen and armour plate. As with the PR Mark IV, the oblique and vertical cameras were all installed in the fuselage, and 'teardrop' transparencies were fitted to each side of the cockpit hood. These modifications signified *the emergence of a coherent approach to photoreconnaissance operations and the development of a distinct type aircraft* which, in the form of the Mark IV and VII, was the backbone of this vital, if often unsung, RAF activity until the end of 1942. Merlin 45 or 46 engine.
    45 built.

The **PR Marks X** and **XI** represented the pressing into PR service of the new, more powerful Merlin 61. This was achieved in 1942 by, essentially, re-engining the Spitfire Mark IX and fitting a retractable tailwheel. Since no air combat was expected of these photoreconnaissance aircraft, the bulletproof windscreen was deleted in favour of a plain, curved one which, in combination with the retracted tailwheel and wing leading edges without gun ports, gave an increase of 5 mph in speed over a comparable Mark IX. The Mark XI appeared before the allocated **PR Mark X** which had a pressurised cockpit and involved longer development of the Mark VII it was based on. It saw limited service from May, 1944.

16 built

The **PR Mark XI** had the Mark IV leading-edge fuel tanks, with a total capacity of 133 gal., in addition to the normal fuselage tankage and external drop tanks. The fitting of the more powerful Merlins resulted in later machines having the pointed, broad-chord rudder which had appeared on the later Mark VIIs. The fuselage camera installations of the PR Mark IV were later augmented by a camera fitted in a blister under each wing, just outboard of the wheel well. As this long-range Mark was also fully tropicalised, it not only served in European theatres but was also employed in the Middle East and the Pacific and, as such, could be regarded as *the most effective photoreconnaissance aircraft of World War Two. It was the main PR variant used by the RAF in Europe and the Far East in the second half of the war.*

It achieved the highest wartime speed of any Spitfire when being tested at Farnborough: on 27 April, 1944, Sqd. Ldr 'Marty' Martindale dived from 40,500 ft to a speed of 606 mph (Mach 0.89) before the engine blew up (he glided safely back to the airfield, despite almost no forward vision owing to a heavily oiled windscreen).

471 built.

The **PR Mark XIII** signified an advance on the PRVII low level reconnaissance aircraft and was, essentially, a Mark V fitted with a specially low-altitude-rated Merlin 32 and a 30 gal. drop tank. Instead of cockpit teardrops, a balloon hood was now fitted behind the bulletproof front windscreen and only four machine guns were installed. It appeared in 1943 – by which time, standard fighters were also being equipped, additionally, with cameras and operating as fighter-reconnaissance (FR) machines.

26 built.

The **PR Mark 19** was the first of the arabic numeral variants to be dedicated to specialist photoreconnaissance roles and, powered by the Griffon 65, represented *the only Griffon-powered PR variant.* Enemy improvements in location and interception of existing photoreconnaissance aircraft had created a need for the greater speed, range and ceiling that the Griffon engined fighters could achieve. A Mark XIV airframe was therefore adapted with the increased wing tankage and camera arrangements of the PR Mark XI.

Later machines were fitted with the Griffon 66 and had pressurised cockpits, now something of a necessity when flights of five to six hours and at an increased height were possible. The leading-edge wing tanks were increased by 20 gallons on each side to give a total internal petrol capacity of 256 gallons (compared with the 84 gallons of the Mark I), often supplemented by a 90 or 170 gal. drop-tank. This airframe, which was produced in May, 1944, was now able to take full advantage of the Griffon engine and resulted in *one of the most outstanding of all the Spitfire variants.*

The PR pilots now had an aircraft that matched the performance of the F Mark XIV, had a greater range than the PR Mark XI, and the cockpit comfort of the PR Mark X. With a top speed of 460 mph, it was one of the fastest piston-engined aircraft of all time; it could cruise at 370 mph and at 40,000 feet, beyond the effective reach even of the German jets at the end of the war – or, indeed, beyond the reach of any jet before the introduction of the swept-wing Sabres and the MiGs of the early 1950s. Indeed, a Flight Lieutenant E. C. Powles claimed a climb to 51,550 feet and an emergency dive, when his cabin pressurisation failed, to a speed of Mach 0.96 This event was incidentally in *the last Mark to embody the classic elliptical wing planform of Mitchell's original design* – it was

The PR 19.

perhaps the 'ultimate Spitfire' and one of the most functionally beautiful aircraft of World War Two. As with the Spitfire F Mark XIV, it was produced in time for the closing stages of the war in Europe and for limited active service in the Pacific and Far Eastern areas. *A meteorological flight of Mark 19s was finally phased out in 1957.*
225 built

Naval Spitfires: the Seafires

At the beginning of the war, naval aviation was underdeveloped – attributed by C. G. Grey to the senior Admirals being 'definitely anti-air-minded' and being 'solid ivory from the jaws up, except for a little hole from ear to ear to let useful knowledge go in and out'. And he attributed the sinking of the *Prince of Wales* and the *Repulse* off the coast of Malaya and the prolonged hunting of the *Bismarck*, *Scharnhorst* and *Prinz Eugen* to lack of effective naval aircraft.

The navy's first all-metal monoplane was the obsolescent and vulnerable Blackburn Skua dive bomber, whose bulk and lack of power resulted in too low a speed; the Roc, a turret fighter variant, was (like the Boulton Paul Defiant mentioned earlier) no match for the new enemy aircraft and the weight and drag of the turret made it even slower than the Skua. The newer Fairey Fulmar, a navalised version of which was designed as an eight-gun fleet defence two-seat fighter, was no great improvement: the Navy had specified a two-seat machine, feeling that a second crew member was necessary for navigating over the open sea but, as a result, the Fulmar was far too large and unwieldy when it came up against the new single-seat opposition now based on the other side of the Channel.

Thus the Admiralty had requested a navalised version of both the Spitfire and the Hurricane in 1939 when the threat of the German navy on the Atlantic seaboard was becoming very real. A carrier deck was marked out on the Eastleigh runway and a Spitfire fitted with an arrester hook. The tests were satisfactory but other more pressing demands for these aircraft (including photoreconnaissance) *delayed the Navy's acquisition of such seagoing fighters until the autumn of 1941*. At this time, proper maritime tests were ordered and, in January 1942, a Mark V Spitfire with an arrester hook landed on HMS *Illustrious*. In view of shortages, some 60 Hurricanes had been put into service as a stopgap but, by mid-1942, mounting losses made it urgent that Spitfires should be finally employed. In September, further trials were ordered to see if the modified Spitfires could operate from smaller carrier decks. HMS *Biter* was chosen and Capt. Eric 'Winkle' Brown made a sideslipping approach (because of the long nose of the Spitfire) and landed successfully – only to discover that the deck had been unmanned and that the wire he had caught with his hook had been flat on the deck – one of Britain's greatest test pilots was obviously very good at accurate three-point landings.

So the first Seafires were, basically, Spitfire Vs fitted with arrester hooks and, because fitted with the B wing, were designated **Seafire IB** and **IIB**.

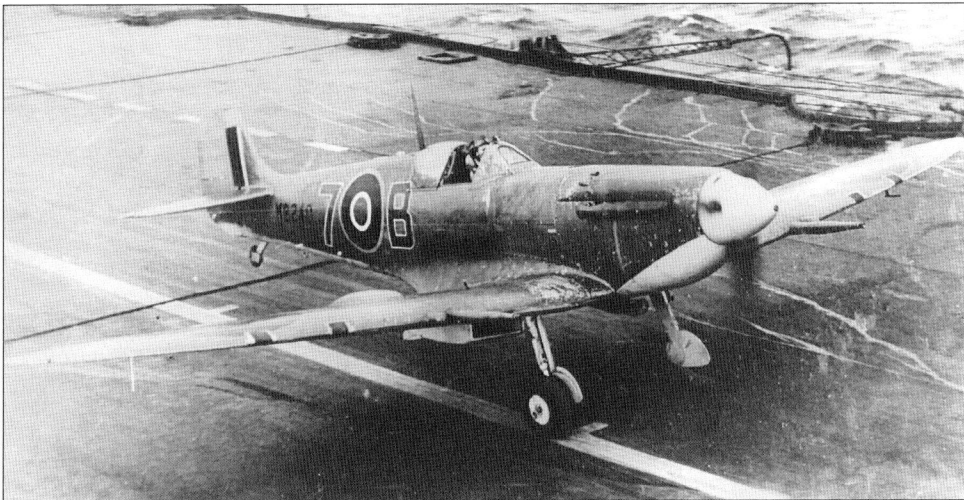

A well weathered Seafire IIC taking off.

The Mark II was additionally equipped with catapult spools (the use of Mark Is being limited to aircraft carriers not fitted with catapults) and, because of the harsher nature of seaborne landings, was now given a strengthened undercarriage. The 'universal' wing fitted to this version resulted in the designation Seafire **IIC**, followed by the **Seafire FRIIC** which signified the installation of cameras. The high priority given to production of improved aircraft for home defence did not permit the more fundamental design changes that were really needed to be incorporated into an aircraft which Mitchell had never contemplated being used for the particular demands of shipborne operations. This aircraft entered service in June, 1942, and its first action was in support of Albacore bombers during Operation Torch.

372 built

(The following extract from Adlan describes some of the problems associated with the early Seafire types; the detail of the Salerno operation (September 1943) makes chilling reading:

> The aircraft had a very limited range, no bomb load and, with the extra weight of the [arrester] hook was not all that much faster than the Wildcat and with lesser firepower....although a beautiful machine just to fly, it was very difficult to deck-land because of its tendency to 'float' over the wires when the engine was cut...

For the Seafire to land on the small deck of an Escort Carrier, even under ideal conditions, calls for considerable skill and experience on the part of the pilot. But at Salerno, the wind conditions were no better than a zephyr breeze and almost a dead calm, conditions entirely to have been expected at that time of the year. Thus the Seafires had to operate with a total wind speed over the deck of only sixteen knots, being the maximum speed of the Escort Carriers, whereas they needed a total wind speed over the dek of at least twenty-eight knots. These were desperately difficult landing conditions for the Seafire pilots; conditions which surely should have been anticipated at the outset when the whole Salerno operation was being planned by Rear Admiral Vian who, despite never having flown an aircraft or having served in an Aircraft Carrier, had been put in charge of this, the first multi Carrier Fleet of the Royal Navy ... After two days the four Escort Carriers had virtually run out of Seafires, no less than forty-eight of which had been written off as the pilots attempted to land in windless conditions.

But see below for a far better report with reference to the Mk XV.)

The **Seafire Mark III** (not to be confused with the Spitfire III prototype) saw *the introduction of specially designed folding wings which allowed its use on a wider variety of vessels.* Another improvement on the F Mark IIC was the improved Merlin 55 M and a low rated version of the engine giving rise to the LF Mark III; this was followed by the FR Mark III with the installation of the Mark II camera provision. Because of the extensive modification to the wings, the Mark III did not become operational until late 1943; *it was still in service at the end of the war with Japan.*
1250 built

The **Seafire Marks XV** and **XVII** were *the first Griffon powered naval versions* (with the Griffon VI). The first of these, the **Seafire Mark XV**, basically, had a Seafire Mark III airframe with folding wings, a Spitfire Mark VIII tail, retractable tailwheel and, on later aircraft, a newly designed spring-loaded 'sting' deck arrester hook which involved a slight modification to the rudder. It finally began to appear in September, 1944.
390 built

(By now the problems with the Seafire mentioned above had now been largely circumvented. Adlam again, with reference to a Seafire XV: 'a beautiful machine to fly. An absolute thoroughbred of an aircraft requiring only the most delicate pressures on the controls for it to respond immediately and perfectly ... On a runway with plenty of space, the simplest of aircraft to land'.)

The **Seafire Mark XVII** was the first naval version to be fitted with the later cut-down fuselage and rear-view canopy. It continued the use of the Mark XV larger rudder and faired sting arrester hook but now had a curved windscreen and the comparative novelty of rocket-assisted take-off. The previous folding wing arrangement was retained but with a strengthened main spar and long-stroke undercarriage. In view of the coming-together of all these improvements, *this*

Mark can therefore be regarded as the first dedicated carrier Spitfire. Although it was produced in September, 1945 – as much as four years after the first improvised naval versions – it was *too late for World War Two but, nevertheless, gave the navy its first 400 mph aircraft.*

233 built

The **Seafire Marks 45** and **46** were produced very soon after the Seafire XVII finally appeared despite their much higher Mark numbers: these had been allocated to allow for future Spitfire developments which, in the event, did not materialise beyond Mark 24. The Mark 45 was to be a naval version of the Spitfire 21 and was capable of another 45 mph above the top speed of the previous type, but the more powerful Griffin 61 engine produced a swing on take-off which did not render it popular for carrier operation. Some later aircraft were fitted with contra-rotating airscrews which, together with increased rudder and elevator areas, made for a considerable improvement in directional stability.

51 built

These modifications, with the rear view canopy, and the Spiteful-type tail constituted the Mark **46** which, like its immediate predecessor, did not have folding wings; there had not been time to develop such an arrangement for the new shape of wing that had come in with the Spitfire Mark 21, on which it was based. (*See* photo p.308.)

25 built

The **Seafire 47**, *the last navy version of the Spitfire*, had the revised tail surfaces, rear-view canopy, and contra-rotating six blade propellers of the later Marks 45 and 46; it was also finally equipped with wing-folding geometry which was now powered (a previous manual arrangement required a five-man ground crew). The air intake filter was now more comprehensively faired into the lower engine cowling with the duct opening positioned just behind the propellers. Although the Griffon engined Seafires were not ready for use in World War Two, the Mark 47 entered front-line squadron service in 1948 and so was *used in the Korean War* and, like the final version of the photoreconnaissance types, had a remarkable performance, including a top speed of 452 mph. Thus, the Seafire 47 can be seen as *the most complete naval revision of the basic Spitfire landplane* and a perfect illustration of how far it had been possible to develop Mitchell's original conception (*see* photo p.285).

90 built.

* * * * *

The website <**www.airhistory.org.uk**> accessed 7 November 2016, supplies the serial number and history of every machine, and gives a grand total of 22,760 Spitfires and Seafires built. As machines were sometimes modified on the production line, assigning numbers to marks when they entered service might easily account for discrepancies and so the above numbers of the various marks might not be entirely accurate. A study of the movement cards at the RAF Museum, comparison between the above web site details and those supplied by Andrews and Morgan and reference to the works by Bruce Robertson and Morgan and Shacklady could be pursued if one is so inclined.

* * * * *

— *Appendix Two* —

Supermarine Aircraft Data

Overleaf is the published data relating to all of R. J. Mitchell's basic designs. It was felt better to collect this information in one place, rather than in the separate sections of the main chapters, as interesting comparisons can thus be easily made of their improving performances; additionally, photographs or the general arrangement drawings (not all to the same scale – for practical reasons) do not readily reveal the relative sizes of his designs.

Meanwhile, tabulated below are the calculation of wing loadings plus indicative extrapolations from variously published data of times taken to climb to a particular height – which also show very clearly the development of aircraft design during the Mitchell years:

Type	Wing loading (lbs per sq ft)	Time to reach 10,000 feet (in minutes)
Sea Eagle (1923)	9.840	
Seagull II (1922)	9.6	25.8
Seamew (1928)	9.5	24.4
Walrus (1933)	11.8	12.5
Southampton I (1925)	10.5	49.5
Air Yacht (1930)	15.8	26
Scapa (1932)	12.4	20.3
Stranraer (1934)	13	10
Type 224 (1934)	16	6.3
Spitfire prototype (1936)	22	4.1

Additionally, the following figures for wing loadings and landing speeds for the Schneider Trophy winners reveal the rapid advance in performance and increased challenge to the pilots over a dramatically short period and also show the difference between dedicated racers and the later fighter:

Type	Wing loading (lbs per sq ft)	Landing speed (mph)
S.4 (1925)	23	85
S.5 (1927)	28	90
S.6 (1929)	39.8	95
S.6B (1931)	42	100
Spitfire (1936)	22	60

Dimensions and Performances of all R. J. Mitchell's Main Types

Type	Span	Length	Wing area (sq ft)	Loaded weight (lb)	Max. speed (mph)	Climb (min)	Service ceiling (ft)
Spitfire (prototype)	36' 10"	29' 11"	242	5,359	364	8.2 (to 20,000ft)	35,400
Type 224	45' 10"	29' 5¼"	295	4,743	228	9.5 (to 15,000ft)	38,800 (absolute)
Air Yacht	92'	66' 6"	1,472	23,348	118	5.3 (to 2,000ft)	6,500
Scapa	75'	53'	1,300	16,080	142	20 (to 9,840ft)	15,500
Walrus	45' 10"	37' 7"	610	7,200	135	12.5 (to 10,000ft)	18,500
Stranraer	85'	54' 10"	1,457	19,000	165	10 (to 10,000ft)	18,500
S.6B	30'	28' 10"	145	6,086	407.5	–	–
S.6	30'	25' 10"	145	5,771	357.7	–	–
S.5	26' 9"	24' 3½"	115	3,242	319.57	–	–
S.4	30' 7½"	26' 7¾"	139	3,191	239	–	–
Seamew	45' 11½"	33' 6⅝"	610	5,800	95	22 (to 9,000ft)	10,950
Southampton	75'	49' 8½"	1,448	15,200	108	29.7 (to 6,000ft)	5,950
Swan	68' 8"	48' 6"	1,264.8	12,832	105	35.75 (to 10,000ft)	10,200
Sea Lion III	28'	28'	360	3,275	175	–	–
Sea Lion II	32'	24' 9"	384	2,850	160	–	–
Sea King II	32'	26' 9"	352	2,850	125	12 (to 10,000ft)	–
Scarab	46'	37'	610	5,750	93	–	–
Sea Eagle	46'	37' 4"	620	6,050	93	20 (to 5,000ft)	–
Seagull II	46'	37' 9"	593	5,691	98	7.7 (to 3,000 ft)	9,150
Seal II	48'	32' 10"	620	5,600	112	17 (to 10,000 ft)	–
Commercial Amphibian	50'	32' 6"	600	5,700	94.4	–	–
Sparrow II	34'	23'	193	1,000	65*	4.1[†] (to 1,000 ft)	–
Sparrow I	33' 4"	23' 6'	256	887	72	–	11,000

* RAF 30 aerofoil   [†] SA12 aerofoil

EARLY AMPHIBIANS

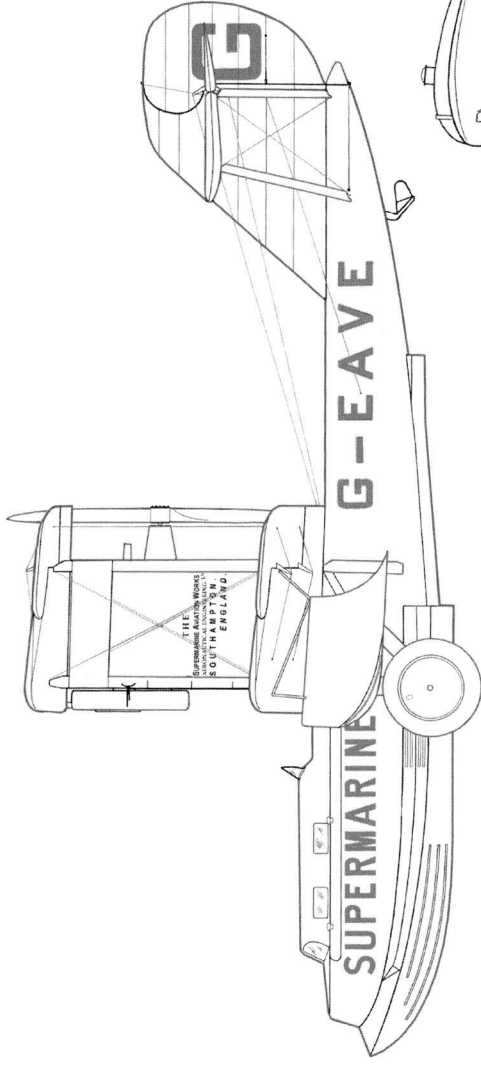

Sea Eagle (1923).

G-EBCR

IMPERIAL AIRWAYS LTD

Commercial Amphibian (1920).

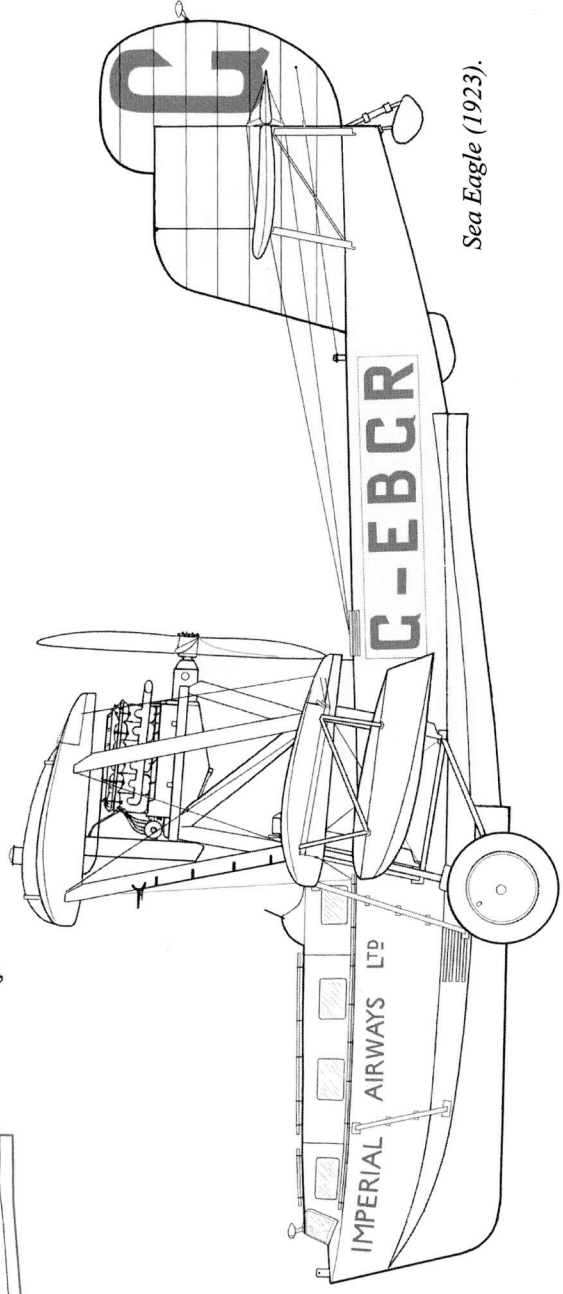

G-EAVE

SUPERMARINE

MEDIUM SIZED NAVAL AMPHIBIANS

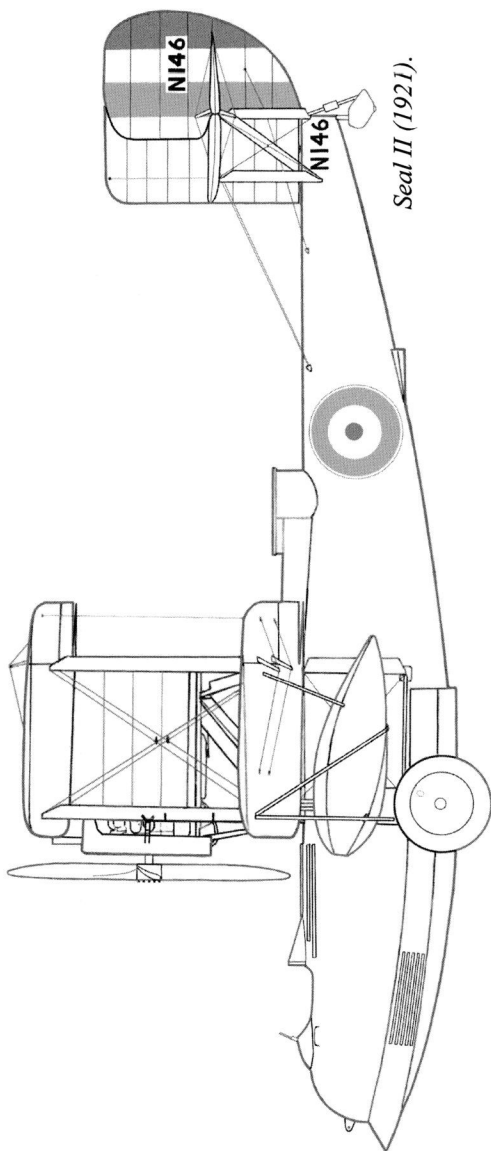

Seal II (1921).

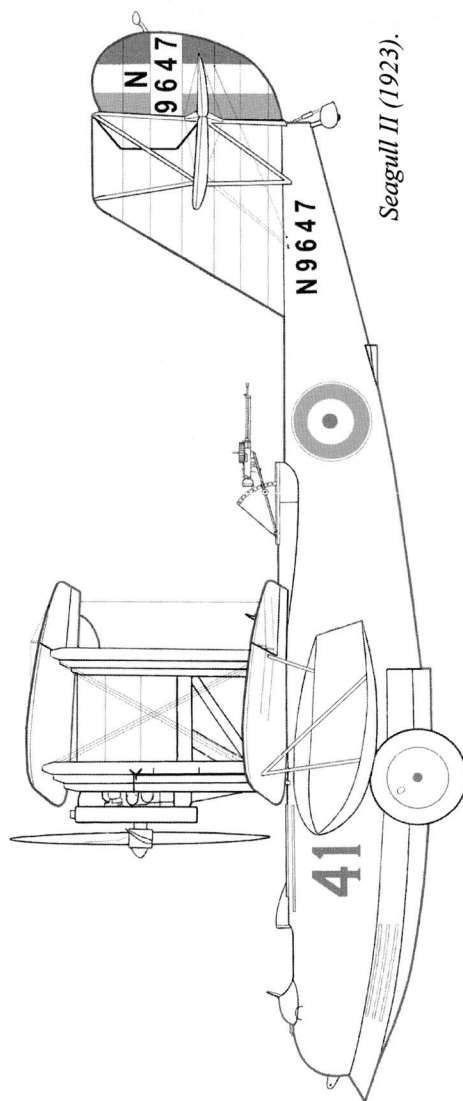

Seagull II (1923).

Note negative lift tailplanes.

L 2190

L 2190

Seagull V/Walrus (1933).

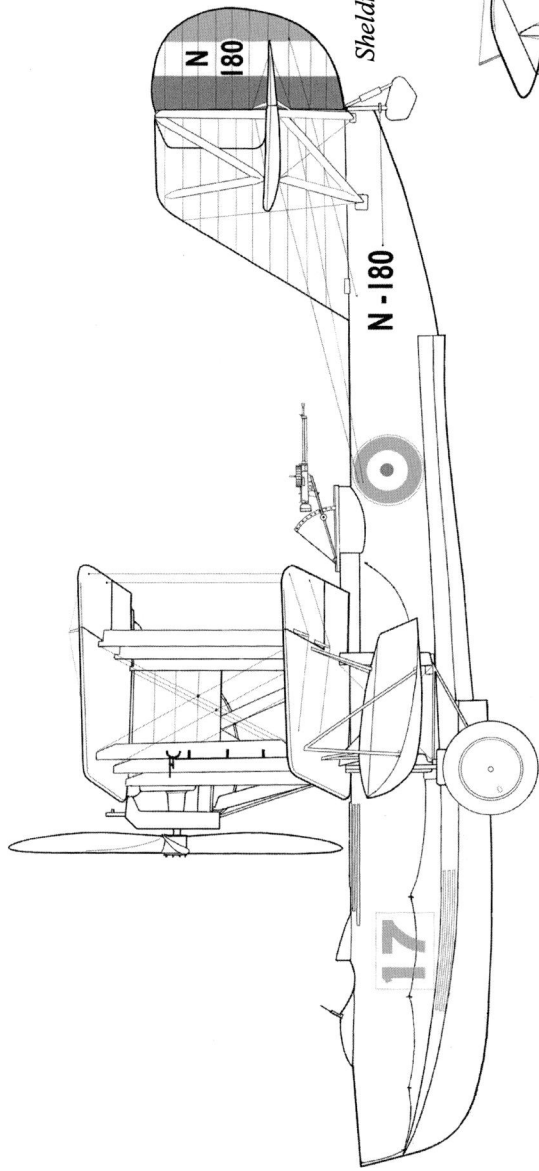

Sheldrake (c. 1925).

N 180

N-180

43

17

Note changing hull, fin and rudder, and tip float profiles.

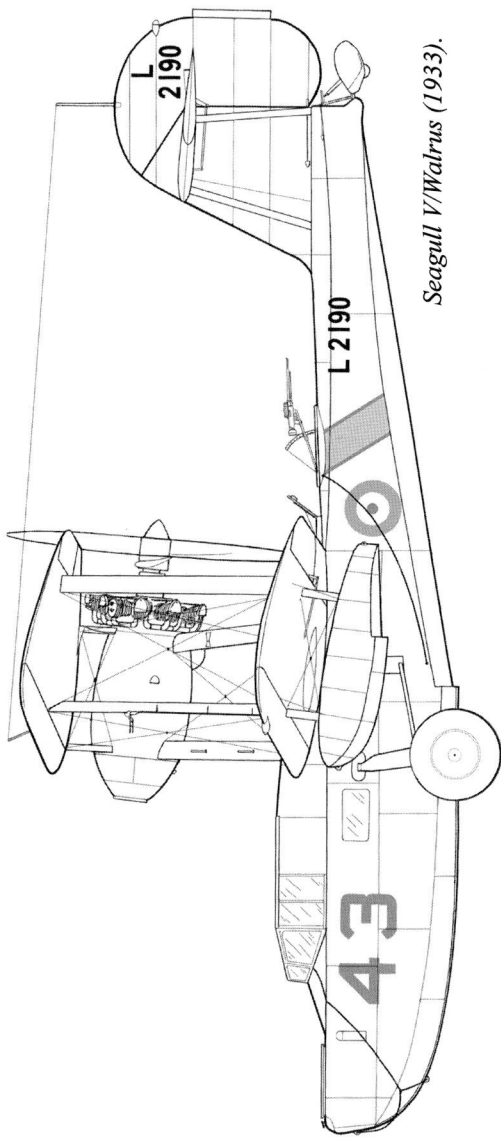

EARLY AMPHIBIANS

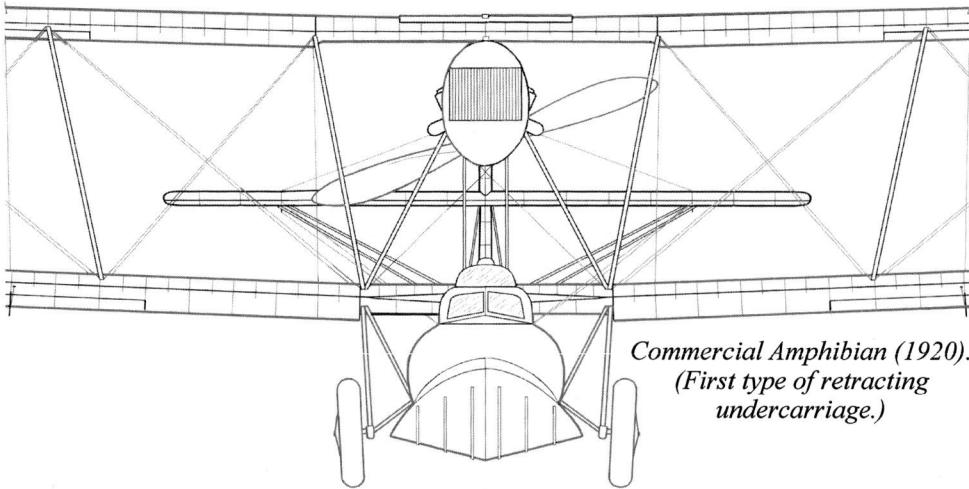

Commercial Amphibian (1920).
(First type of retracting
undercarriage.)

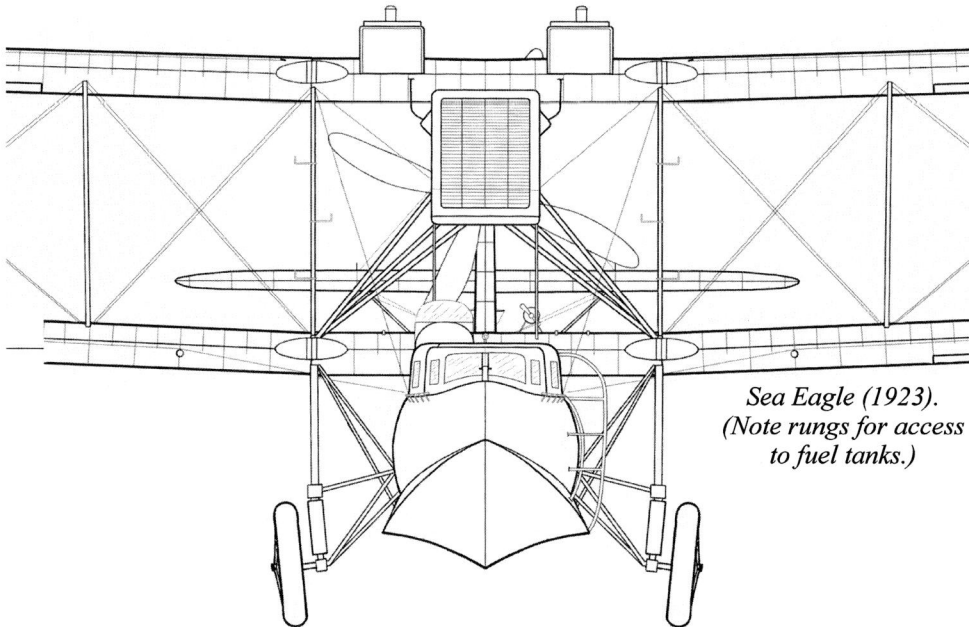

Sea Eagle (1923).
(Note rungs for access
to fuel tanks.)

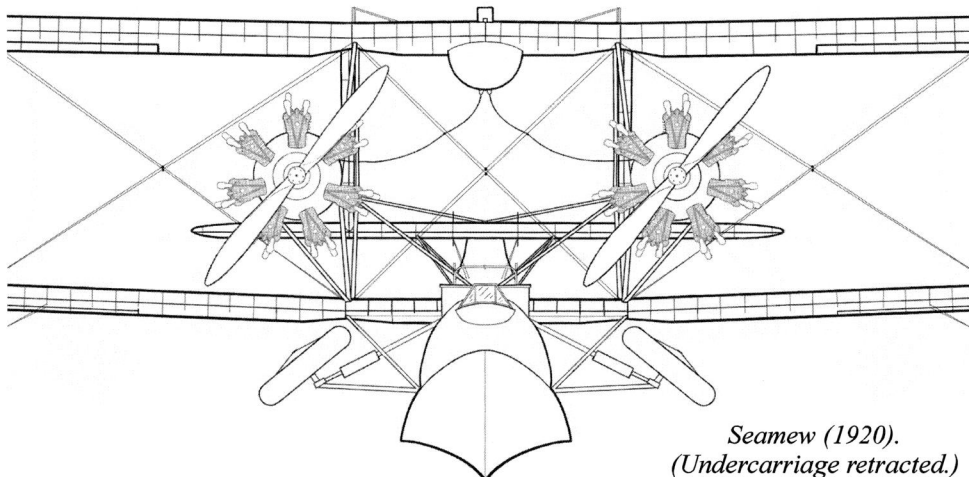

Seamew (1920).
(Undercarriage retracted.)

NAVAL FLYING BOATS

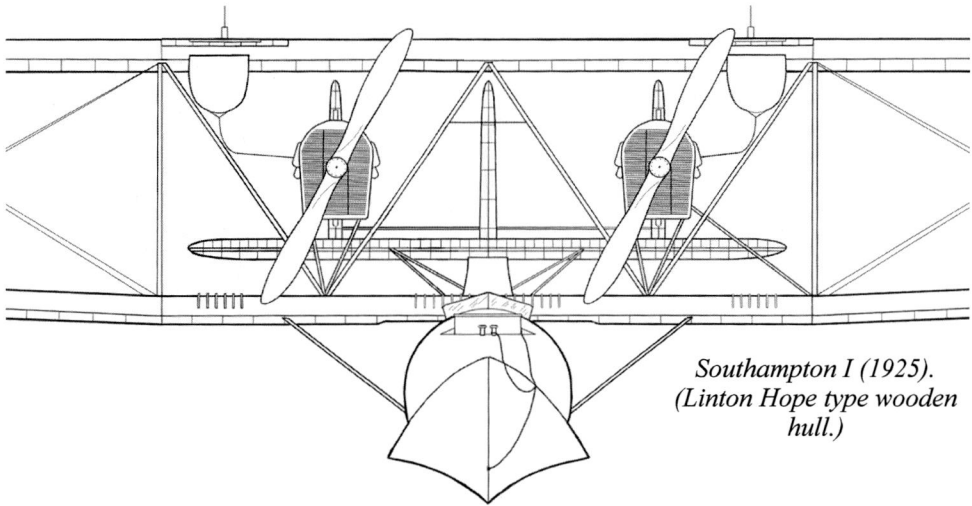

Southampton I (1925).
(Linton Hope type wooden
hull.)

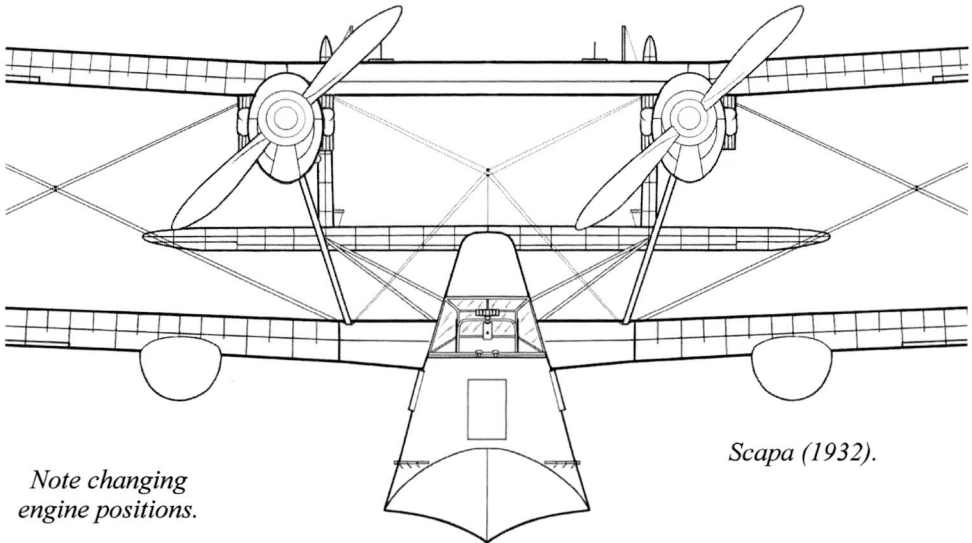

Note changing
engine positions.

Scapa (1932).

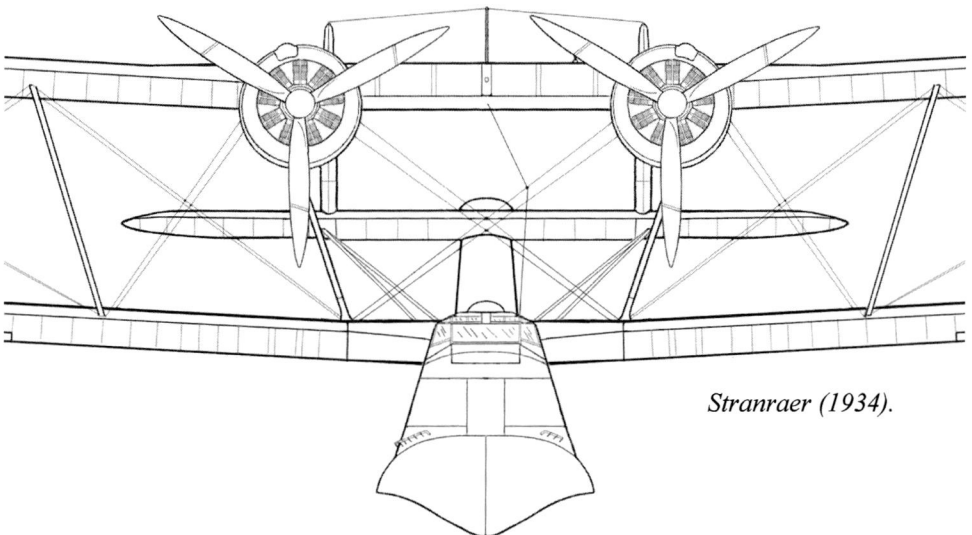

Stranraer (1934).

S.6B (1931).
(2,300 hp Rolls-
Royce 'R' engine.)

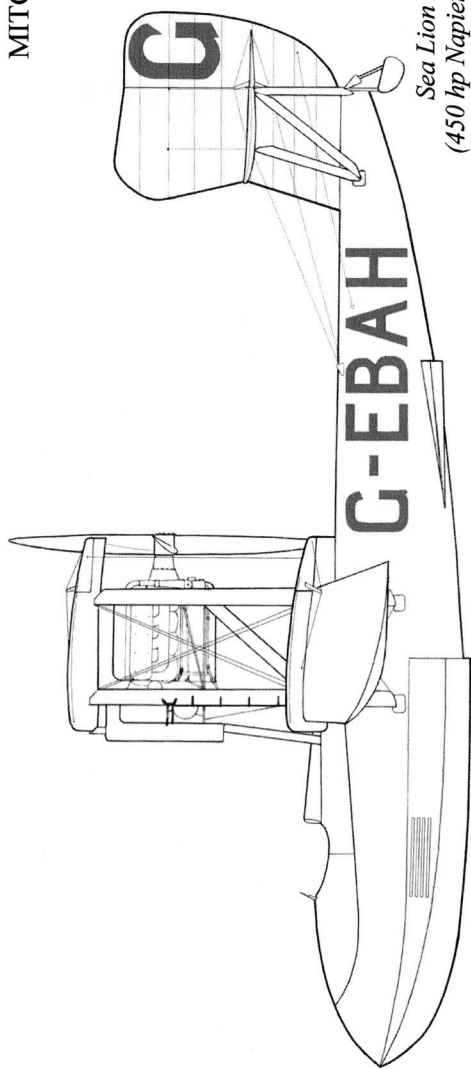

Sea Lion II (1922).
(450 hp Napier Lion engine.)

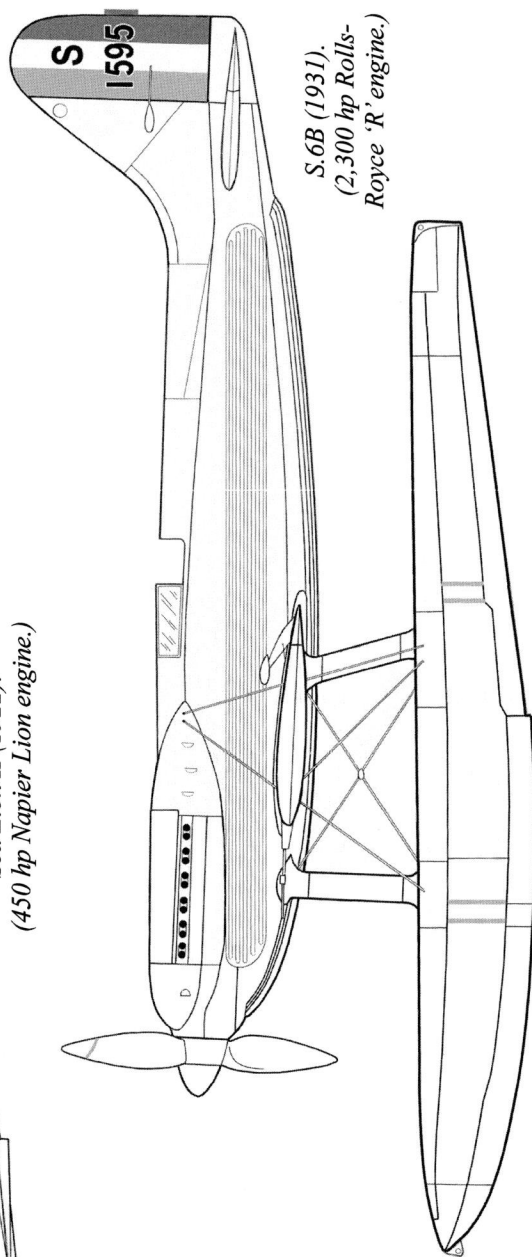

S
I595

G-EBAH

G

— *Appendix Three* —

Notes on R. J. Mitchell's wooden hulls

One of the fortuitous circumstances in R. J. Mitchell's career was joining a firm which had adopted a method of flying-boat hull construction of considerable potential while ever wooden hulls were in vogue. The Pemberton Billing N.1B hull set the precedent for the employment of sound boat-building techniques in the design of flying-boat hulls, followed by the advice of Linton Hope and Admiralty staff. (*See* p. 29.)

The standard flying boat of World War One was the Felistowe, in its various versions derived from American models, but, although it was successfully employed, it had structural and hydrodynamic deficiencies. The Air Ministry was concerned to see if the Linton-Hope type of hull could be adopted on aircraft of the large Felixstowe-Porte size. Thus, in 1917, Specifications N.3B and N.4 had been issued for this purpose. Capt. David Nicolson, in two articles for *Flight*, described these and strongly advocated the Linton Hope approach they embodied:

> The P.5 and N.4 types patented by Major Linton Hope are entirely different in design and construction ... Being of circular cross-section, with fair and easy lines, they offer much less air resistance, consequently with the same horsepower are driven at higher speeds; they are much stronger weight for weight than the F type, more seaworthy, and generally show the impress of the trained naval architect's hand.

However, these types could not be built immediately by specialist manufacturers of seaplanes as these companies were fully committed to the wartime production of standard service machines. Supermarine and Mitchell's good fortune was that the building of two smaller prototype Linton Hope machines had meantime been contracted out to the then Pemberton Billing firm and, although Hope did not entirely get his own way, as he indicated in another *Flight* article, the less than perfect machine embodied the alternative and much more promising approach to flying-boat hull design:

> these boats were very difficult to get off the water ... and with later experience it was obvious that the main step was too far aft and the rear step much too far forward. In spite of these faults in design, the AD boats showed the great strength of the flexible construction, and some bending and crushing tests carried out by the RAE works at Farnborough show what they were able to resist.

Structural strength was a feature of Mitchell's hulls and the eye of the yacht designer also imparted a relative sleekness to his hulls which were far less obvious in most contemporary land-based aircraft; but the immediate essential rightness of the Linton Hope approach was its flexibility, whereby a flying boat could absorb the shocks of sea landings and take-offs when there could be no recourse to the normal devices of aircraft undercarriages. Mitchell, as a young Chief Designer with a background in locomotive engineering and from the land-locked Potteries, was thus fortunate to inherit considerable sophisticated marine know-how and also a much more advantageous approach than that of the slab-sided Felixstowe flying boats of World War One, particularly because of their shock absorption. As C. G. Grey said of these hulls, 'they were almost basket-like in their flexibility, and so got through the water without that jarring shock which was common to most high-speed motor boats.' This approach and its method of construction had also been spoken of approvingly in a lecture to the Royal Aeronautical Society given by Capt. D. Nicholson, who had been involved with the alternative flat-sided hulls of the Felixstowe flying-boats:

> Construction is such that the structure is capable of resilient distortion, so that when alighting it can spring, reducing the shock. The hull cross-section is egg-shaped, very light, possesses great strength, and is built of longitudinal stringers with bent hoop timbers inside and light frames outside the stringers, skinned with double planking, through-fastened together. No web frames or cross-bracings are required, and the hull is a continuous structure with steps externally added.
>
> With a hull of the conventional [Felixstowe] F.5 type, such as the Cromarty, you start by criticising it as a commercial proposition, for you run into such items as turnbuckles, bolts and nuts, wires, cables, sheet metal, steel tubing ... You must employ not one trade but a number, such as boatbuilders, carpenters, sheet metal workers, fitters, machine hands, riggers – and are immediately in the midst of demarcation troubles in arranging the working squads. With a Linton Hope hull, you need only one class of labour – boatbuilders; a small number of men and boys can be placed on the job, and if pieceworked under supervision, the chances of holdups are small; there is no complication, and they carry straight through and finish their job. A standard Supermarine [Channel] four-seat hull, 31 ft long, takes 3 men and 2 boys on an average 5.5 weeks to build, working a 47 hour week.

Mitchell seized upon the basic principle of mounting necessarily rigid flying surfaces and engine mountings upon the Linton Hope flexible and robust hull and this approach bore fruit in the durable Sea Eagle, Seagull II/III, and Scarab fleets – and especially in the Southampton I which a, by then, desperate Air Ministry was relieved to order, unusually, straight from the drawing-board. (A small example of attention to the interface between rigidity and flexibility is mentioned in a *Flight* article about the Sea King: the pilot's 'controls are mounted on the triangular tubular frame so well known in all Supermarine boats, and whose function it is to allow the circular hull to flex and 'give' in a seaway, without interfering with the smooth working of the controls.')

Webb, during his apprenticeship in the hull building section, also noted how the brass nails and screws that held the final planking in place had to be fixed precisely in line and how the hull was then finally sanded down by hand and varnished until it had a surface 'akin to the best kept dining room table'. Cozens' description of the fate of the unsuccessful 1914 PB1 machine also indicates a similar concern with good workmanship and finish: 'A [Southampton] *Echo* 'Letterbox' contributor wrote to say that his father, who was working at Super-marine was told to fetch an axe and Pemberton Billing, after looking at the beautiful machine for a long while, broke it up'. He went on: 'All through the lifetime of the wooden flying boats the air of sturdy solidarity was due to the beautiful diagonal mahogany or Red Cedar planking of the hulls, covered by four coats of Copal varnish, which gave it a look of a piece of well polished furniture'. Unfortunately, photographs of the earlier aircraft do not show details of planking and there is evidence that the earlier machines, which were smaller and lighter than the Southampton, were built with 'the top side being of single-skin planking, covered with fabric treated with a tropical doping scheme.' As Webb joined

Supermarine in 1926, it would appear that he was speaking of work on the later Southampton flying boats and it is also possible that Cozen's account is influenced by the memories of the later machines' finish. But, at least, readers today are now able to see the magnificent reconstruction of a Southampton hull at the RAF Museum, and confirm for themselves the carefully aligned brass fixings and the 'luxury yacht' finish.

As mentioned in Chapter Two, the Linton Hope hull was double planked although the positioning of the outer layer at about 90 degrees to the inner, diagonal, one was usually superceded by the former being laid longitudinally. Between the two layers was a layer of varnished fabric to aid waterproofing. What is less certain is whether the final finish was varnished wood or a doped-on fabric, also varnished.

Whilst Cozens wrote of Supermarine aircraft having an 'air of sturdy solidarity ... due to the beautiful diagonal mahogany or Red Cedar planking of the hulls, covered by four coats of Copal varnish', Supermarine give a variant description of the construction with reference to their Sea King II: 'the hull is of circular construction with built-on steps, which can be replaced in case of damage. The steps are divided into watertight compartments, the top side being of *single-skin planking, covered with fabric* treated with a tropical doping scheme' Also a *Flight* description of the machine says that 'the boat hull is of the typical Supermarine type, boat-built and through fastened, with copper or brass fixings throughout. The mahogany *single-skin* planking is riveted to rock elm timbers and frames, and *covered externally with fabric* suitably treated with pigmented dope.' Further, the Sea Lion III is described as having 'mahogany *single-skin planking* ... covered externally with fabric suitably treated with pigmented dope.' One is reminded of Cozens' report that when the Sea Lion (Mark II – as it then was) was first started up, the vibration at the tail was so great that the pilot refused to fly it until the rear fuselage had been stiffened up – by wrapping and gluing canvas [extra?]around it. [My italics above.]

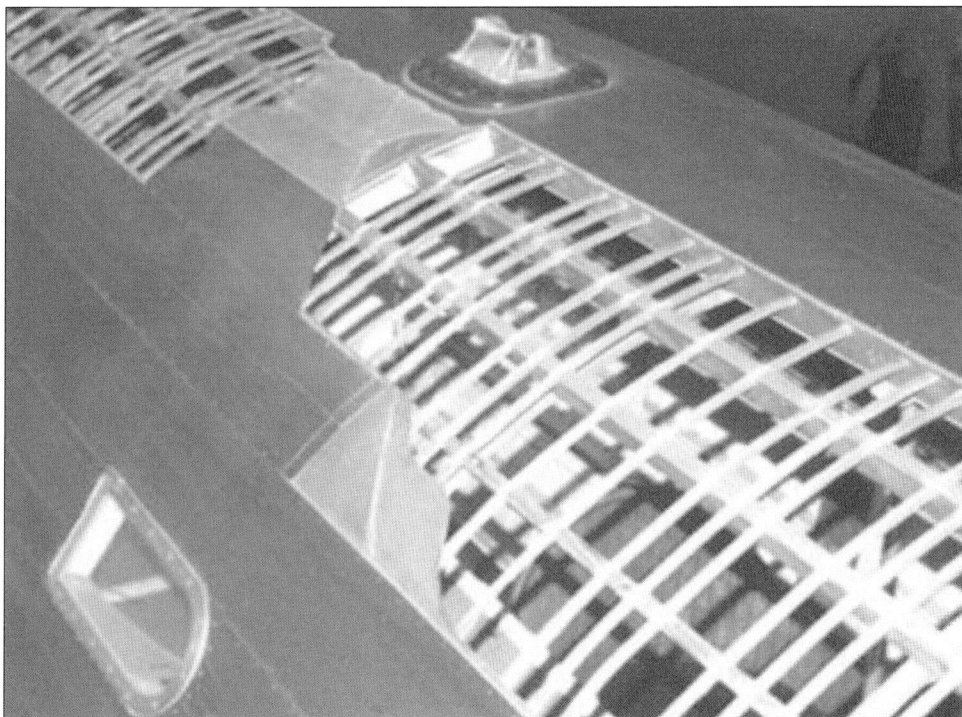

Detail of restored Southampton hull, showing the close-spaced hoops, the lengthwise stringers and the outer skin of thin mahogany strips. No canvas outer covering.

It should be noted that the Sea King and the Sea Lion shared a common ancestry and probably the same fuselage with the N.1B Baby – hence the similarity in the above extracts –and it would seem that Mitchell had thus inherited an aircraft significantly lightened in order to achieve the very sprightly performance it undoubtedly had. And specific centre of gravity considerations or design requirements might very well have led to Mitchell requiring variations in planking and finishing in other of the firm's aircraft at about this time – for example a Supermarine patent allows for planking to be omitted where external steps are to be added:

> If it is desired to reduce the weight of the hull to the greatest possible extent, the skin-planking on the hull proper may be omitted where side wings or other projections cover that portion of the hull. Where this planking is omitted it is preferred to use a fabric covering for the hull proper so that it is maintained watertight, even although the wing or other projection may be perforated. The close spacing of the bent timbers and stringers provide sufficient support for the fabric to be a satisfactory watertight skin in cases of emergency.

Thus, when we consider Mitchell's first complete designs, the Commercial Amphibian, Sea Eagle, Seagull II/III, Seal II, and Scarab/Sheldrake, which can be regarded as coming from a common stable, some or all may also have canvas exteriors – indeed, *Flight* describes the Seal II as 'boat-built of planking over a light skeleton of timbers and stringers, and covered in fabric *on the outside*. [My italics.]' The machines in this group are, however, all about 11 feet longer than the previously mentioned machines and would probably require the stiffening of double planking, especially bearing in mind the need here to design more staid, robust designs – with or without fabric doped on. There is, however, an intriguing report in *Flight* of the visit of HRH the Prince of Wales to Supermarine where it is said that 'the building of the Seagull flying boat hulls was greatly admired by His Royal Highness': perhaps the company saw an advantage in producing hulls that were, like the Southampton, fine examples of the boatbuilder's art, bearing in mind their British, Australian and Spanish naval customers. But it could be that the prince merely saw well finished planking awaiting a protective layer of fabric to be doped on.

Incidentally, it is worth considering that wherever there was double planking, with the usual layer of canvas in between, fabric might also have been applied externally to protect the woodwork from splitting in the sun as well as to prevent the soaking up of water – one thinks particularly of the 'tropicalised' Seagull IIIs for Australia. The commonly held view that Supermarine hulls were a mahogany colour might be because of a varnished wood finish or because the doped fabric would allow the colour of the timber to show through; on the other hand, the 'pigmented dope' finish might be none too opaque and mahogany in hue, given that Supermarine might very well have been strongly attached to reminders of their boating heritage.

— *Appendix Four* —

Jacques Schneider

Jacques Schneider was born near Paris on 25th January, 1879. He was the son of the owner of the Schneider armaments factory at Le Creusôt and trained as a mining engineer. But his sights were soon set well above the ground. Ballooning had been a French passion ever since the 1780s but there was an especial interest created in 1906 with the advent of the James Gordon Bennett balloon distance competitions. And so Schneider qualified as a free balloon pilot (licence No. 181) and, in 1913, he gained and held for a long time the French altitude record of 10,081 metres and also made a cross-country flight from France to the Black Sea in his balloon *Icare*. (Charles Rolls was also a balloonist, with 170 ascents from 1901 until his death when the tail of his Wright Flyer broke off.)

Hydroplaning was another sport for the rich that had been developing at the beginning of the century, notably at Monaco, led by the Marquis Charles de Lambert who made a breakthrough when he utilised an Antoinette aircraft internal combustion engine and an aircraft propeller. He began to reach speeds approaching 50 kilometres per hour and this also attracted the attention of Jacques Schneider who later drove one of these machines from Cairo to Khartoum with four passengers, including Lord Kitchener.

This event took place in 1914, four years after an accident in a hydroplane had caused multiple fractures of an arm and had prevented him from continuing with his third passion, flying. Schneider (who was at that time France's Under Minister for Air) had met Wilbur Wright in 1908 when on a visit to Le Mans to demonstrate his aircraft. Schneider had also become a close friend of Louis Blériot; and he joined the Aero Club of France in 1910 and was awarded his pilot's brevet (No. 409) in the March of the following year.

Because of his hydroplane crash, his name might have disappeared into the dusty annals of early aviation. However, his accident by no means diminished his passion for flight and, as race referee at a Monaco aviation meeting in 1912, he noticed that seaplane design was lagging far behind that of other types of aircraft. As aerodromes, obviously, did not exist, he foresaw that the best solution for long-range aerial passenger service between nations could be seaplanes – hybrid vehicles that had attributes of both hydroplanes and flying machines. Schneider thought that a seaplane competition would not only combine these two great loves of his but would also be the means by which such 'hydro-aeroplanes' might improve more quickly. By using his wealth to endow such a competition, he left his mark on the aeronautical world and his vision turned out to influence aircraft design for many years to come.

On November 5, 1912, at the banquet following the fourth Gordon Bennett Aviation Cup race for landplanes, he announced his trophy for a seaplane competition. The prize turned out to be a silver plated Art Nouveau classic designed by E. Garbard, a work of art costing 25,000 francs (about 67,000 euros in 2001). and proposed a course of at least 150 nautical miles. This competition was known under various names: Schneider Trophy, Schneider Cup, the Flying Flirt and the Hat Rack; in Italy it was the 'Coppa Schneider': the official French name was 'Coupe d'Aviation Maritime Jacques Schneider'. It required certain conditions to be fulfilled: (i) if a sponsoring aero club won three races in five years, it would retain the cup and the winning pilot would receive 75,000 francs. (ii) the annual race was to be hosted by the previous winning club and the races were to be supervised by the Fédération Aéronautique Internationale. (iii) each club would be permitted to enter up to three competitors with an equal number of substitutes available.

In 1921 the course was increased from 150 to 212 nautical miles, after a 2.5 nautical mile water navigation test. After 1921, an additional requirement was added: the competing aircraft had to remain moored to a buoy for six hours without crew aid.

There soon developed a keen public interest in this type of competition and crowds in excess of 250,000 began to gather to watch what were popularly known as the 'Schneider Cup races'. Whilst the commercial flying boat did eventually develop into a very effective and 'civilised' form of intercontinental travel, the Schneider Trophy is mainly remembered as a competition between nations for the prestige of merely flying faster than anyone else. However, as we have seen in previous chapters, the necessary concerns with streamlining and engine development in this endeavour had a far-reaching influence on the progress of aviation in general. Jacques Schneider did not live to see the full flowering of maritime aviation in the 1930s but, at least, he was able witness the modern streamlined competition monoplanes of 1925 and 1926. He was not able to attend the 1927 Trophy event in Venice, due to ill health after an appendicitis operation and, not long afterwards, on May 1st, 1928 he died in reduced circumstance, aged 49 – at Beaulieu-sur-Mer, near Nice, not far from the Monaco course where the first Schneider competition had been held.

— *Appendix Five* —
Kinkead's S.5 Crash.

The accident was reported in *Flight* on March 15, three days after the crash:

> The machine, for reasons yet unexplained, suddenly dived into the So-lent from an altitude of about 100 ft while Flight-Lt Kinkead was flying over the measured course, and immediately sank without leaving a trace of man or machine, other than just a few small pieces of wreckage.
>
> Flight-Lt Kinkead had been waiting at Calshot some ten days for an opportunity to beat the 297 mph set up last November by the Italian pilot, Maj. De Bernardi. Weather conditions, however, were by no means favourable most of the time, and only once did he get a chance to make a trial flight, which was entirely successful.
>
> On the morning of March 12 conditions seemed favourable for a flight, but an oil leak developed, and by the time this was set aright, the weather had broken up. All through the afternoon a snowstorm rendered a start hopeless. Then, shortly after 4 p.m., conditions changed entirely, and it was calm and sunny. Lt Kinkead therefore decided to make an attempt and the S.5 was brought out once more. After a preliminary run of the 875 hp Napier 'Lion' engine, the pilot packed himself away in the tiny cockpit and the seaplane was launched into the water.
>
> Besides the official timekeepers, a number of spectators were pres-ent to witness the attempt. Amongst these were several foreign Air Attachés, Capt. H. C. Biard, Mr R. J. Mitchell, designer of the machine, and many other prominent figures in aeronautics. With little if any wind to help him, Flight-Lt Kinkead experienced some difficulty in getting the S.5 off the water, but once in the air the machine flew well at terrific speed. After a short circle round, Flight-Lt Kinkead brought the ma-chine down and made an absolutely perfect landing, thus complying with the regulations which demand that the machine attempting the record must make two alightings to prove its seaworthiness – the first landing was made the previous day.
>
> Although, by this time, a mist was forming over the water, Lt Kinkead immediately took off – again with some difficulty – once more, this time obviously with the intention of making the attempt on the record. The S.5 soon disappeared from view in the direction of the Isle of Wight, and then after a short interval was observed as a tiny speck rapidly approaching Calshot, the scream of its engine getting louder and louder.
>
> When about two miles from the station, near Calshot Lightship [the start of the 3 km record course], all who were intently watching were horrified to see the machine plunge suddenly, nose first into the sea. A huge column of water rose into the air as the machine struck, and when this subsided, not a trace of the S.5 was visible. To add to the terrible effect of this disaster, the scream of the engine came across the water for what appeared to be an appreciable time after the machine disappeared from view.

The account of the death of Fl. Lt Samuel Kinkead in Chapter Six was necessarily brief but the recently published book, *Racing Ace: the Fights and Flights of 'Kink' Kinkead* by Julian Lewis, to which we are indebted, has brought to light certain eyewitness accounts which suggest that the cause of the fatal accident might very well have been structural failure – contrary to earlier accounts and to the findings of the internal RAF Court of Inquiry and the Southampton Coroner's

Inquest, both of which gave the cause of death as 'stalling'. I would now like to consider various options:

(i) The possibility of a high speed stall caused by violent manœuvres can surely be ruled out as there is nothing in the accounts of Kinkead's flight pattern immediately prior to his accident to suggest that this highly experienced pilot and instructor, with the press watching, as well as foreign dignitaries, senior RAF personnel, Biard the Supermarine test pilot, and R. J. Mitchell, and with a world record for the taking, would be doing anything other than professionally descending steadily to the required height of no more than 50 metres above the surface of the water, to be achieved 500 metres before the start of the 3 km record course, and doing nothing more than making minor corrections to his altitude or direction on the fatal approach run.

(ii) The possibility of a stall whilst having to make a landing approach because the visibility had deteriorated must take into account the fact that most of Kinkead's many decorations had been gained during World War One and afterwards, whilst often flying very low. Dr Lewis' book provides an invaluable service in describing these exploits and so one must pay due regard to Kinkead's airmanship in this respect even though he was now flying with the very reduced view from the cockpit that was an especial feature of the current batch of Schneider Trophy machines. In particular, this necessitated the ability to feel one's way down to the water surface virtually blind – described by another Schneider pilot, Fl. Lt H. M. Schofield, as 'a game of great patience'. Nothing Schofield wrote suggested that Kinkead would have adopted a different landing approach, particularly in misty conditions, and this contention is borne out by the *Times*' description of his skillful landing of the S.5 on the day before the record attempt: 'he chose an angle of glide *which almost imperceptibly brought the floats nearer and nearer the water until the monoplane was just skimming the surface* [my italics]– a grey insect over a grey sea. Then a tiny feather of white foam broke behind the floats and rapidly grew bigger. The machine seemed about to settle, but lifted just once in a long hop and then, touching again, threw up a spurt of spray, and gently floated on the water'.

(iii) Even if Kinkead had made a rare landing misjudgement, there seems to have been no mention of a gradual decrease in engine sound before his crash, consistent with slowing down from high speed (expected to be over 300 mph) to the necessary landing approach speed (about 100 mph); on the contrary, spectators were impressed by the sound of an approaching high-revving engine which suddenly ceased and two commentators even described his impact with the water like that of a shell. Harold Perrin, the Secretary of the Royal Aero Club, who had had a good view from the side, reported that the reappearance of fog, coupled with glare, prevented Kinkead from judging his height and that he flew straight in.

(iv) This last explanation of the crash would seem more plausible except for the fact that most witnesses reported an unexpected change of attitude; indeed, the coroner at the inquest is reported as 'trying to find out if there is a possible explanation of the sudden dive of the seaplane' and a Napier engineer's testimony was that 'the machine soon appeared in sight, travelling towards me from Cowes at about 250 feet above the water, and going very fast … The machine appeared to be intentionally descending and, after a few seconds, nose dived very suddenly into the sea from about 50 feet up'. This account is echoed in the *Morning Post* headline: 'Vertical Nosedive from 100 feet' or the *Times* report that Kinkead 'dived straight into the Solent from a height of between 100 feet and 50 feet' and that 'no man could have survived such a terrific impact with the water' as the machine had 'dived rather than flown into the sea'. Dr Lewis also records similar accounts of a sudden dive at high speed in the *Daily Sketch*, the *Daily Express* and in the local paper, the *Southern Daily Echo*.

(iv) It is, of course, well known how unreliable and interpretative are eye-witness accounts of unexpected events but, given the similarity of these particular reports and numerous testimonials to Kinkead's skill as a pilot, a mechanical failure rather than pilot error must thus be given particular attention – and it would foreshadow the alarming and near fatal oscillation of the S.6B rudder, three years later, which resulted in stretched control wires and had required rear fuselage strengthening and anti-flutter balance weights on control surfaces. Lewis cites newspaper reports from other observers with a view from the side of 'abnormal movement of the tail unit and fin' (the *Times*) and of the aircraft 'going at a very good speed' when its tail developed 'a pronounced flutter' (the *Daily Express*). Biard, who was standing next to Mitchell at the time of the accident, is also quoted as being quite certain that the crash was a result of structural failure and his likely reaction at the time might have contributed to Mitchell's reported distress for some time afterwards.

It should be acknowledged that the other two similar S.5 machines had successfully completed the Venice Schneider Trophy course, which involved the taking of two extremely sharp corners on each of the required seven laps whilst returning an average course speed of about 88% of the top speed available to them – and there is no record of any Supermarine 'never exceed' stipulations at that time. Kinkead's speed record S.5, N221, had not been extensively tested, having been held in reserve and consequently unused in the Schneider Contest a few months earlier, but the preliminary flight on the Sunday before the fatal crash had been without incident. The second mandatory flight, flown on the day of the accident, also went off without any problem (although heavy spray on the tail unit during porpoising prior to take-off was afterwards considered to have perhaps weakened the machine's structure) but its specially tuned engine and light fuel load might have allowed Kinkead to achieve a hitherto untested speed that brought with it unexpected structural problems.

(v) On the other hand, some reports of a sudden engine roar at the very end must be given due consideration. No evidence of a decoupling of engine and propeller was given at the inquest and so there thus remains the possibility that Kinkead had discovered a problem and had begun to try to get down, that he had throttled back but almost immediately poured on the power in an attempt to correct some worsening situation. As the speed of sound is slower than the speed of light, these last sounds would not be synchronous with what could be (imperfectly) seen by the observers in the misty conditions and so this aspect remains problematic.

(vi) But most problematic, in view of the various considerations above, are the findings of the RAF Court of Inquiry that the accident was 'due to stalling of the machine' and the verdict of the Southampton Coroner's inquest that Kinkead's fatal injuries were caused by diving into the sea 'owing to lack of speed while attempting to alight'. In the latter case, it was also maintained that no evidence had been found in the wreckage to point to any physical causes for the accident although how this could be upheld in view of the extensive damage caused by impact and by the subsequent salvaging operation might perhaps only be explained by a wish to draw a discreet veil over the whole matter. (One also reads in Lewis that a file concerning the private RAF Court of Inquiry, referred to in Kinkead's Service Record, has not survived.) Hugh Trenchard, Chief of the Air Staff, had been a reluctant participant in the recent Schneider Trophy preparations as he was still very protective of his new Air Force; he saw involvement in the competitions as peripheral at best to the main functions of his command, as these events employed extremely specialised and minimalist machines flying dangerously low at otherwise unattainable speeds. From another perspective, certain senior Air Ministry officials were unlikely to be happy with findings of mechanical failure as, in their case, they were in favour of continuing RAF involvement in the Schneider contests for the sake of research and development and of the prestige accruing to the British aviation industry via this blue riband event.

It is thus interesting that, at the inquest, the Inspector of Accidents, appointed by the Secretary of State for Air, felt able to assert that 'no part of the aircraft structure or controls broke, or failed to function normally during the flight' and that, whilst agreeing that 'the rudder or tail of the machine was seen to be fluttering', said that it was 'probably the reflection of the sunlight from the rudder that gave the impression' when the machine turned. Although we have only Southampton's *Southern Daily Echo* reports of the inquest to respond to (inquest transcripts are routinely destroyed after a period of time), this is the only mention here or in Lewis of the machine turning. Had a turn been necessary, it would surely have been made well before the crash site, in order to maximise the entry speed into the actual speed course, and one would in any case expect only slight movements of the rudder at the speed that Kinkead was going – producing equally slight, progressive, changes of colour tone to the rudder (but not with reference to the tail as a whole – as reported above by witnesses). And one wonders how bright was the sunlight if Kinkead had actually been attempting to land because of the worsening visibility.

Whilst a stall due to 'pilot error' remains a possible reason for the death of Kinkead, there is, to say the least, no overwhelmingly supporting evidence and, indeed, R. W. Owen, a fitter who also looked after Kinkead's car, is reported as recording in his diary his incredulity at the inquest finding. He had also noted earlier that the aircraft had 'crashed from 50 ft full out ab[out] 340 mph (?) flying towards the sun'. It would seem that he might have been disposed to the Perrin view (iii above) but his queried note on Kinkead's possible speed is especially interesting. This mention of a speed well above what had been possible in the Venice S.5s is borne out in more detail in an RAF press release:

> Kinkead had preserved complete reticence as to the exact speed at which
> he travelled in his trial flight on Sunday morning [March 11], and no one
> had realised that he had reached, on his air speed indicator, a rate of no
> less than 330 miles per hour ... Kinkead was so enthusiastic after his
> trial flight that he said he believed he could attain probably 350 miles
> an hour. It should be realised that this type of monoplane had never
> before been flown, probably, at more than 300 miles an hour, and ...
> no one could be certain that stresses, which were within the capabil-

ity of aircraft engine and propeller at 300 miles an hour, might not rise to an unexpected magnitude when the speed was increased to 350 miles an hour ... Kinkead had been urged to content himself with beating the existing record by the requisite 5 miles an hour, or at least to keep within a speed of 310 miles, but he, as a pilot, took a risk which designers and aircraft instructors would have rightly hesitated to permit in existing machines.

It is worth noting that this report only says that 'designers [unspecified] would have hesitated [not "had hesitated"] to permit' a flat out speed. The *New York Times* of March 14 printed the following version:

> The theory advanced as to the accident now is that in endeavouring to reach the higher speed the machine, despite all human skill and foresight could do to make it function perfectly, broke down under the strain.
>
> It was stated tonight that Lieutenant Kinkead had been urged to content himself with beating existing records and to keep within a speed of 310 miles per hour, as the designers of the machine thought it risky to exceed that limit.

This last version now asserts that advice to fly circumspectly came from 'the designers of the machine'. Mitchell (unnamed) was, of course, the Chief Designer of the S.5 and one wonders if Kinkead had mentioned to him anything unusual during his earlier test flight – as Biard had done, prior to the S.4 going to the 1925 Baltimore competition. If there had been a report of something not quite right, Mitchell would certainly have wished to avoid a repeat accident – given his well known concern for the safety of his pilots.

Perhaps one is here straying too far into the realms of speculation but it would not have been out of character for Kinkead to have taken upon himself the risk, in order to set a speed record that might not be broken by another country's machine for some considerable time. It is thus a great pity that it is not possible to assert unequivocally that Kinkead's death was the direct result of a desire to achieve the very best speed for the RAF and the Empire (Kinkead was born in South Africa).

As it stands, the findings of the RAF and of the inquest can be seen as the least worst outcome for all parties except the pilot. This is an especial pity as little mention is made of the Schneider Trophy pilots when credits are given for winning the Battle of Britain and all that followed. If these pilots had not successfully flown their, frankly, dangerous aircraft, the Spitfire might not have been ready in time for the outbreak of World War Two – one notes at least D'Arcy Greig's dedication in *My Golden Flying Years* 'to all those involved with the Schneider Trophy races that helped so much in the development of the Spitfire in later years'.

* * * * *

It is surely worth a moment to consider, via Kinkead and his fellow pilots, what was involved in Schneider Trophy flying. In 1927, the fastest RAF fighter they might have flown was the 155 mph Gloster Gamecock; and the Fairey Flycatcher floatplane made available for practice by the High Speed Flight had a top speed of about 125 mph. The pilots then moved up to the previous Schneider Trophy contender, the Gloster III, 100 mph faster and unstable when cornering. One has to bear in mind that many practice days were lost because of weather or sea conditions unsuitable for these sensitive machines and it was only a few months (in effect therefore, weeks) before the High Speed Flight received their more stable but much faster Gloster IV and Supermarine S.5 competition floatplanes – a total increase in top speed of about 170 mph since converting onto the Flycatcher floatplane. In the ten years between the advent of the Game-cock and of the Gladiator, the top speeds of RAF fighters had risen by an average of about 10 mph per year and these landplanes were far more straightforward to fly than the Schneider floatplanes, which drenched and blinded their pilots on taking off, apart from their overall visibility problems in the air. Yet James (*Gloster Aircraft since 1917*) records that the Gloster IV machine (introduced to Schofield as 'the blind wonder') which Kinkead flew in 1927 was shipped to Venice after only fifty minutes of flying and that he flew it there, very low, after only 105 minutes on the type recorded in his log book. And D'Arcy Greig states that suitable conditions resulted in his own total flying time in these specialist aircraft, between 25 July 1928 and 7 September 1929, as only eleven hours, twenty-three minutes. It should also be added that, whilst parachutes were now becoming standard military equipment for the RAF, Kinkead and his fellow pilots would have had no chance of using them at the height they flew; nor would extricating themselves from their cramped cockpits have been easy whatever the circumstances. Blacking out from G forces was also a new phenomenon to contend with.

It is well recognised that Mitchell's Spitfire and its Rolls-Royce Merlin engine were ultimately a by-product of the Schneider Trophy competitions, where manufacturers were free of inhibiting Air Ministry requirements, but not enough has been said in this context about the skills of the High Speed Flight pilots who took part in this early test flying. No doubt others would have come along and accepted the risks involved but it was these pilots who led the way with their successive wins in 1927, 1929 and 1931. Of these, Kinkead's 1927 group can be seen as the one which took the primary and most dramatic leap into the unknown. Their Flight Leader, Sqd. Ldr L. H. Slatter spoke after the event of the skill of all his pilots but he singled out Kinkead for putting up the 'most extraordinary show'. Thereafter Kinkead was put in command of the nucleus of the High Speed Flight retained at Felixstowe and, had he not died, he would have been in charge of the full High Speed Flight for the next competition.

It is hoped that he will be remembered in this light rather than as a pilot who was said to have died because of an error of judgement.

— *Appendix Six* —

Lady Lucy Houston, DBE

The life story of Lady Houston presents a particular insight into the bygone late Victorian and Edwardian ages when wealth and privilege was confined to a far smaller few, when the gap between them and the poor was much more clearly defined, and when the British Empire was at its greatest extent and influence. She came from the underprivileged class but rose to become the richest woman in Great Britain and a champion, however extreme and eccentric, of the status quo and, especially, of Britain and the British Empire; Winston Churchill described her as a 'modern Boadicea' and the historian A. J. P. Taylor wrote that 'the Battle of Britain was won by Chamberlain, or perhaps Lady Houston'.

Born in 1857, Fanny Lucy Radmall, claimed to be the seventh child of a seventh child (and always claimed to be much younger than she really was). Her father, Thomas Radmall, was a woolen draper who had moved to live over his warehouse near St Paul's Churchyard and she therefore considered herself to be pure Cockney as well as the descendant of Sussex yeoman stock – 'before William the Conqueror came over to mess the place up'.

At the age of sixteen she was described by the playwright Arthur Wing Pinero as 'a small-part actress' and, in those days of 'mashers' at every stage door, was almost immediately taken up by the 34-year-old Frederick Gretton who, in 1867, had become a partner in the well-known brewery, Bass, Ratcliff & Gretton in Burton on Trent. Over the next five years, his main involvement with the firm was in the malting department but then he met Fanny and he eloped with the 16-year-old to Paris where they lived together as man and wife. At the atelier of the artist Edouard Détaille she met Madame de Polés, whose friends included archdukes and princes as well as the future King Edward VII – which led to her becoming a royalist of the most romantic sort as well as learning the ways of the world, society manners, and the art of dressing. She also seemed to have copied Madame de Polés' habit of carrying a large, shabby handbag into which she randomly stuffed often large sums of money, expensive jewellery, and important documents.

After some seven years, Fred Gretton died, leaving her the then considerable sum of £7,000 per year for life. She took a house in Portland Place, and surrounded herself with all the necessary servants: butler, coachman, lady's maid, footmen and maids. Half a century later, she told a friend: 'Today is a very sad day to me. Someone I loved dearly died on this day 50 years ago. He worshipped me and said I was the apple of his eye. Now wasn't that a sweet thing to say?'

In about 1882, she met the eldest son of Sir Theodore Brinckman who was twenty-one when they married in the following year – believing Lucy to be younger than her twenty-six years. They lived the social life typical of a son of the aristocracy, but she divorced Theodore in 1895 on the grounds of his adultery. Then, in 1901, she married the ninth Lord Byron of Rochdale and went to live in an old Georgian house in Hampstead, which she named Byron Cottage, and was now able to wear robes, ermine and a tiara at the resplendent coronation of George V. Byron Cottage now became a base for her developing interest in politics which began with her championing the cause of Votes for Women and which, on one occasion, involved an open letter to Christobel Pankhurst, with the proposal to buy 615 parrots (in red, white and blue cages) and to try to teach them to screech 'Votes for Women' in unison. Not surprisingly, this was a failure but at least she was more practical in her helping out of Emily Pankhurst with what was believed to be thousands of pounds when the famous suffragette ran into severe financial trouble.

She was also effective in her contribution to the 1914–18 war effort by setting up a much needed rest home for overworked nurses and, as a result, became the fifth lady to be created a 'Dame of the British Empire'. Among later donations was £30,000 to the Miners' Relief Fund, £30,000 to the Lord Mayor's Distress Fund, £10,000 to the Navy League Fighting Fund, £40,000 to King George's Jubilee Fund, £10,000 to the Liverpool Cathedral Fund, and £10,000 to the Maternity and Child Welfare Fund. Much of her charitable work, however, was completely unsung, such as seeking out and giving money to tramps on Hampstead Heath.

The £100,000 given to subsidise the final British Schneider Trophy entry is, of course, the reason for her being the subject of this Appendix and there was also a similar sum donated later to enable the first flight over Everest (*see* below). Such vast sums (for the 1930s) were possible, thanks to her subsequent marriage to Lord Houston. Lord Byron had died in 1917, after their marriage of 17 years, and in 1922, on the yacht of Sir Thomas Lipton (the 'boating grocer'), she met Sir Robert Paterson Houston, the owner of the Houston line of steamships. He also owned a yacht, the 1600 ton steam yacht, *Liberty*, on which she met members of the Russian royal family and learned at first hand of the brutalities of the Bolsheviks, which further increased her developing dread of the spread of their influence in world affairs. They were married in December 1924 and so she thus became Lady Houston although not before she had refused to choose from a selection of jewellery as a birthday gift: on being asked what she considered was appropriate, she indicated a string of black pearls but informed her future husband she thought he would consider them too expensive at £50,000 – they arrived the next day. Later, on being shown his will, which left her a million pounds, she tore it up, declaring that if that was all she was worth, she wanted nothing. The will was remade and, when Lord Houston died in 1926, he left her money and property amounting to six million pounds, including the 304 ft yacht. A very bad attack of jaundice and her extreme reaction to Lord Houston's death led to a curator being appointed to administer her vast estate in Jersey. She sent for seven of the most eminent psychiatrists and doctors from Great Britain and France who declared her perfectly capable of conducting her own affairs.

Whilst Lord Houston had been domiciled in Jersey and his estate was not subject to British taxation, his widow was contacted by the Treasury when she took up residence again at Byron Cottage. Having ascertained that it might very well cost her £20,000 in legal fees to settle, she went herself to see Winston Churchill, who was then Chancellor of the Exchequer, to negotiate. At a second visit, she presented Churchill with a cheque for £1.5 million by way of a final settlement. Most of the time, however, she spent in her bedroom, suffering from very low blood pressure and the house took on a distinct air of neglect; a contributory factor was the high turnover of servants, due to her demanding nature and irascibility.

In her 70s, she spent much of her time on *Liberty*, cruising off the coast of the Riviera, at anchor in the harbour at Cannes or on the Seine, between Paris and the coast. The main deck of Liberty was off-limits to her crew at certain hours as she took daily exercise, whatever the weather, with no clothes on. More publicly, her tirades against Russia and her championship of the British Empire had by now become extreme, for she saw the Labour governments of Ramsay MacDonald in 1924 and 1929 as threatening to erode the domination of the British Empire and the social order, in which she had led such a charmed life. She published *Potted Biographies, a Dictionary of anti-National Biography*, in which she gave chapter and verse of alleged anti-war activities or speeches by Labour MPs which appeared to belittle Britain, and she intervened in numerous by-elections with poster campaigns which she financed herself. And at one time she anchored off various ports on the south coast of England with *Liberty* decked out with six foot lettering, lit up at night, as *Time* reported: 'this week irrepressible Dame ("Fanny") Lucy was at it again on her yacht, the *Liberty* once owned by not-quite-so-rich and eccentric Joseph Pulitzer. From the *Liberty*, on which Lady Houston lives with steam constantly up, blazes at her whim an electric sign DOWN WITH MACDONALD, THE TRAITOR'. Unfortunately, from her point of view, the top-hamper of her yacht prevented her reaching moorings off the Houses of Parliament with her message.

Also in 1933, she had bought as her mouthpiece the *Saturday Review* outside whose offices she was wont to draw up and dictate copy to her long-suffering manager; *Time* gives an instance of his problems:

> cringingly he told her that the leading wholesale newsdealers of Great Britain, on advice of their solicitors, had refused to distribute the next copy of the *Saturday Review* if it should contain, as planned, Lady Houston's personally penned opinion of the Prime Minister ... that he was 'squandering millions on peace conferences' while he let the Empire's defence forces go to ruin. This was only to be expected, she slashed, from a man who, like Scot MacDonald, urged British munitions workers to strike during the War at a time when British soldiers at the front were short of shells. 'How can you be secure?' Dame Lucy planned to query the readers of the *Saturday Review*. 'How can you be sure your dear ones will not be sacrificed through the treachery of this traitor?'

The vehemence of her published views thus made good copy in the press and it was by no means ill-founded as Baldwin's third premiership now attracts most blame for British unpreparedness for war in 1939 – as we have seen reflected in the tribulations of the aircraft industry. Notably, eight years earlier, Lady Houston had sent her secretary-companion to the private residence of the Chancellor of the Exchequer, at the time Neville Chamberlain, with a £200,000 cheque to finance a squadron of fighters for the defence of London against Germany. It then became front page news when the government refused the offer on the grounds that it could not accept money when accompanied with conditions as to how it would be spent. (It may be that the Prime Minister, smarting from her many attacks upon himself, did not want to provide reasons why she might be given any higher honours than her DBE.)

It is thus possible to see how Lady Houston's funding of the 1931 Schneider Trophy entry was in large measure influenced by a wish to embarrass Ramsay

MacDonald and his socialist party which had refused to put government money into the project. And her opposition to their moves to give India a measure of self-government were also behind her financing (again, to the tune of £100,000) the successful British attempt to fly over Everest for the first time in 1933; but, whilst it can be seen as an imperialist gesture in support of the British Raj, Lady Houston was also concerned with the wider view that any withdrawal from India would leave a vacuum that Russia would seek to fill.

At the Everest Flight celebratory luncheon, the extreme jingoism of her 'message' from her yacht was not read out, despite her financial backing of the project. The following extract, in characteristic 'syntax', explains why she had supported the flight over Everest: 'some great deed of heroism might rouse India and make them remember that though they are a different Race – they are British Subjects – under the King of England – who is Emperor of India – and what more can they want? ... this is surely a proof to them that pluck and courage are not dead in our Race and perhaps – who can tell? – this may make them remember all the advantages and privileges they have enjoyed under English Rule ...' (A less sophisticated expression of the Air Ministry's purpose in 'showing the flag' with the Supermarine Far East Flight of 1927/8.) A recompense for being snubbed at the Everest Flight luncheon came when a lake that had been discovered on the southern slopes of Mount Everest in the course of the flight was named Parvati Lal, the Lady of the Mountain. And a film of the flight was released in the following year which, after some altercations, reinstated an introduction by Lady Houston.

But despite, or because of, her many stunts, despite her munificent gestures, her pamphlets and her articles in the *Saturday Review*, her influence with the political establishment was not commensurate with her wealth or her acquaintance with the great and the good; whilst many at the time shared her admiration for such 'strong men' as Hitler and Mussolini, they shrank from being associated with her unrestrained attacks upon the British government.

A final example of her inability to influence events was her public concerns over the burning issue of the abdication in the late 1930s. One might have expected that Lady Houston would have been violently opposed to the idea of an American becoming the queen of England; however, she had met Edward VII in her Paris days and his grandson had even visited Byron Cottage; so her personal feelings for him took pride of place, especially as Baldwin (who had succeeded Ramsay Macdonald but who was no better in Lady Houston's eyes) was opposed to the King's proposed marriage. She exchanged letters with the King and with Queen Mary but the abdication took place on December 10th, 1936, nevertheless.

Now eighty years old, Lady Houston printed in her *Review* a letter she had sent to his successor, King George VI, which repeated some of her views about the present government but expressed an expectation [forlorn hope?] that he would maintain the scepticism of their policies that his departed brother had sometimes appeared to voice. There had thus been much to convince her that her fears for the demise of the world she believed in were well-founded and that there was little to show that she had been able to halt the work of those in power whom she considered, with some justification, to be, at best, inept. Food now had no interest for her and she died on the night of December the 29th, 1936.

[Full details of her eccentricities, her wide-ranging and generous gifts, as well as her perspicacity concerning the way that world events were tending, can now be found in a well-researched book, just published by Miles Macnair – *see* Bibliography.]

Mitchell being introduced to Lady Houston at Calshot on the occasion of the 1931 Schneider Trophy Contest.

* * * * *

Bibliography

Authors quoted from, or referred to, in the main text are documented below. The Bibliography also provides details of material for wider reading – grouped according to specific areas of interest.

1/. Supermarine Aviation Works
The Aeroplane / The Aeroplane Monthly.
Andrews, C. F. and Morgan, E. B., *Supermarine Aircraft since 1914* (Putnam, 1981).
Cozens, G. A., 'Concerning the Aircraft Industry in South Hampshire', Unpublished MS, Solent Sky Museum.
Flight / Flight International.
Griffiths, Harry, *Testing Times: Memoirs of a Spitfire Boffin* (United Writers Publications Ltd, Cornwall, 1992).
Honeysett, Jon, 'Death in the Afternoon: The Bombing of Vickers Supermarine Works, September 15, 24 and 26th', Unpublished MS, Solent Sky Museum.
Jane's All the World's Aircraft (Sampson Low, 1921–1937).
King, H. F., 'Sires of the Swift: A Forty Year Record of Supermarine Achievement', in Flight (October, 2006).
Pegram, Ralph, *Beyond the Spitfire: the Unseen Designs of R. J. Mitchell* (The History Press, 2016).
Rance, Adrian, *Fast Boats and Flying Boats: a Biography of Hubert Scott-Paine* (Ensign Publications, 2004).
Roussell, Mike, *Spitfire's Forgotten Designer: the Career of Supermarine's Joe Smith* (The History Press, 2013).
Russell, C. R., *Spitfire Odyssey* (Kingfisher Railway Productions, 1985).
Scott, J. D., *Vickers: a History* (Weidenfeld and Nicholson, 1962).
Shelton, John K., *From Nighthawk to Spitfire: the Aircraft of R. J.Mitchell* (The History Press, 2015).
——*Schneider Trophy to Spitfire: the Design Career of R. J.Mitchell* (First Edn, Haynes, 2008).
Smithies, E., *Aces, Erks and Backroom Boys* (Orion Books, 1990).
Stoney, Barbara, *Twentieth Century Maverick: the Life of Noel Pemberton Billing* (Bank House Books, 1989).
Webb, Denis Le P., *Never a Dull Moment at Supermarine: a personal history* (J. and K. H. Publishing, 2001).

2/. The Schneider Trophy
Air Ministry Aeronautical Research Committee, 'Reports on the Schneider Trophy Contests for the years 1927 and 1931', HMSO.
Banks, F. R., 'Memories of the Last Schneider Trophy Contests', in Journal of the Royal Aeronautical Society, January, 1966.
Barker, R., *The Schneider Trophy Races* (Chatto and Windus, 1971).
Bazzocchi, Dr E., 'Technical Aspects of the Schneider Trophy and the World Speed Record for Seaplanes', in Journal of the Royal Aeronautical Society (February, 1972).
Buchanan, J. S., 'The Schneider Cup Race, 1925', in Proceedings of the Tenth Meeting, 61st Session of the Royal Aeronautical Society.
Design Staff of Supermarine under R. J. Mitchell, 'The Schneider Trophy Seaplane – Some Notes on the Special Features of the S.6B', in Aircraft Engineering, October, 1932.
Eves, E., *The Schneider Trophy Story* (Airlife Publishing, 2001).
Gouge, M., 'Doolittle wins in Baltimore', in Airpower (November, 2005).
Hawkes, E., *British Seaplanes Triumph in the International Schneider Trophy Contest, 1913–1932* (Real Photographs, 1945).

Hirsch, R., *Schneider Trophy Racers* (Motorbooks International, 1993).

James, D. N., *Schneider Trophy Aircraft, 1913–1931* (Putnam, 1981).

Mitchell, R. J., 'Racing Seaplanes and their Influence on Design', in Aeronautical Engineering Supplement to The Aeroplane, 25th December, 1929.

——'Schneider Trophy Machine Design, 1927', in Proceedings of the Third Meeting, 63rd Session of the Royal Aeronautical Society.

Mondey, D., *The Schneider Trophy* (R. Hale, 1975).

Orlebar, Wing Commander A. H., *The Schneider Trophy* (A. F. C. Seeley Service and Co., 1933).

Pegram, Ralph, *The Schneider Trophy Seaplanes and Flying Boats: Victors, Vanquished and Visions* (Fonthill Media, 2012).

Shelton, John K., *Schneider Trophy to Spitfire: the Design Career of R. J.Mitchell* (First Edn, Haynes, 2008).

The Supermarine S.4–S.6B, Profile Publications, No. 39.

Waghorn, Fl. Lt R. D. H., 'The Schneider Trophy, 1929', Royal Aeronautical Society, May, 1930, Yeovil Branch.

3/. The Spitfire and Other Supermarine Aircraft

Alcorn, J., 'Battle of Britain Top Guns', in *The Aeroplane Monthly* (September 1996).

——'Battle of Britain Top Guns Update', in *The Aeroplane Monthly* (July 2000).

Caygill, P., *The Darlington Spitfire* (Airlife, 1999). See also Roger Darlington World [blog] <http://www.rogerdarlington.me.uk/Spitfire.html> accessed 30 Oct. 2016.

McKinstry, L., *Spitfire: Portrait of a Legend* (Murray, 2007).

Morgan, E. B. and Shacklady, E., *Spitfire: the History* (Kay Publishing, 1987).

Neil, Wing Commander Tom, DFC & bar, AFC, *The Silver Spitfire* (Weidenfeld and Nicolson, 2013).

Price, A., *The Spitfire Story* (Jane's, 1982).

——*Supermarine Spitfire* (Chevron Publishing Ltd, 2010).

Rendall, Ivan, *Spitfire: Icon of a Nation* (W&N, 2009).

Robertson, B., *Spitfire: the Story of a Famous Fighter* (Harleyford, 1960).

Royal Aeronautical Society, Southampton Branch, 'Forty Years of the Spitfire', Proceedings of the R. J. Mitchell Memorial Symposium (6th March, 1976).

Sarkar, D., *How the Spitfire Won the Battle of Britain* (Amberely, 2010).

'Spitfire 70', a *FlyPast Special* (Key Publishing, 2006).

The Supermarine Spitfire I and II, Profile Publications, No. 41.

Doyle, N., *From Sea-Eagle to Flamingo* (The Self Publishing Association Ltd, Upton-upon-Severn, 1991).

Nicholl, G. W. R., *The Supermarine Walrus: the Story of a Unique Aircraft* (G. T. Foulis and Co., 1966).

The Supermarine Walrus and Seagull Variants, Profile Publications, No. 224.

Twentyfirst Profile, Vol.1 No.6 'the B.12/36 Bomber' and Vol.1 No.9 'the Air Yacht' (21st Profile Ltd, ISSN 0961-8120).

4/. R. J. Mitchell

Black, A., 'R. J. Mitchell: Designer of Aircraft', in The Reginald Mitchell County Primary School Commemorative Brochure, 1959.

Mitchell, Dr Gordon, 'R. J. Mitchell: My Father', in *Aeroplane Monthly* (March 1986).

——*Reginald Mitchell, 1895–1937* (RAF Souvenir Book, 1966).

——[editor and in part author of] *R. J. Mitchell: Schooldays to Spitfire* (Nelson and Saunders, 1986; reprinted by Tempus Publishing, 2006).

Smith, Joe, 'R. J. Mitchell, Aircraft Designer', in *The Aeroplane*, 29th January, 1954.

——'The First Mitchell Memorial Lecture', in Journal of the Royal Aeronautical Society, 58, 1954.

5/. Pilots' Accounts

Adlam, Hank, *On and Off the Flight Deck* (Pen and Sword, 2007).

Battle, H. F. V., OBE, DFC, *Line!* (Newbury, 1984).

Biard, H. C., *Wings* (Hurst and Blackett, 1935).

Greig, Air Commodore D'Arcy A., *My Golden Flying Years* (Grub St., 2010).

Henshaw, A., *Sigh for a Merlin* (Blackett, 1977; reprinted by Crecy Publishing, Ltd, 2007).

Lewis, Julian, *Racing Ace: the Fights and Flights of Samuel 'Kink' Kinkead* (Pen and Sword, 2011).

Livock, G. E., *To the Ends of the Air* (HMSO 1973).

Pudney, J., *A Pride of Unicorns: Richard and David Atcherley of the RAF* (Oldbourne Book Co. Ltd, 1960).

Quill, J., *Spitfire: a Test Pilot's Story* (Murray,1983; reprinted by Crecy Publishing Ltd, 1998).

Schofield, Fl. Lt H. M., *The High Speed and Other Flights* (John Hamilton Ltd, 1932).

Snaith, Grp Capt. L. C., 'Schneider Trophy Flying', 1968 Lecture to the Royal Aeronautical Society Historical Group.

Wellum, G., *First Light* (Penguin Books, 2003).

6/. Aviation in General

Anderson, J. D., *A History of Aerodynamics and its Impact on Flying Machines* (Cambridge University Press, 1997).

Baker, A., *From Biplane to Spitfire: The Life of Air Chief Marshall Sir Geoffrey Salmond* (Pen and Sword, 2003).

Boyle, A., *Trenchard* (Collins, 1962).

Godwin, J, *Early Aeronautics in Staffordshire* (Staffordshire Libraries, Arts and Archives, 1986).

Goodall, M. H., *The Norman Thompson File* (Air-Britain, 1995).

King, Allan J., *Wings on Windermere: The History of the Lake District's Forgotten Flying Boat Factory* (Mushroom Model Publications, 2009).

Master of Sempill, Colonel *The, Air and the Plain Man* (Elkin, Matthews and Marrot, 1931).

Penrose, H., *British Aviation: the Adventuring Years, 1920–1929* (Putnam, 1973).

——*British Aviation: the Great War and Armistice, 1915–1919* (Putnam, 1969).

——*British Aviation: Widening Horizons, 1930–1934* (Putnam,1979).

Stroud, J., *Annals of British and Commonwealth Air Transport 1919–1960* (Putnam, 1962).

Templewood, Viscount [Sir Samuel Hoare], *Empire of the Air: The Advent of the Air Age, 1922–1929* (Collins, 1957).

The Centenary Journal of the Royal Aeronautical Society, 1866–1966.

Jarrett, P., ed., *Biplane to Monoplane: Aircraft Development, 1919–1933* (Putnam, 1997).

Nayler, J. L. and Ower, E., *Aviation: Its Technical Development* (Owen, 1965).

Mason, T., *British Flight Testing: Martlesham Heath, 1920–1939* (Putnam, 1993).

Meekoms, K. J. and Morgan, E. B., *The British Aircraft Specifications File, 1920–1949* (Air-Britain, 1994).

Sinnot, C., *The Royal Air Force and Aircraft Design, 1923–1939* (Frank Cass, 2001).

Smith, C. B., *Testing Time* (Cassell, 1961).

Foxworth, T., *The Speed Seekers* (Doubleday, 1974).

Lewis, P., *British Racing and Record-Breaking Aircraft* (Putnam, 1970).

Duval, G. R., *British Flying Boats and Amphibians, 1909–1952* (Putnam, 1966).

Grey, C. G., *Sea Flyers* (Faber and Faber, 1942).

Hendrie, Andrew, *The Cinderella Service: RAF Coastal Command, 1939–1945* (Pen and Sword, 2006).

Killen, J., *A History of Marine Aviation* (Muller, 1969).

King, H. F., *Aeromarine Origins* (Putnam,1966).

London, P., *British Flying Boats* (Sutton Publishing Co., 2003).

Munson, K., *Flying Boats and Seaplanes since 1910* (MacMillan, 1971).

Penny, R. E., 'Seaplane Development' in *Journal of the Royal Aeronautical Society*, September 1927. [Contribution by R. J. Mitchell.]

Shenstone, B. S., 'Transport Flying Boats: Life and Death' in *Journal of the Royal Aeronautical Society*, December, 1969.

7/. Aero Engines

Banks, Air Commodore F. R., 'Fifty Years of Engineering Learning', in *Journal of the Royal Aeronautical Society*, March, 1968.

——*I Kept No Diary* (Airlife publications, 1978).

Bulman, G. P., 'An Account of Partnership: Industry, Government and the Aero Engine', 2002, Rolls-Royce Heritage Trust, HS31.

Harker, R. R., *Rolls Royce from the Wings* (Oxford Illustrated Press, 1976).

Lovesey, A. C., 'Milestones and Memories from Fifty Years of Aero Engine Development', 11th Sir Henry Royce Memorial Lecture, Royal Aeronautical Society, 7th November, 1966.

8/. Lady Houston

Allen, Warner, *Lady Houston DBE* (Constable, 1947).

Macnaie, Miles, *Lady Lucy Houston DBE; Aviation Champion and Mother of the Spitfire* (Pen and Sword, 2016).

Wentworth Day, J., *Lady Houston DBE* (Allan Wingate, 1958).

9/. Other Aircraft

Andrews, C. F. and Morgan, E. B., *Vickers Aircraft since 1908* (Putnam, 1974).

Barnes, C. H., *Bristol Aircraft since 1910* (Putnam, 1964).

——*Shorts Aircraft since 1900* (Putnam, 1967).

Jackson, A. J., *Blackburn Aircraft since 1909* (Putnam, 1968).

James, D. N., *Gloster Aircraft since 1917* (Putnam, 1971).

London, P., *Hawker and Saro Aircraft since 1917* (Putnam, 1974).

Mason, F. K., *Saunders Aircraft since 1920* (Putnam, 1961).

Taylor, H. A., *Fairey Aircraft since 1915* (Putnam, 1974).

'Hurricane Special' edition of *The Aeroplane*, October, 2007.

Profile Publications:

No. 3, *The Focke-Wulf Fw 190A*.

No. 11, *The Handley Page Halifax BIII, VI, VII*.

No. 40, *The Messerschmitt Bf 109E*.

No. 44, *The Fairey IIIA*.

No. 57, *The Hawker Fury*.

No. 65, *The Aero Lancaster I*.

No. 76, *The Junkers Ju 87A and B*.

No. 81, *The Hawker Typhoon*.

No. 84, *The Short Empire Boats*.

No. 98, *The Gloster Gladiator*.

No. 111, *The Hawker Hurricane I*.

No. 142, *The Short Stirling*.

No. 183, *The Consolidated PBY Catalina*.

No. 197, *The Hawker Tempest*.

Index

(Page references in *italic* indicate photographs or illustrations.)

index

To Carole Michael Beatty
with best wishes
John Shull